COLLIDERS AND COLLIDER PHYSICS AT THE HIGHEST ENERGIES

Related Titles from AIP Conference Proceedings

537 Physics Potential and Development of Muon Colliders and Neutrino Factories: Fifth International Conference on $\mu^+\mu^-$ Colliders
Edited by David B. Cline, September 2000, 1-56396- [to come]

533 Next Generation Nucleon Decay and Neutrino Detector: NNN99
Edited by Milind V. Diwan and Chang Kee Jung, August 2000, 1-56396-956-4

531 Particles and Fields: Seventh Mexican Workshop
Edited by Alejandro Ayala, Guillermo Contreras, and Gerardo Herrera, July 2000, 1-56396-954-8

512 Nuclear Physics at Storage Rings: Fourth International Conference: STORI99
Edited by Hans-Otto Meyer and Peter Schwandt, June 2000, 1-56396-928-9

508 Hadron Physics: Effective Theories of Low Energy QCD
Edited by A. H. Blin, B. Hiller, M. C. Ruivo, C. A. Sousa, and E. van Beveren, March 2000, 1-56396-927-0

496 Workshop on Instabilities of High Intensity Hadron Beams in Rings
Edited by T. Roser and S. Y. Zhang, December 1999, 1-56396-910-6

490 Particles and Fields: Eighth Mexican School
Edited by Juan Carlos D'Olivo, Gabriel López Castro, and Myriam Mondragón, November 1999, 1-56396-895-9

488 High Energy Physics at the Millennium: MRST'99
Edited by Pat Kalyniak, Stephen Godfrey, and B. Kamal, October 1999, 1-56396-902-5

444 Particle Physics and Cosmology: First Tropical Workshop/High Energy Physics: Second Latin American Symposium
Edited by José F. Nieves, September 1998, 1-56396-775-8

To learn more about these titles, or the AIP Conference Proceedings Series, please visit the webpage **http://www.aip.org/catalog/aboutconf.html**

COLLIDERS AND COLLIDER PHYSICS AT THE HIGHEST ENERGIES

Muon Colliders at 10 TeV to 100 TeV
HEMC'99 Workshop

Montauk, New York   27 September–1 October 1999

EDITOR
Bruce J. King
Brookhaven National Laboratory

Melville, New York, 2000
AIP CONFERENCE PROCEEDINGS ■ VOLUME 530

Editor:

Bruce J. King
Brookhaven National Laboratory
Building 901A
Upton, NY 11973-5000
USA

E-mail: bking@bnl.gov

The articles on pp. 1–12, 13–31, 48–65, 86–121, 122–141, 142–164, 165–180, 208–216, 239–248, 249–259, 308–310, and 333–338 were authored by U. S. Government employees and are not covered by the below mentioned copyright.

Authorization to photocopy items for internal or personal use, beyond the free copying permitted under the 1978 U.S. Copyright Law (see statement below), is granted by the American Institute of Physics for users registered with the Copyright Clearance Center (CCC) Transactional Reporting Service, provided that the base fee of $17.00 per copy is paid directly to CCC, 222 Rosewood Drive, Danvers, MA 01923. For those organizations that have been granted a photocopy license by CCC, a separate system of payment has been arranged. The fee code for users of the Transactional Reporting Service is: 1-56396-953-X/00/$17.00.

© 2000 American Institute of Physics

Individual readers of this volume and nonprofit libraries, acting for them, are permitted to make fair use of the material in it, such as copying an article for use in teaching or research. Permission is granted to quote from this volume in scientific work with the customary acknowledgment of the source. To reprint a figure, table, or other excerpt requires the consent of one of the original authors and notification to AIP. Republication or systematic or multiple reproduction of any material in this volume is permitted only under license from AIP. Address inquiries to Office of Rights and Permissions, Suite 1NO1, 2 Huntington Quadrangle, Melville, N.Y. 11747-4502; phone: 516-576-2268; fax: 516-576-2450; e-mail: rights@aip.org.

L.C. Catalog Card No. 00-106072
ISBN 1-56396-953-X
ISSN 0094-243X
Printed in the United States of America

CONTENTS

Preface vii

Detector Backgrounds for a High Energy Muon Collider 1
    O. Benary, S. Kahn, and I. Stumer

Acceleration for a High Energy Muon Collider 13
    J. S. Berg

Muon Collider Physics at Very High Energies 32
    M. S. Berger

Fringe Field Effects in Muon Rings 38
    M. Berz, K. Makino, and B. Erdélyi

Physics in 2006 48
    S. Dawson

A Scaling Radial-Sector FFAG Lattice for a Muon Accelerator 66
    A. Garren

Physics Opportunities with a High-Energy Collider of Same-Sign Muons .. 68
    C. A. Heusch

Collective Effects in High-Energy Muon Colliders 80
    E. Keil

Prospects for Colliders and Collider Physics to the 1 PeV Energy Scale 86
    B. J. King

Parameter Sets for 10 TeV and 100 TeV Muon Colliders and their Study at the HEMC'99 Workshop 122
    B. J. King

Mighty MURINEs: Neutrino Physics at Very High Energy Muon Colliders 142
    B. J. King

Neutrino Radiation Challenges and Proposed Solutions for Many-TeV Muon Colliders 165
    B. J. King

Technicolor Signatures at the High Energy Muon Collider 181
    K. Lane

Comments on Frictional Cooling and the Zero Energy Options for Cooling Intense Muon Beams 190
    P. Lebrun

Constraints on Plasma Compensation of Beam-Beam Effects in Muon Colliders 201
    K. V. Lotov

New Physics and the Post-LHC Era 208
    J. D. Lykken

Effects of Kinematic Correction on the Dynamics in Muon Rings 217
    K. Makino and M. Berz

Very Low-Energy Cooling Possibilities Towards Muon Colliders and Neutrino Factory 228
    K. Nagamine

The Limitations of Muon Beam Density at Ionization Cooling 236
    V. V. Parkhomchuk
High Energy Physics Potential at Muon Colliders 239
    Z. Parsa
Focusing and Acceleration of Bunched Beams 249
    Z. Parsa and V. Zadorozhny
Detector Challenges for $\mu^+\mu^-$ Colliders in the 10–100 TeV Range 260
    P. Rehak, D. Cline, E. Gatti, C. Heusch, S. Kahn, B. King, T. Kirk,
    P. Norton, V. Radeka, N. Samios, V. Tcherniatine, and W. Willis
Kaluza-Klein Physics at Muon Colliders 290
    T. G. Rizzo
Might These Machines be Affordable 308
    N. P. Samios
Remarks on High Energy Muon Collider................................. 311
    A. Skrinsky
Limit on a Horizontal Emittance in High Energy Muon Colliders
due to Synchrotron Radiation 316
    V. Telnov
Problems and Stoppers for $\gamma\gamma$, $\gamma\mu$, μp Colliders
Using Very High Energy Muons 318
    V. Telnov
Some Problems in Plasma Suppression of Beam-Beam Interactions
at Muon Colliders 324
    V. Telnov
Coherent e^+e^- Pair Creation at High Energy Muon Colliders............... 330
    V. Telnov
FFAG Lattice Without Opposite Bends 333
    D. Trbojevic, E. D. Courant, and A. Garren
Muon Collider Workshop Summary..................................... 339
    W. Willis
Final Focus Challenges for Muon Colliders at Highest Energies 347
    F. Zimmermann

List of Participants... 367
Author Index.. 371

Preface

The workshop "Studies on Colliders and Collider Physics at the Highest Energies: Muon Colliders at 10 TeV to 100 TeV" (HEMC'99) was held in Montauk, NY, U.S.A., from 27 September to 1 October, 1999. Its goal was to provide a first assessment of the long-term potential of muon colliders to explore the basic building blocks of the natural Universe. Broad-based in its outlook, it included discussion on the challenges and feasibility of accelerator and detector technologies as well as theoretical speculations and classification schemes for the physics processes that such future colliders might uncover. It was highly appropriate that the historic Montauk lighthouse gave us the imagery of a guiding beacon to inspire us in such forward-looking studies!

Forty-six accelerator and high energy physicists from around the world attended the 5 days of the workshop. (The participants are listed at the back of these proceedings.) Approximately 1/3 of the program consisted of technical evaluations in working group sessions and 2/3 was for plenary sessions giving non-technical presentations that could be understood across the sub-fields. The program and other information on the workshop can be found on the internet at http://pubweb.bnl.gov/people/bking/heshop/ .

The Brookhaven National Laboratory Center for Accelerator Physics (BNL CAP) and the U.S. Department of Energy (DOE) were co-sponsors for the workshop and we thank them for their enthusiastic support. The workshop really only become more than a glint in the organizers' eyes when Bob Palmer, the Director of BNL CAP, took an immediate liking to the concept and agreed to back it. Financial support for some overseas participants from the DOE office of David Sutter was particularly appreciated, and these participants added much to the workshop.

Bob Palmer's one major reservation concerning the workshop was the considerable additional workload it would entail for the CAP secretariat – Kathleen Tuohy and Patricia Tuttle – who, besides their normal workload, were already burdened with the organization of another meeting. It is a credit to their dedication and "can-do" helpfulness that Kathy and Pat very graciously agreed to do the workshop anyway, despite their conscious knowledge of the stressful months of extra work ahead. Their talent and their experience enabled them to pull it off wonderfully, to rave reviews from the participants, and we are deeply grateful to them for their most essential contributions to the success of the workshop.

We very much appreciated the help and encouragement of the advisory committee: Alain Blondel (Ecole Polytechnique), John Ellis (CERN), Shouxian Fang (IHEP, Beijing), Eberhard Keil (CERN), Kanetada Nagamine (KEK/RIKEN), Jose Nieves (U. Puerto Rico), Robert Palmer (BNL), Ken Peach (Rutherford Lab.), Chris Quigg (FNAL), Heidi Schellman (Northwestern U.), Andrew Sessler (LBNL),

Robert Siemann (SLAC), Alexander Skrinsky (INP, Novosibirsk), Alvin Tollestrup (FNAL) and Edward Witten (Princeton, Inst. Advanced Study). Thanks are also due to Juan Gallardo of BNL CAP, for technical support.

In many ways, September of 1999 was an inauspicious time to hold a workshop, coming at the end of a financial year that was hard hit by all manner of travel restrictions, budget cuts and curbs on meetings. We and the secretariat wish to thank several of the BNL staff for steering us through the maze, including Christine Ronick of Staff Services, Michael Goldman of the Office of General Counsel, and Greg Ogeka and his staff in the Finance and Administration office of the BNL Directorate.

It is a shame that the special circumstances of the time led to many eager would-be attendees missing out on obtaining permission and/or funding to participate in the workshop. However, and to paraphrase Nietzsche, what didn't kill the workshop may even have made it stronger. The unanimous enthusiasm and productivity of those who were able to attend conspired to turn the workshop into a resounding success and a small but valuable step towards an awareness of our ultimate prospects for experimental high energy physics.

It was a real pleasure to be part of an atmosphere where everyone was engaged, thinking about and contributing to novel topics that had not been previously considered in any depth. The quality of the articles in these proceedings testifies to the ingenuity and commitment of the participants in addressing these issues. We hope the reader finds them as enlightening and thought-provoking as we do.

Bruce King, Colin Johnson and Joe Lykken
HEMC'99 Organizers

Detector Backgrounds for a High Energy Muon Collider

O. Benary[*], S. Kahn[§], and I. Stumer[§]

[§]*Brookhaven National Laboratory, Upton, NY 11973*
[*]*Tel-Aviv University, Ramat-Aviv, Tel-Aviv 69978, Isreal*

Abstract. The potentially large background produced from the decay products of muons could affect the quality of the physics in a high energy muon collider. This paper examines the kinds of background present and their expected rate. The results are based on a simulation of muon colliders with center of mass energies of 100 GeV, 500 GeV and 4 TeV.

INTRODUCTION

How well the background can be controlled will determine the kind of physics that can be done in a muon collider. Most backgrounds are associated with the products of the decaying muons that get into the detector region. A 2 TeV/c muon beam with 2×10^{12} μ per bunch will produce 2×10^5 decays per meter. The number of decays per length scales with $1/\gamma$ because of the Lorentz contraction. The electrons produced from muon decays will not have the designed momentum of the collider ring and will either interact with the wall of the beam chamber producing electromagnetic showers or produce bremsstrahlung in regions of large transverse magnetic field. The design of a detector for a muon collider will be constrained by the necessity to reduce the electromagnetic background.

Detector backgrounds have been studied for a potential muon collider at three different energies:
- A 50 × 50 GeV/c Higgs factory collider.
- A 250 ×250 GeV/c muon collider at the same energy as the phase 1 NLC.
- A 2 × 2 TeV/c muon collider at an energy not accessible to an electron collider.

Results of these studies have been reported elsewhere [1-2]. No explicit investigation has been done for backgrounds in muon colliders with $\sqrt{s}$ between 10 TeV and 100 TeV. This paper will review what has been learned about muon collider background in the studies that have been previously performed. Extrapolation to higher energies may not be straight forward since the shielding geometry and the beam final focus elements have been optimized at each energy to reduce backgrounds. The effort to minimize the backgrounds will have a strong influence on both the design of the detector and the design of the magnets in the *final focus* of the intersection region (IR).

BACKGROUNDS

The following classes of backgrounds have been investigated for a muon collider:
- Muon Decay Background.
- Beam Halo Background.
- Beam-Beam Interactions.

Most of the work to date has been devoted to the backgrounds associated to muon decays. The backgrounds associated to muon decays include:
- Electron showers.
- Synchrotron Radiation and pair production.
- Photonuclear interactions producing hadrons.
- Bethe-Heitler muon production.

These will be discussed thoroughly in the following sections. Beam halo backgrounds come from accelerator sources. The beam scraping system will be designed to reduce halo, however even a small residual muon halo can produce a large background in the detector.

The backgrounds associated with the muon decay have been studied with a GEANT simulation of a detector that would be appropriate for a muon collider. Figure 1 shows a sketch of the geometry of such a detector along with the final focus magnets of the collider ring. The detector geometry includes the following features:
- Conical tungsten shield over the beam extending to 20° in both forward and backward directions. The shield prevents most of the electromagnetic showers from penetrating into the detector region.
- Expanding shield inner cone beyond minimum aperture position (1.1 meters from IP for 2×2 TeV collider) is 4σ of the beam size.
- Between IP and minimum aperture point the inner radius is an inverse cone with an angle equal to but opposite sign of the previously mentioned angle of expansion after the minimum aperture point. These cones are designed so that the detector does not *see* any surface where incident decay electrons could have interacted.
- The open space between the IP and the beginning of the tungsten shield is small. It is only 3 cm for the 2×2 TeV collider.
- The inner surface of each shield is shaped in a sawtooth manner. These steps and slopes collimate the electrons in the beam and maximize the absorption of the electromagnetic showers from the electrons that graze the cone surface. This reduces the funneling of low energy electrons down the beam pipe. Figure 2 shows the geometry of the inner surface of the tungsten shield.
- An 8 T dipole magnet with collimator inside is placed upstream of the first quadrupole magnet to sweep decay electrons away before the final collimation.

Table 1: Parameters describing the geometry of the intersection region for the muon colliders studied at three energies.

Parameter	50 ×50 GeV	250 ×250 GeV	2 ×2 TeV
Shield Angle	20°	20°	20°
Open Space to IP	6 cm	3 cm	3 cm
Min Aperture Point from IP	80 cm	1.1 m	1.1 m
R_{iris} at Min Aperture	0.8 cm	0.5 cm	0.5 cm
Distance to 1st Quad	7 m	8 m	6.5 m

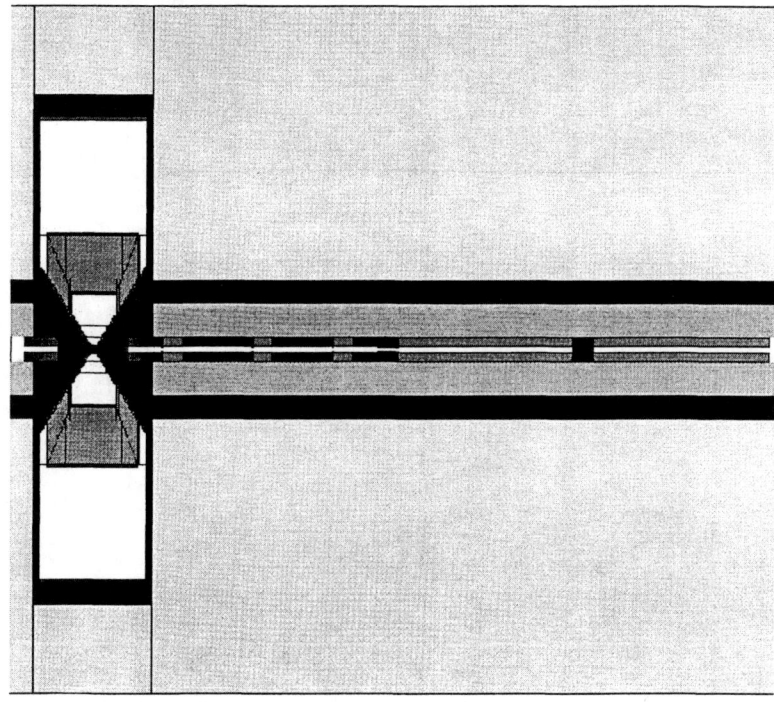

FIGURE 1. Sketch of the IP region and 130 meters of the final focus magnet system. This sketch shows the geometry of the detector used in the GEANT simulation of the 2 × 2 TeV Muon Collider.

Muon Decay Background

The electromagnetic showers produced from the muon decay electrons interacting with the beam chamber wall are mostly contained in the tungsten shielding. A small residual flux of soft photons does reach the detector region. Electrons in the showers are trapped at small radius by the 2-4 T detector solenoid field. This is shown in table 2. Table 2 also shows fluences for protons, neutrons, and pions. Hadrons are produced in ~1% of the interactions in an electromagnetic shower. These hadrons are produced as a result of photonuclear processes:

- Giant Dipole Resonance in the energy region $5 < E_\gamma < 30$ MeV.

- Quasi-Deuteron Region in the energy region $30 < E_\gamma < 30$ MeV.
- Baryon Resonance Production in the energy region 150 MeV $< E_\gamma < 2$ GeV.
- Vector Dominance in the energy region $E_\gamma > 2$ GeV.

The GEANT program used in these simulations has been modified to include these processes. Hadron production is also possible at the highest energies through leptoproduction. This process has been ignored in the studies so far, however they certainly would be important at very high energy muon colliders. The MICAP option in GEANT for hadron showering package was used since it handles neutrons down to thermal energies. Low energy neutrons can be bothersome since they live a long time. They can bounce around entering the same detector element several times giving extraneous signals. In order to alleviate this problem, borated-polyethelene is placed where possible around the tungsten shielding and the calorimeter to soak up as many neutrons as possible. This reduces the neutron flux substantially.

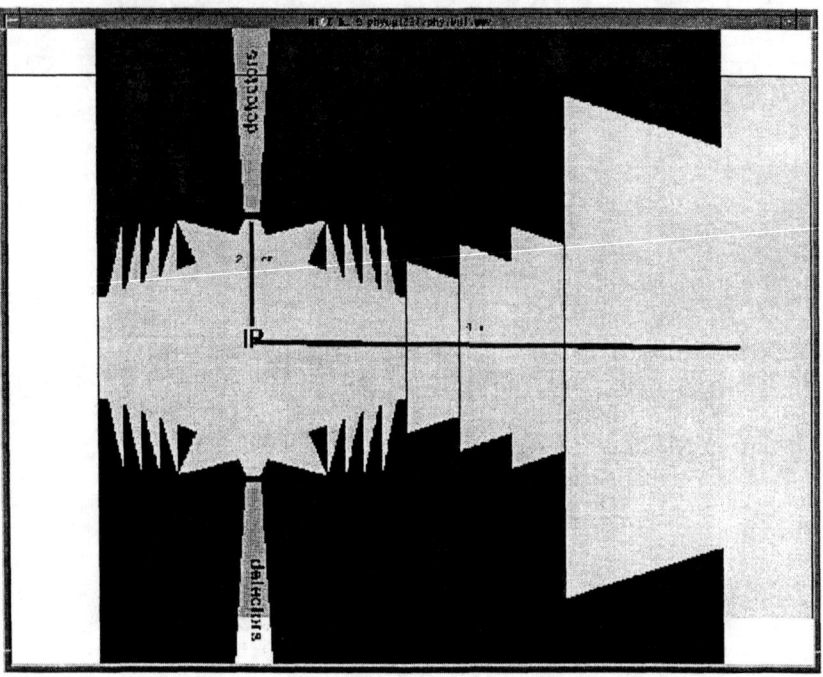

FIGURE 2: Absorber and collimation system on the inside surface of the tungsten shield. This system is designed to reduce electrons in the beam pipe in the vicinity of the IP.

Shown in Table 2a, b are the radial particle fluxes in $cm^{-2}/crossing$ for two bunches of 10^{12} μ each at different radii for the 50×50 GeV and 2×2 TeV machines. The rates are shown for γ, n, p, π, e, and μ. The energy thresholds used to define the particle fluxes are $E_\gamma > 25$ keV, $E_n > 40$ keV, $E_p > 10$ MeV, $E_\pi > 10$ MeV. The fluxes between 5 cm and 20 cm are what a vertex detector would see. A central tracker might be located at 50-100 cm and the calorimeter would start at 150 cm. The fluxes shown in Table 2 can be interpreted as occupancy in a 300 μm × 300 μm silicon pad detector that might be used as a vertex detector. If one assumes that the probability that low energy photons and neutrons interact in the silicon pad detector is 0.003 and 0.0001, respectively, hits and occupancy rates can be calculated. Table 3A and B show the background hits and occupancy for the 50 × 50 GeV and 2 × 2 TeV colliders, respectively. Figure 3 shows the occupancy for all three energies studied. The left-hand part of Figure 3 shows the occupancies for all particles while the right-hand part shows the occupancies for the charge particles only. The total occupancy is above 1% at small radii, mostly due to photon conversions. The charge particle occupancy is equal or below 1% at all the radii in the figure. One can reduce the occupancy rates by using smaller pixel sizes or using new innovated detector ideas.

TABLE 2A. Flux per crossing for two 50 GeV beams of 10^{12} μ per bunch.

Radius	γ	n	p	π	e	μ
5 cm	4300	32			3.8	0.15
10 cm	1100	36		0.24	0.3	0.07
15 cm	480	75		0.11		0.03
20 cm	270	98		0.09		0.007
50 cm	40	37	0.05	0.015		0.0004
100 cm	9	18	0.005			0.0002
150 cm	4	9	0.02			2.1×10^{-5}

TABLE 2B. Flux per crossing for two 2 TeV beams of 10^{12} μ per bunch.

Radius	γ	n	p	π	e	μ
5 cm	2700	120	0.05	0.9	2.3	1.7
10 cm	750	110	0.20	0.4	−	0.7
15 cm	350	100	0.13	0.4	−	0.4
20 cm	210	100	0.13	0.3	−	0.1
50 cm	70	120	0.08	0.05	−	0.02
100 cm	31	50	0.04	0.003	−	0.008
150 cm						0.003

TABLE 3A. Background hits and occupancy for 50 × 50 GeV μ collider. *Hit* rates are quoted for two beams with bunches of 10^{12} μ's each. *Occupancy* is quoted for two beams with 4×10^{12} μ's each and that the silicon detector has 300 × 300 μm pads.

Radius cm	γ cm^{-2}	n cm^{-2}	Charged cm^{-2}	Hits cm^{-2}	Total Occupancy	Charged Occupancy
5	13	0.03	4	17	6.1%	1.4%
10	3.3	0.04	0.6	4	1.4%	0.2%
15	1.4	0.07	0.14	1.6	0.6%	0.05%
20	0.8	0.1	0.10	1.0	0.4%	0.04%
50	0.1	0.03	0.065	0.2	0.07%	0.02%
100	0.03	0.02	0.005	0.06	0.02%	0.002%
150	0.01	0.01	0.02	0.04	0.014%	0.007%

TABLE 3B. Background hits and occupancy for 2× 2 TeV µ collider. *Hit* rates are quoted for two beams with bunches of 10^{12} µ's each. *Occupancy* is quoted for two beams with 2×10^{12} µ's each and that the silicon detector has 300×300 µm pads.

Radius cm	γ cm^{-2}	n cm^{-2}	charged cm^{-2}	Hits cm^{-2}	Total Occupancy	Charged Occupancy
5	8.1	0.012	5.0	13.1	2.4%	0.9%
10	2.3	0.011	1.3	3.6	0.65%	0.23%
15	1.1	0.010	0.93	2.0	0.36%	0.17%
20	0.6	0.010	0.53	1.1	0.20%	0.10%
50	0.2	0.012	0.15	0.36	0.06%	0.03%
100	0.1	0.005	0.05	0.16	0.03%	0.01%

The lifetime of a vertex detector in a high radiation environment is of concern. The radiation damage to a silicon vertex detector at 5 cm from the beam can be estimated. Table 4 shows the hits/cm^2/yr (1 year = 10^7 sec) and expected lifetime of the device under normal operating conditions. An acceptable number of hits for the silicon vertex detector lifetime is 1.5×10^{14}.

TABLE 4. Radiation damage by neutrons on a silicon vertex detector situated at 5 cm from the beam. It is assumed that the machine is operating 10^7 sec/year at 15 Hz with 1000 turns per machine fill. The neutrons counted are those with $E_{kin} > 100$ keV.

Energy	µ/bunch	Neutrons/cm^2 Per crossing	Hits/cm^2/year	Lifetime (years)
50×50 GeV	4×10^{12}	30	1.8×10^{13}	8
250×250 GeV	4×10^{12}	50	3×10^{13}	5
2×2 TeV	2×10^{12}	100	3×10^{13}	5

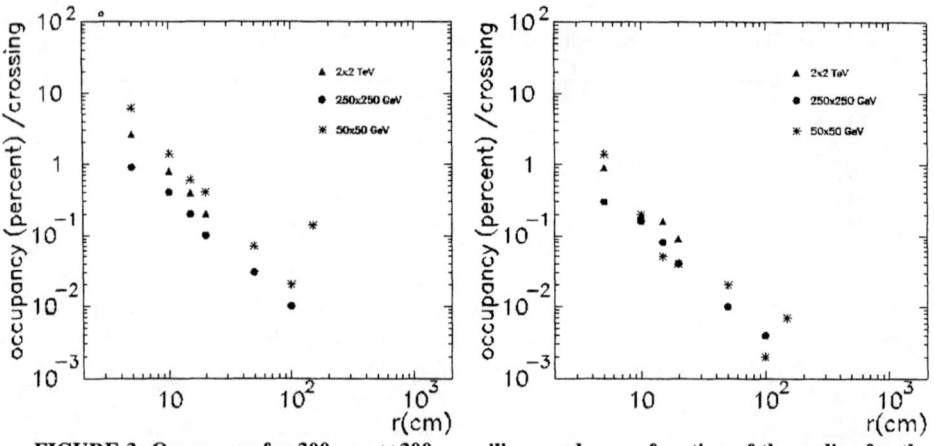

FIGURE 3: Occupancy for 300 µm × 300 µm silicon pads, as a function of the radius for the three energies studied. Left-hand figure shows the total occupancy while the right-hand figure shows the occupancy for charged particles only.

Bethe-Heitler Muons

A significant flux of muons with high energy come from μ pair production in electromagnetic showers. These Bethe-Heitler muons will be a concern at high energy colliders. Figure 4 shows the spectrum of Bethe-Heitler muons in the final focus region for the 2×2 TeV collider. This figure is obtained by convoluting the energy spectrum from the decay of 2 TeV muons ($<E_{electron}>$ = 670 GeV) with the μ pair production cross section. Figure 5 displays the trajectories of typical Bethe-Heitler muons from the last 130 m section of beam pipe before the detector in the 2×2 TeV collider. The most serious effect of these muons is that they may make deeply inelastic collisions in the electromagnetic or hadronic calorimeter. A large deposit of energy in a single cell in the vicinity of other jets can cause severe fluctuations in the global parameters such as transverse energy or missing transverse energy. This is a more serious problem for the higher energy muon colliders.

Table 5 gives the attributes of the Bethe-Heitler muons found at the three energies. In the 250×250 GeV and 2×2 TeV cases there is massive lead shielding around the final focus quadrupoles to reduce the Bethe-Heitler muons as much as possible. In the higher energy cases ≤1% of these muons reach the calorimeter. In the 2×2 TeV case the B-H muons have $\langle p_\mu \rangle$ = 15.4 GeV/c and deposit 2.9 GeV in the calorimeter on average. This sits on top of a uniform pedestal of 100 GeV to 100 TeV of deposited energy from all other sources. The size of the pedestal is dependent on the calorimeter technology used. Since calorimeters based on scintillator technology will have a higher pedestal of deposited energy, they may not be a desirable choice. The high neutron background in the calorimeter favors the use of liquid argon as the active calorimeter medium. The B-H muons cause large local fluctuations in the energy deposited in a cell. These energy spikes can (1) cause false triggers when event selection is based on transverse or overall energy balance, (2) generate false jets, and (3) give incorrect energy of real jets. False jet generation can be eliminated with longitudinal energy balance. Errors in energy of real jets can be a more serious problem. They can be reduced to some extent with radial energy distribution cuts.

Figure 6 shows the energy deposition for B-H muons in calorimeter cells segmented in $\Delta\phi$ and $\Delta\cos\theta$ for the 2×2 TeV muon collider. In this study the lead shield around the final focus quadrupoles is not included. The left-hand figure has no timing cut, where as the right-hand plot has a 1 ns timing in synch with the particles from the beam IP. The figure shows that B-H muons in the central (barrel) part of the calorimeter are eliminated. Figure 7 illustrates the paths of muons from the IP and B-H muons arriving at the central and the forward calorimeters. A B-H muon, which was in time with the muon beam at its production point at 100 m away, would go directly to the calorimeter, whereas a muon heading to the IP would travel 100 m *plus* an additional time for the transit of a particle created at the IP to reach the calorimeter.

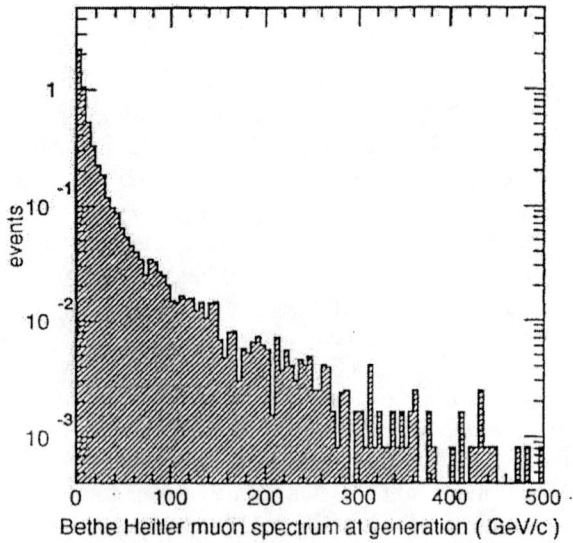

FIGURE 4. Bethe-Heitler muon spectrum at generation.

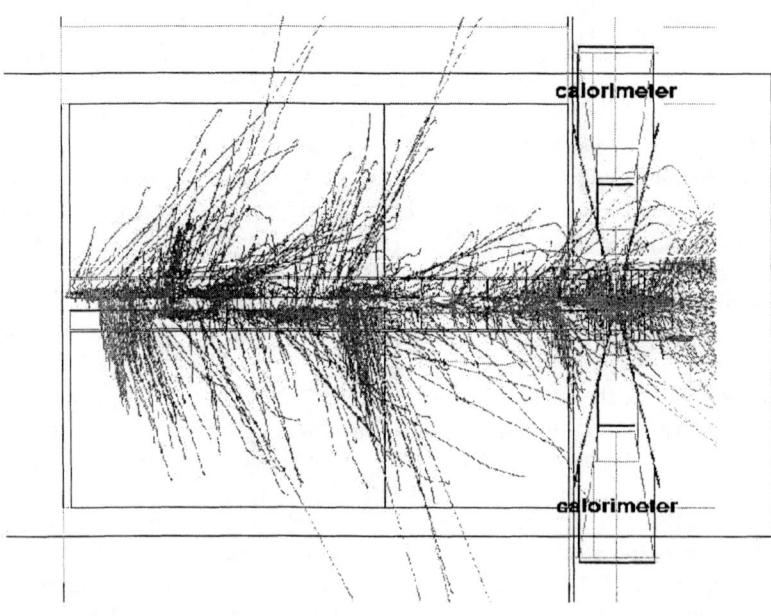

FIGURE 5. Trajectories of Bethe-Heitler muons from sources in the shielding around the beampipe in the last 130 m from the IP for the 2×2 TeV collider. The outer edge of the calorimeter is approximately 4 m

TABLE 5: Attributes of Bethe-Heitler muons at the energies of the three muon colliders studied. This table gives the parameters at the source of the muon pairs and in the calorimeter. It is assumed that lead shielding surrounds the final focus quadrupoles for the 250×250 GeV and 2×2 TeV colliders.

Collider Energy	50×50 GeV	250×250 GeV	2×2 TeV
Source Length	20 m	33 m	130 m
μ (p>1GeV/c) per electron	9.6×10^{-6}	8.3×10^{-5}	5.4×10^{-4}
Beam μ per bunch	4×10^{12}	2×10^{12}	2×10^{12}
B-H μ per bunch crossing	6100	17500	28000
$\langle p_\mu \rangle_{initial}$ GeV/c	4.4	9.5	22
μ's entering calorimeter	25	160	220
$\langle p_\mu \rangle$ in calorimeter, GeV/c	1.8	6.3	15.4
$\langle E_{dep} \rangle$, GeV	0.4	1.3	2.9
Total E_{dep} by all μ's	10	210	640
E_{dep} pedestal, GeV	1	25	50
Fluctuation in E_{dep}, GeV	1	15	55
$E_\perp$ pedestal, GeV	0.5	15	15
Fluctuation in $E_\perp$, GeV	0.5	8	40

The difference in time-of-flight between signal muon path and the B-H muon path would exceed 1 ns which can be used to remove the background. The situation is not the same for the forward and backward calorimeters. The time difference between the path of a forward-going B-H muon reaching the forward calorimeter and path for a beam muon *plus* produced particle to reach the forward calorimeter is very small. The timing cut will only eliminate the B-H muons in the backward calorimeter in this case. Thus the timing cut can only eliminate half of the B-H muon background in the forward and backward calorimeters.

Pair Production

Although coherent beam-beam electron pair production (beamsstrahlung) is small [3] incoherent pair production from processes like $\mu^+\mu^- \rightarrow \mu^+\mu^- e^+e^-$ is significant in the 2×2 TeV collider. The cross section for such processes is estimated to be 10 mb [4]

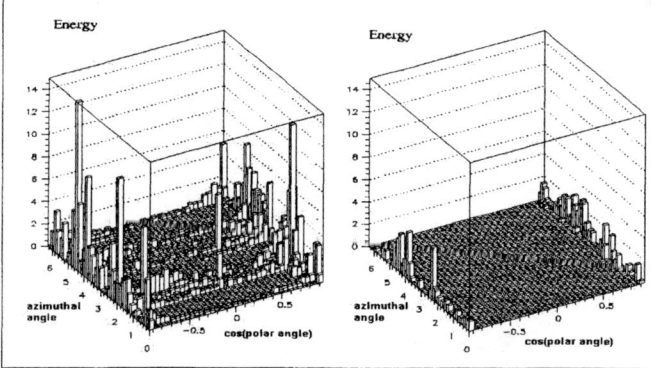

FIGURE 6. Energy deposition from Bethe-Heitler muons into a calorimeter at a 2 × 2 TeV muon collider. The calorimeter cells are segmented in cells with $\Delta\phi = 0.1$ and $\Delta\cos\vartheta \approx 0.05$. The right-hand plot has a timing cut of 1 ns.

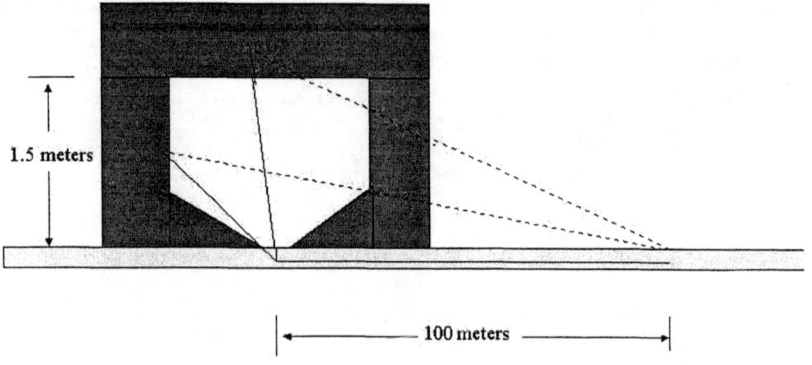

FIGURE 7. Sketch illustrating the paths of signal muons from the IP and Bethe-Heitler muons generated by an electromagnetic shower in the beam chamber wall 100 meters upstream. The difference in path length between the signal muons and the B-H muons is large enough to distinguish by a 1 ns difference in time-of-flight in the central calorimeter. In the forward calorimeter a 1ns difference in time-of-flight is not sufficient to be distinguished the signal muon from the B-H background. This sketch is *not* to scale.

which would mean approximately 3×10^4 electron pair per beam crossing. Although these electrons do not have significant transverse momentum initially they can be deflected into the detector by the oncoming beam. This potential background was examined with a simple tracking program to follow electrons created on the axis (the worse case) as they are deflected away from the opposing bunch. After they are past that bunch the electrons are trapped in spiral orbits in the detector magnetic field. Figure 8 shows the trajectories of the electron tracks for initial momentum in the range of 3.8 to 3000 MeV/c in a 4 tesla solenoid field. The study was done for a 2 tesla and 4 tesla detector solenoid field. In the 2 tesla case electrons with $P_{init} < 30$ MeV/c do not make it out to the 10 cm detector plane, while those with $P_{init} > 100$ MeV/c have an initial angle that keeps them within the nose shield. Approximately 10% of the electron tracks are between these limits and pass through the detector plane at 10 cm. This gives an electron track fluence < 10 tracks/cm^2. For a detector plane at 5 cm there would be an electron track fluence of 30 tracks/cm^2. If the detector solenoid field is 4 tesla, no electron tracks reach 10 cm and the fluence at 5 cm is reduced by a factor of 2.

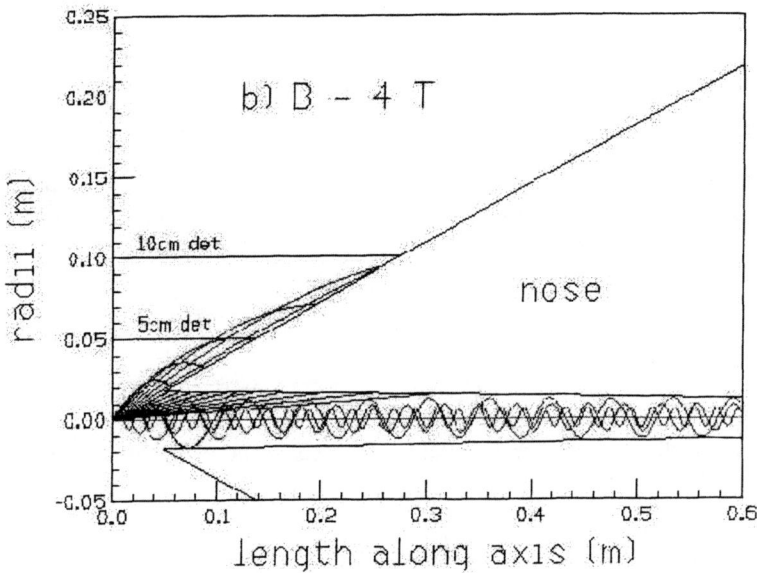

FIGURE 8. Trajectories of inelastic electron pair tracks with initial momentum between 3.8 and 3000 MeV/c. These tracks are in a 4 T solenoid field.

Beam Halo Background

Beam halo background is the background from the tails of the beam that can get into the detector. Since muons can traverse long distances without interacting, any source of beam loss in the ring can potentially deflect some muons into detector. This is generally not as serious a problem in hadron or electron colliding rings. Since each beam has 2×10^{12} muons, a small beam halo that could get into the detector would be very troublesome. Halo calculations are hard to do since they depend on a detailed model of the ring, the injection system and the injected beam profile. In addition imperfections in the building of the accelerator can contribute. Beam halo can have two components: (1) that which is associated to the initial filling of the ring which will die out in a small number of turns and (2) that which is generated or regenerated in the ring and can persist throughout the fill. The former is not of as great concern since the physics data from the first few turns can be ignored.

A collimation system[5] in the ring is designed to scrape the beam at a position in the ring 180° away from the IP. This system would scrape the beam at 3 σ in x and y. Studies with very limited statistics have been performed tracking muons with x, x', y, or y' greater than 3.5, 4, 4.5, 5 beam σ around a 2 TeV lattice. In this study muons were tracked through finite sized magnets, not lumped elements. Muons were allowed pass through material with multiple scattering corrections. With samples of 200 μ's no beam halo tracks appeared in the detector region. These samples are too small to make significant conclusions. Much more work in this area needs to be done.

CONCLUSIONS

The highest energy muon collider that physics backgrounds have been studied is the 2×2 TeV machine. It is not easy to make meaningful extrapolations to 100 TeV com muon colliders. It is likely that certain backgrounds such as hadron production and Bethe-Heitler muons will increase with energy. It can be assumed that a very high energy muon collider in the range of 10-100 TeV in the center of mass will not be the first machine built. One will learn much better how to estimate and control these physics backgrounds by building lower energy machines such as the 4 TeV center of mass machine that was studied at SNOWMASS '96.

ACKNOWLEDGMENTS

This research was supported by the U. S. Department of Energy under Contract No. DE-ACO2-98CH10886.

REFERENCES

1. $\mu^+\mu^-$ *Collider: A Feasibility Study.* BNL-52503, FNAL Conf 96/092, LBNL-38946 (1996).
2. Ankenbrandt, C. M. *et al.*, "Status of Muon Collider Research and Development and Future Plans", *Phys. Rev. Special Topics –AB 081001* (1999).
3. P. Chen, *Beam-Beam Interactions in Muon Colliders,* Nucl. Phys. B (Proc. Suppl.) **51A**, 179 (1996).
4. I. Ginzburg, *The e^+e^- Pair Production at $\mu^+\mu^-$ Collider,* Nucl. Phys. B (Proc. Suppl.) **51A**, 186, (1996).
5. A. Drozhdin *et al.*, *Scraping Beam Halo in $\mu^+\mu^-$ Colliders,* AIP Conf. Proc. **441**, p 242 (1998), FERMILAB-Conf-98/042.

Acceleration for a High Energy Muon Collider

J. Scott Berg

Brookhaven National Laboratory; Building 901A; PO Box 5000; Upton, NY 11973-5000

Abstract. We describe a method for designing the acceleration systems for a muon collider, with particular application and examples for a high energy muon collider. This paper will primarily concentrate on design considerations coming from longitudinal motion, but some transverse issues will be briefly discussed.

INTRODUCTION

The cost of a high energy muon collider will be clearly be dominated by the cost of accelerating the muons to their maximum energy. It is thus important to study possible techniques for acceleration, the advantages and disadvantages of these techniques, and how to perform a cost optimization of a final design.

The acceleration of muons poses various challenges that that are not present in the acceleration of other types of charged particles. Since muons have a finite lifetime (approximately 1000 turns), one cannot take a long time to accelerate them, so traditional accelerating synchrotrons such as are used for protons cannot be used. Due to the difficulty of cooling the muon beam, the longitudinal emittances tend to be large (as high as 0.047 eV-s), and this ends up making the acceleration significantly more challenging, especially at lower energies.

The larger mass of the muons (as compared to electrons) prevents them from emitting substantial amounts of synchrotron radiation, and thus there is nothing preventing them from being bent in an arc. While a conventional synchrotron will not work, a recirculating accelerator is certainly possible. An additional advantage of a recirculating accelerator is that it potentially gives high RF-to-beam power efficiencies, since the same linac can be used for multiple passes through the system.

Design Parameters

Table 1 gives parameters for high energy muon colliders that are relevant to the design of the acceleration systems [1, 2]. In that table, $p_{\min}$ is the momentum at which acceleration begins, $p_{\max}$ is the momentum in the collider, ϵ_L is the longitu-

TABLE 1. Parameters for various high energy muon colliders.

p_{min} MeV/c	p_{max} TeV/c	ϵ_L meV-s	ϵ_n μm	N 10^{12}
186	0.5	24	50	4
186	10	21	38	3
186	100	47	8.7	0.8

dinal emittance (normalized), ϵ_n is the normalized transverse emittance, and N is the number of particles per bunch.

We will assume that maximum accelerating gradients and the r_s/Q of the cavities scale with frequency according to

$$v = 30 \text{ MV/m} \sqrt{\frac{f}{800 \text{ MHz}}} \qquad \frac{r_s}{Q} = 1000 \text{ }\Omega/\text{m} \frac{f}{800 \text{ MHz}}, \qquad (1)$$

where f is the RF frequency.

METHODS FOR COMPUTING PARAMETERS

This paper will primarily address how to compute longitudinal parameters, but will include very rough estimates of transverse parameters for the design. There are really two types of systems that will be considered: straight linacs and recirculating accelerators. Straight linacs are used for acceleration from the lowest energies, since the relative energy spreads at lower energies will be impossible to get through a conventional arc. Once the beam reaches a sufficient energy, however, recirculating accelerators will be used.

Straight Linacs

At the lowest energies, the velocity of the particles cannot be considered to be constant, and thus it is probably best not to accelerate on-crest; instead, one should allow particles to undergo synchrotron oscillations, with the time-of-flight variation coming purely from the velocity variation with energy.

As a first approximation to the behavior of synchrotron motion in the linacs, we can take an adiabatic approximation wherein the RF bucket is determined by the sinusoidal RF fields and the velocity variation of the muons with energy. The smallness parameter in the adiabatic approximation is the quantity

$$\frac{1}{k_s p} \frac{dp}{ds}, \qquad (2)$$

where k_s is the synchrotron wave number (computed later), p is the muon momentum for the reference particle, and s is the distance along the reference orbit. It will turn out that this quantity is not in fact very small; however, lacking a better analytic description, one can nonetheless use the results under the adiabatic approximation to give a first (probably optimistic) guess as to what the parameters may be.

One could consider running on-crest at some point during the acceleration process, particularly once the adiabatic approximation becomes particularly bad. This will first of all necessarily lead to some longitudinal emittance growth. Furthermore, it will be necessary to introduce some momentum compaction to counteract the effect of the velocity variation with energy, but doing that for a beam with the relative energy spreads that this beam will have would be highly nontrivial.

Hamiltonian Description

The Hamiltonian describing longitudinal particle motion in the linac is

$$-\frac{1}{c}\sqrt{[E_0(s) + \Delta]^2 - (mc^2)^2} + \frac{E_0(s)\Delta}{p_0(s)c^2} + p_0(s)$$
$$+ \frac{qv(s, t_0(s) + \tau)}{\omega}\left[\sin\left(\omega[t_0(s) + \tau] + \phi(s, t_0(s) + \tau)\right)\right.$$
$$\left. - \omega\tau\cos\left(\omega t_0(s) + \phi(s, t_0(s) + \tau)\right) - \sin\left(\omega t_0(s) + \phi(s, t_0(s) + \tau)\right)\right], \quad (3)$$

where $E_0(s)$ is the energy of the reference particle, $p_0(s) = \sqrt{E_0^2(s) - (mc^2)^2}$ is its momentum, and t_0 is its arrival time at longitudinal position s in the linac. The particles have mass m and charge q. The gradient and phase of the fundamental mode with frequency ω at longitudinal position s and time t are $v(s,t)$ and $\phi(s,t)$ respectively. These quantities are related through

$$\frac{dt_0}{ds} = \frac{E_0(s)}{p_0(s)c} \qquad \frac{dE_0}{ds} = qv(s, t_0(s))\cos\left(\omega t_0(s) + \phi(s, t_0(s))\right) \quad (4)$$

The canonical coordinate is τ for the deviation of the arrival time at a given longitudinal position s of a particle from that of a reference particle, and the canonical momentum is $-\Delta$, where Δ is the deviation of the energy of a particle from that of the reference particle. The Hamiltonian ignores effects of beam loading, and only considers the effect of the fundamental mode in the linac.

The reason for putting time dependence in v and ϕ is to take into account power input, wall losses, and in principle beam loading (although the latter really requires a self-consistent solution, but the average effect could be put in). Except for beam loading, these time variations occur over time scales which are long compared to the time that it takes the bunch to pass by a given point in the linac. Thus, for the purposes of this discussion, the time dependence will be ignored.

It will be convenient to define an effective phase of $\psi(s) \equiv \omega t_0(s) + \phi(s)$. The adiabatic approximation is really an assumption that $v(s)$ and $\psi(s)$ change very slowly. "Very slowly" must be in relation to some other time scale, and it turns out that this time scale is really defined by the quantity (2) being small (this will not be demonstrated here). The adiabatic approximation is computed by performing calculations as if $v(s)$ and $\psi(s)$ were in fact constant. The idea is that synchrotron oscillations are occurring so rapidly the $v(s)$ and $\psi(s)$ really are nearly constant over the period of a synchrotron oscillation.

There are of course variations in $v(s)$ due to the fact that the fields amplitudes not constant over the length of a cavity cell. For these variations, one can make another adiabaticity argument: as long as the variation in particle energy and arrival time offset is small over the length of a cavity cell, the average value of $v(s)$ can be used.

Thus, these two adiabaticity arguments allow us to remove the s dependence from v and ψ in subsequent discussions, and the results will be valid to the extent that the adiabaticity arguments are valid.

Linearization

There are two periodic sets of fixed points of the above Hamiltonian: one set with $\Delta = 0$ and $\omega\tau = 2\pi m$, the other set at $\Delta = 0$ and $\omega\tau = 2\pi m - 2\psi$. As long as $\psi < 0$, the fixed point at $\tau = 0$ is stable. Linearizing the Hamiltonian about that fixed point, we get

$$\frac{1}{2}\frac{(mc^2)^2}{c(p_0c)^3}\Delta^2 - \frac{1}{2}qv\omega\sin\psi\,\tau^2 \tag{5}$$

From this, we can compute the square of the aspect ratio of a matched beam to be

$$\frac{\sigma_\Delta^2}{\sigma_\tau^2} = -\frac{qcv\omega(p_0c)^3\sin\psi}{(mc^2)^2} \tag{6}$$

and the square of the synchrotron wave number k_s (2π divided by the synchrotron wavelength) to be

$$k_s^2 = -\frac{qv\omega(mc^2)^2\sin\psi}{c(p_0c)^3}. \tag{7}$$

Here σ_Δ is the RMS energy spread, and σ_τ is the RMS bunch length in arrival time units.

Bucket Area

The Hamiltonian (3) can be used to find an equation describing the separatrix of the RF bucket. The separatrix contains the unstable fixed point at $\omega\tau = -2\psi$,

and the Hamiltonian has a value of
$$\frac{2qv}{\omega}(\psi\cos\psi - \sin\psi) \tag{8}$$
at that point and therefore along the separatrix. The separatrix is at τ has the energy deviation values
$$\Delta = E_0 w \pm p_0 c\sqrt{2w + w^2} \tag{9}$$
$$w = \frac{qvp_0c^2}{(mc^2)^2\omega}[(2\psi + \omega\tau)\cos\psi - \sin(\psi + \omega\tau) - \sin\psi] \tag{10}$$

From this, the half-width of the bucket can be computed to be
$$2p_0 c\sqrt{\frac{qvp_0c^2}{(mc^2)^2\omega}}\sqrt{\psi\cos\psi - \sin\psi}\sqrt{1 + \frac{qvp_0c^2}{(mc^2)^2\omega}(\psi\cos\psi - \sin\psi)}. \tag{11}$$

Now one might want to ask how large a bunch the bucket can hold. This should really be computed by computing the area of the RF bucket directly. However, a simpler method can be used which is approximately correct. Take the bucket half-width to be $k\sigma_\Delta$, where k is an arbitrary factor indicating how full you would like the bucket to be. This gives an expression for σ_Δ. Next, use the aspect ratio (6) to compute σ_τ in terms of σ_Δ. The product of σ_τ and σ_Δ is the longitudinal emittance ϵ_L, and thus we have another expression for σ_Δ. Equating these two expressions for σ_Δ, we get a relationship between the longitudinal emittance that the bucket will hold and the bucket and beam parameters. The result is

$$\frac{4}{k^2}\frac{p_0 c}{\omega}\sqrt{\frac{qvp_0c^2}{(mc^2)^2\omega}}\frac{\psi\cos\psi - \sin\psi}{\sqrt{-\sin\psi}}\left[1 + \frac{qvp_0c^2}{(mc^2)^2\omega}(\psi\cos\psi - \sin\psi)\right] = \epsilon_L. \tag{12}$$

Decay Losses

It is well known that for decaying particles at constant velocity, if they travel a distance s, the number of particles N at the end of that distance is related to the number N_0 at the beginning of that distance by
$$N - N_0 e^{-sm/p\tau} \tag{13}$$
where τ is the lifetime if the particles in their rest frame and p is the particles momentum.

When the particles are undergoing constant acceleration parallel to their momentum, a calculation is necessary. Integrating
$$\frac{dN}{ds} = -\frac{m}{p\tau} \tag{14}$$

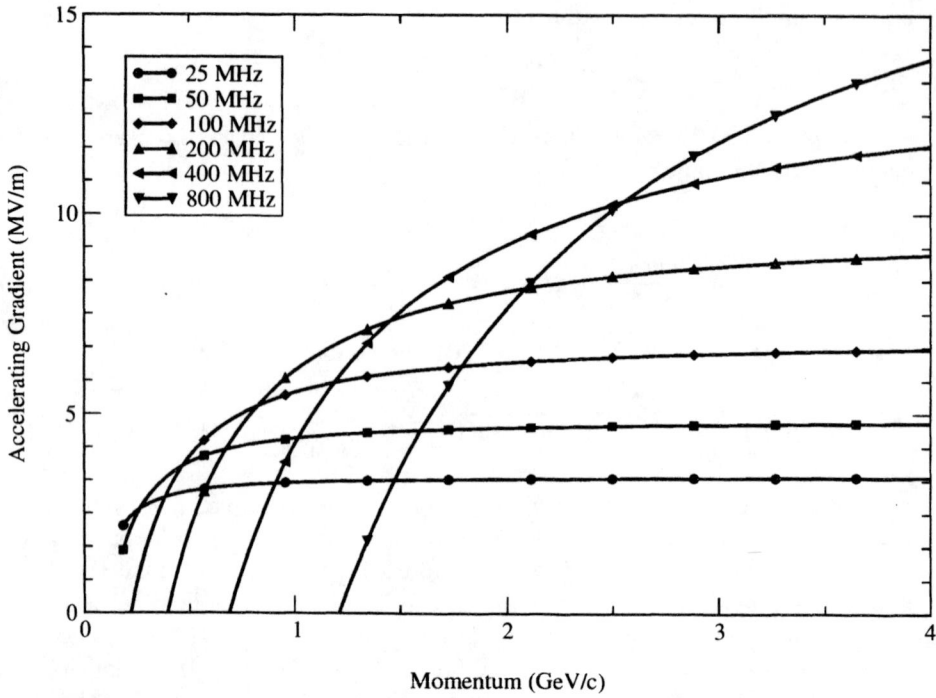

FIGURE 1. Accelerating gradient as a function of momenta for various RF frequencies, using 0.5 TeV/c parameters.

when the energy varies according to

$$\frac{dE}{ds} = qv \cos \psi, \qquad (15)$$

the result is

$$N = N_0 \left(\frac{E + pc}{E_0 + p_0 c} \right)^{-\frac{mc}{\tau q v \cos \psi}}, \qquad (16)$$

where E is the final energy of the particles, p is their final momentum, and the 0 subscripted numbers are the initial quantities.

In principle there may be corrections that come about from the finite energy spread in the distribution, but these will not be treated here.

Linac Designs

To apply this to the design of the initial linac for a muon collider, one can take the longitudinal emittance of the beam, choose a value for k, choose a frequency

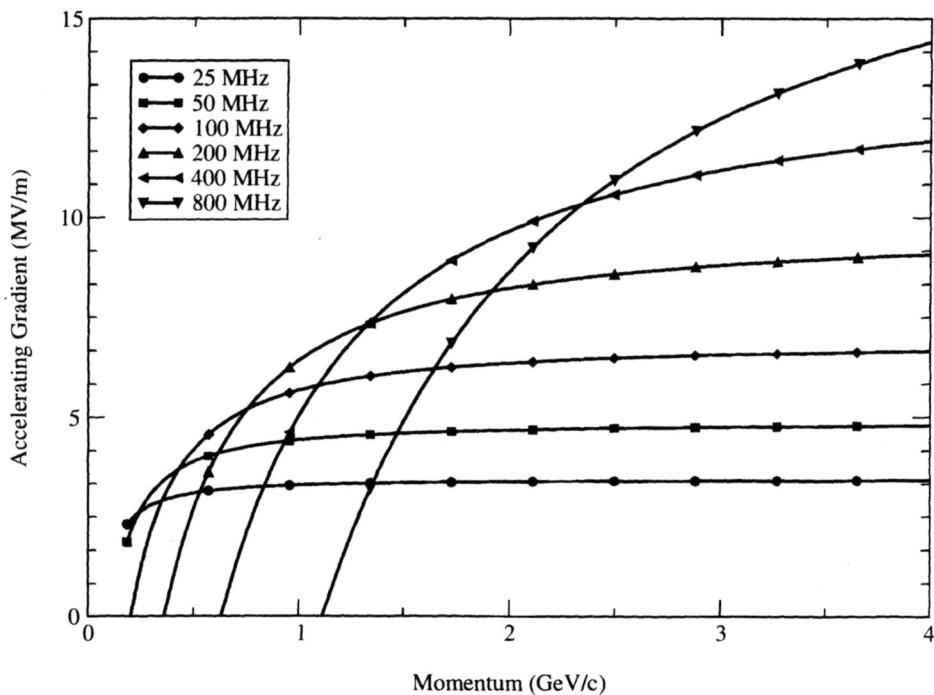

FIGURE 2. Accelerating gradient as a function of momenta for various RF frequencies, using 10 TeV/c parameters.

ω and its corresponding gradient v, and use (12) to solve for ψ as a function of p_0. Since the synchronous phase gives the effective accelerating gradient, one thus has the maximum accelerating gradient one can achieve for a given beam emittance and linac parameters as a function of beam momentum. One can plot this for various RF frequencies, and the results are shown in Figs. 1–3. For those figures, a value of 4 was chosen for k, and a linac filling factor was assumed to be 0.65 (thus, the maximum average accelerating gradient is really 0.65 times the value from (1)).

Using these plots, one can come up with a scheme for accelerating the bunch in a linac. Assuming that the bunch shape adiabatically follows the bucket, and that one varies the phase of the RF along the linac to keep the bucket area constant, then Figs. 1–3 really do show the gradient in the linac as a function of reference momentum. To minimize decay losses, one wants to have the highest gradient possible for a given momentum. Thus, one should switch from one frequency linac to the next when the reference momentum reaches the value where lines for adjacent frequencies cross. One would like to minimize the number of different frequency RF systems used; examining the graph, this suggests that maybe one should choose to jump in frequency by a factor of 4 from one linac to the next.

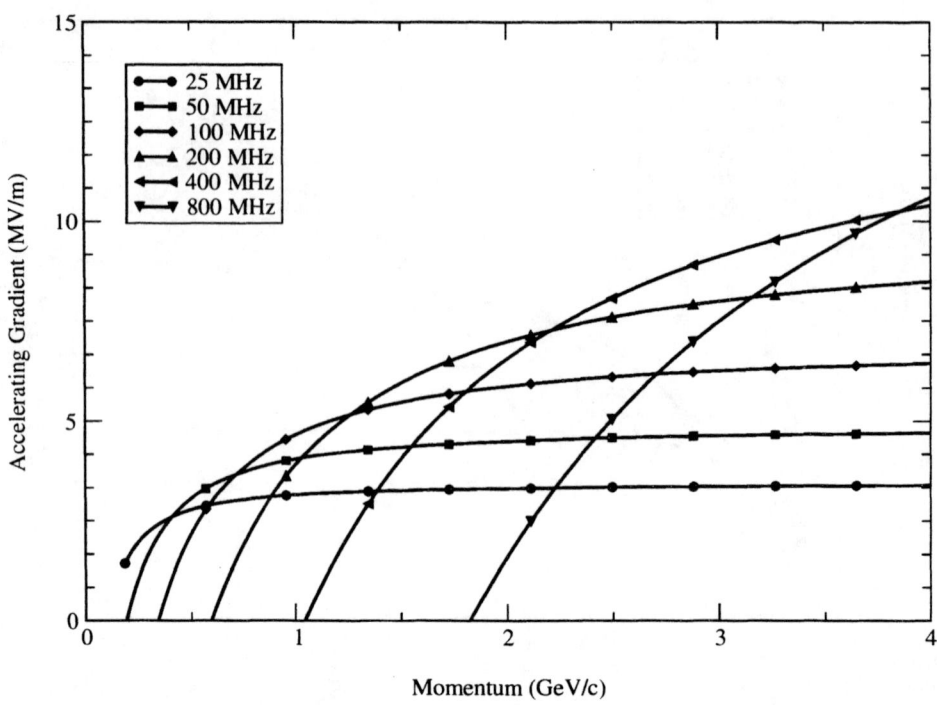

FIGURE 3. Accelerating gradient as a function of momenta for various RF frequencies, using 100 TeV/c parameters.

TABLE 2. Parameters for linacs accelerating to 4 GeV.

p_{min} GeV/c	p_{max} GeV/c	f MHz	L m	$\sigma_{\tau,in}$ ps	$\sigma_{\tau,out}$ ps	$\sigma_{\Delta,in}$ MeV	$\sigma_{\Delta,out}$ MeV	Decay %
\multicolumn{9}{c}{0.5 TeV/c Parameters}
0.186	0.68	50	147	1232	536	19	45	6.5
0.68	2.10	200	211	280	136	86	176	2.8
2.10	4.0	800	166	71	47	336	511	0.9
\multicolumn{9}{c}{10 TeV/c Parameters}
0.186	0.62	50	125	1159	537	18	39	5.7
0.62	1.92	200	194	281	136	75	154	2.8
1.92	4.0	800	177	71	44	294	473	1.0
\multicolumn{9}{c}{100 TeV/c Parameters}
0.186	0.59	25	162	2257	1080	21	44	7.5
0.59	1.82	100	259	565	275	83	171	4.0
1.82	4.0	400	257	144	87	326	542	1.5

Using the graphs, the schemes suggested in Tab. 2 seem optimal.

The choice of a maximum energy for the linacs of 4 GeV is based on the fact that the arcs for a recirculating accelerator are particularly difficult to construct for energies below this, primarily because of the large energy spread and the requirements on the momentum compaction over the energy range that the arc accepts

[3].

An analysis of the results from Tab. 2 suggests the following:

- The larger longitudinal emittances gives require substantially longer linacs, and require those linacs to be at lower frequencies. This will substantially effect the cost of such systems. The systems become more efficient and less costly if the longitudinal emittance is reduced.
- There must be a longitudinal matching section from one linac to the next. This matching section can in principle use the lower frequency linac itself. However, these matching sections are potentially very long, and may require momentum compaction to be generated using some sort of arc, which would be difficult with these energy spreads.

Furthermore, it turns out that due to the fact that the adiabatic approximation is not very good in this case, the bunch does not in fact re-orient itself in phase space according to what was given in Tab. 2. The linear matching issues can be corrected for, but it is important to study the phase space dynamics to determine what the effective "bucket" is in this case. In addition, the asymmetric shape of the bucket causes problems with matching in the tails of the distribution.

Recirculating Accelerators

Once the beam can be reasonably expected to pass through an arc, it becomes more efficient to use a recirculating accelerator to accelerate the beam. A recirculating accelerator consists of two (or more) linacs, connected by one or several arcs. The beam makes several passes through the linacs. This makes more efficient use of the linacs and the RF power, at the cost of more decays and potential complexity in the arc design.

Drift-Kick Map for Linac

In our coordinate system $(\tau, -\Delta)$, we represent the linear map for the arcs (plus the drift behavior in the linacs) as

$$\begin{bmatrix} 1 & -D \\ 0 & 1 \end{bmatrix}, \tag{17}$$

and the linear map for the energy kick from the linac as

$$\begin{bmatrix} 1 & 0 \\ qvL_{\text{lin}}\omega \sin\phi & \end{bmatrix} \tag{18}$$

where L_{lin} is the length of the linac, and ϕ is the RF phase of the reference particle (same convention as previously). D is a parameter which will be discussed and computed later. Thus, the synchrotron tune for the drift-kick pair will be given implicitly by

$$\sin \pi \nu_s = \frac{1}{2}\sqrt{qvL_{\text{lin}}\omega D \sin \phi} \qquad (19)$$

The minimum (at the center of the arc) RMS bunch length (in arrival time units) will be

$$\sqrt{\frac{2 \sin \pi \nu_s}{qvL_{\text{lin}}\omega \sin \phi}}\sqrt{\epsilon_L \cos \pi \nu_s}, \qquad (20)$$

and the maximum length (center of linac) will be

$$\sqrt{\frac{2 \sin \pi \nu_s}{qvL_{\text{lin}}\omega \sin \phi}}\sqrt{\epsilon_L \sec \pi \nu_s}. \qquad (21)$$

Similarly, the minimum RMS energy spread (center of linac) will be

$$\sqrt{\frac{qvL_{\text{lin}}\omega \sin \phi}{2 \sin \pi \nu_s}}\sqrt{\epsilon_L \cos \pi \nu_s}, \qquad (22)$$

and the maximum energy spread (center of arc) will be

$$\sqrt{\frac{qvL_{\text{lin}}\omega \sin \phi}{2 \sin \pi \nu_s}}\sqrt{\epsilon_L \sec \pi \nu_s}. \qquad (23)$$

Hamiltonian Description

To obtain an RF bucket, one must construct an averaged s-independent Hamiltonian which behaves as if the accelerating gradient and the time-of-flight variation are occurring simultaneously instead of sequentially. This Hamiltonian should have the correct linear tune (as computed for the linear transfer map described above), and correctly represent the nonlinearity in the RF. It can only correctly give the matched beam ellipse at one point, since the matched beam ellipse varies with position in the ring, while a time-independent Hamiltonian has the same matched ellipse everywhere. Thus we choose an "averaged" matched ellipse for the Hamiltonian to represent: this is most easily chosen by replacing the $\cos \pi \nu_s$ and $\sec \pi \nu_s$ in (20)–(23) with 1. The resulting Hamiltonian is

$$-\frac{1}{2}\frac{\pi \nu_s}{\sin \pi \nu_s}\frac{D}{L_{\text{tot}}}\Delta^2 + \frac{\pi \nu_s}{\sin \pi \nu_s}\frac{cvL_{\text{lin}}}{\omega L_{\text{tot}}}[\sin(\omega \tau + \phi) - \sin \phi - \omega \tau \cos \phi], \qquad (24)$$

where L_{tot} is L_{lin} plus the length of the arc (which we will call L_{arc}).

Following the same sort of procedure as we did for the linac, we find that the longitudinal emittance accepted by the bucket is

$$\epsilon_L = \frac{2}{k^2}\frac{evL_{\text{lin}}|\sin \phi - \phi \cos \phi|}{\omega \sin \pi \nu_s}. \qquad (25)$$

Linac Contribution to D

There are two contributions to D: one from the arcs, and another from the linacs. The contribution from the arcs is well known and is described by the momentum compaction α_C. The trick is to characterize the contribution from the linacs. It would be convenient if one could lump the contribution from the linacs in with the contribution from the arcs. This is certainly feasible: if a Hamiltonian can be written as $H_\Delta(\Delta) + H_\tau(\tau)$, a well-known technique in symplectic integration to get a second-order accurate map is to integrate H_Δ for half a length step, followed by H_τ for a full length step, followed by H_Δ for a half step [4].

Thus, we can use the linac Hamiltonian (3), and integrate only the part depending on Δ. Using (4) and taking v and ϕ to be constant, we find that after linearizing in Δ,

$$\tau_1 - \tau_0 = \frac{1}{qvc\cos\phi}\left(\frac{1}{\beta_1} - \frac{1}{\beta_0}\right)\Delta, \tag{26}$$

where the subscript 0 refers to the beginning of the integration, and the subscript 1 refers to the end. β is the speed of the reference particle divided by c. There will thus be two contributions from the linacs to D: one from the linac before the arc, where the initial condition in the above integration will be the center of that linac and the final condition will be the end of the linac. Added to that will be a second contribution from the linac after the arc, where the initial condition is the beginning of the linac, and the final condition is the center of that linac. The net result is that there will be a contribution to D which is

$$\frac{1}{qvc\cos\phi}\left(\frac{1}{\beta_2} - \frac{1}{\beta_0}\right) \tag{27}$$

where the subscript 0 refers to the center of the linac before the arc in question, and the subscript 2 refers to the center of the linac that follows the arc.

Arc Parameters

Given longitudinal design parameters, we now have what we need to specify some basic arc parameters. From the previous discussions, D can be written as

$$D = \frac{4\sin^2\pi\nu_s}{qvL_{\text{lin}}\omega\sin\phi}. \tag{28}$$

But from its basic definition, the definition of α_C, and the above discussion, it can also be written as

$$\frac{L_{\text{arc}}}{\beta_1^2 p_1 c^2}\left(\alpha_C - \frac{1}{\gamma_1^2}\right) + \frac{1}{qvc\cos\phi}\left(\frac{1}{\beta_2} - \frac{1}{\beta_0}\right). \tag{29}$$

Here the subscript 1 refers to the value in the arc itself, and $\gamma_1 = 1/\sqrt{1-\beta_1^2}$. As a result, we have an expression for α_C in terms of longitudinal design parameters:

$$\alpha_C = \frac{1}{\gamma_1^2} + \frac{\beta_1^2 p_1 c^2}{L_{\text{arc}}} \left[\frac{4\sin^2 \pi \nu_s}{qvL_{\text{lin}}\omega \sin\phi} + \frac{1}{qvc\cos\phi}\left(\frac{1}{\beta_0} - \frac{1}{\beta_2}\right) \right]. \quad (30)$$

The vertical RMS beam size is given by

$$\sigma_y = \sqrt{\frac{\beta_y \epsilon_y}{\beta\gamma}}, \quad (31)$$

where β_y is the vertical beta-function and ϵ_y is the normalized vertical emittance. Similarly

$$\sigma_x = \sqrt{\frac{\beta_x \epsilon_x}{\beta\gamma} + \left(\frac{D_x \sigma_\Delta}{\beta pc}\right)^2}, \quad (32)$$

where D_x is the horizontal dispersion function. Generally, to compute these values, a lattice needs to be laid out. But one can get lower bounds by assuming a constant focusing and bending channel, which would give

$$\beta_x = \beta_y = \rho\sqrt{\alpha_C} \qquad\qquad D_x = \rho\alpha_C, \quad (33)$$

where ρ is the radius of curvature of the arc.

Supplying RF Power

In this paper, we will assume that power is supplied to the linacs in the recirculator in such a way as to precisely replace the energy removed from the linac by the beam. Such a scheme has the advantage that the longitudinal phase space for the beam can remain matched irrespective of the beam current, assuming that sufficient power is available to make this scheme work for the highest expected beam current. Other schemes, such as one where the stored energy in the linacs is allowed to droop, potentially require that the arcs have different momentum compactions depending on the current in the beam to achieve longitudinal matching, potentially making it difficult to run at a current other than the maximum design current.

This paper will make some simple assumptions about how RF power is supplied: there are assumed to be no losses, either through the walls or into loads which are put in for "matching" purposes. The RF simply stores energy into the linacs, and that energy is extracted by the beam. This will necessarily produce the most optimistic values for peak power requirements and efficiencies. More realistic scenarios should be computed at some point.

Peak power requirements are simple to compute: the beam extracts a certain amount of energy, and that energy must be resupplied by time the beam comes around again. One must take into account the fact that there are actually two beams.

Efficiency is important for high energy machines, and this can be computed as follows: the energy supplied to a beam of N particles over n turns through a linac is $nNqv\cos\phi$ per unit length. The energy stored in the linac initially was $v^2/\omega(r_s/Q)$ per unit length, and we supplied $Nqv\cos\phi$ per unit length $(n-1)$ more times, so the total energy supplied is $v^2/\omega(r_s/Q)+(n-1)Nqv\cos\phi$. The maximum possible efficiency is therefore

$$\varepsilon = \frac{nNqv\cos\phi}{(n-1)Nqv\cos\phi + \dfrac{v^2}{\omega(r_s/Q)}}. \tag{34}$$

The real efficiency will of course be less than this, due to wall losses and loads, plus efficiencies of the devices supplying the RF power.

Recirculator Designs

If we specify

- The synchrotron tune
- The total energy gain in the recirculator
- The gradient and frequency of the RF
- The number of turns in the recirculator
- The quantity k
- The longitudinal emittance

the above description tells us how to compute the phase at which we should run the RF. While the synchrotron tune may seem like an odd quantity to specify, it in fact makes sense to do so. A high synchrotron tune is advantageous for several reasons:

- It gives a smaller energy spread in the beam (important for simplifying arc design)
- It can minimize collective instabilities
- It can prevent degradation of polarization [5]

However, there is a maximum value for the synchrotron tune, which is about 0.15 per linac-arc pair. The reason for this is that the motion is in fact described by a s-dependent Hamiltonian, and not the s-independent Hamiltonian (24). The bucket computed for that Hamiltonian is only correct in the limit of small synchrotron tune for the linac-arc pair. For a larger synchrotron tune, the edges of the bucket will degrade until the bucket completely disappears at a synchrotron tune of 0.5.

The source of this degradation is the nonlinearity from the RF. This suggests that we try a synchrotron tune of 0.15 per linac-arc pair. For a racetrack design, this corresponds to a one-turn synchrotron tune of 0.3. The racetrack design seems most efficient in terms of avoid the overhead necessary at the entrance and exit of each linac, but in principle a design with more sides would allow even larger synchrotron tunes.

Using these constraints, we can come up with designs for the recirculators. One can imagine that a collider will be built up by upgrading the machines over time, essentially adding recirculating stages. The various machines might have single beam energies of 70 GeV, 500 GeV, 10 TeV, and 100 TeV. Thus, the recirculator stages will have maximum energies at these points.

Because any time spent in arcs is essentially lost (and gives excess decays), one does not want to create a recirculator which is unnecessarily long. If one makes a recirculator which works from 4 to 70 GeV, for instance, the arcs at 4 GeV will be nearly as long as the arcs at 70 GeV, and a substantial number of excess decays will occur. Thus, it is important to create even more recirculator stages. Around a factor of 4 in energy per recirculator seems like a good compromise between decay losses and excess hardware. Thus, a good set of cutoff energies for the recirculators are starting at 4 GeV, then 17 GeV, 70 GeV, 190 GeV, 500 GeV, 2.2 TeV, 10 TeV, 32 TeV, and finally up to 100 TeV.

The question now becomes how to choose the appropriate RF frequency and number of turns for the recirculators. For low energy recirculators, the length of the recirculator is so short that a kicker to switch from one arc to another would be at best very difficult. Therefore, the low energy recirculators tend to have their number of turns limited by the requirement that the energy jump should be greater than a few (8 is the choice made here) times the RMS energy spread in the beam. This ceases to be an issue in higher energy recirculators.

These issues can be avoided completely if one goes to an Fixed Field Alternating Gradient (FFAG) type of scheme, where a single arc is used for all passes [3]. There are many problems with this type of arc:

- Making the bunch arrive at the right phase of the RF for each pass.
- Achieving a decent dynamic aperture and avoid emittance blowup.
- Creating the complex and often large magnets that are required.
- Creating the required momentum compaction as a function of energy.

The first problem, making sure the particles arrive at the correct RF phase, is probably the most difficult of these problems. Possible solutions are to simply supply the required RF power (which may be prohibitively large), to add some ferrite or similar material which can cause the resonant frequency of the cavity to be changed (which may give significant problems with losses and heating, particularly in a superconducting environment), or to use other schemes to vary the resonant frequency of the cavity.

The next issue becomes the energy spread in the beam, which will turn out to be very large. It turns out that in the low energy recirculators, the energy spread is so large as to require FFAG-like arcs even when individual arcs are used for each pass [3]. These arcs won't have many of the problems of a single-arc design, but are still very complex. It is clearly advantageous to reduce these energy spreads if at all possible, if for no other reason than to reduce likely emittance blowup caused by the energy spread. This requirement will tend to push you toward lower frequency RF and more turns in the recirculator.

However, when one tries to design the arcs, one runs into another problem: the momentum compactions required can easily become too large. Once the momentum compaction becomes above around 0.03 or so, the arcs get very difficult to design [3]. Lowering the momentum compaction tends to push you toward higher frequency RF and fewer turns in the recirculator.

Higher frequency RF tends to reduce decays (higher gradient), tends to be more efficient (less stored energy), is easier to create power for, and is in general less expensive, but has higher wakefields (which can be a significant problem considering the high beam currents under consideration here). Going to more turns will give more decays, requires more arcs in a multiple arc design and therefore is more expensive, but will generally be more efficient both in terms of average power and in terms of linac usage.

Table 3 contains values for recirculator parameters for these schemes. The arcs are assumed to have 2 T average bending fields. In reality, the arcs may have higher average bending fields in cases where the relative energy spreads are lower; it would be nice to take this into account somehow in the computations, but it is unclear how to do so. This is particularly important at higher energies where the arcs get prohibitively long with 2 T average bend fields (the 2 T average field is kept nonetheless for comparison purposes only). L_{arc} is the length of 180° of arc, and L_{linac} is the length of one of the two linac in the recirculator. P_{peak} is the power that must be supplied to replace the energy extracted by the beam at the same rate the beam is extracting it.

These values were arrived at by various compromises. The values for the lower energy recirculators are often forced. Going to lower frequencies requires momentum compactions that are too high. Going to higher frequencies gives energy spreads which are so large that they don't even allow multiple passes. Generally the number of turns is chosen to be the maximum allowable for passive switching between arcs. Note the large relative energy spreads in these recirculators.

For the higher energy recirculators, there are more choices to be made, and this is reflected in putting multiple lines in the table for a given energy range. The highest frequency given is generally the maximum frequency possible, and the number of turns is the maximum for that frequency. The relative energy spreads are generally decreasing as we go up in energy, but it might be nice to further decrease the relative energy spread so as to make the arcs easier.

TABLE 3. Parameters for recirculators.

p_{min} GeV/c	p_{max} GeV/c	f MHz	n	L_{arc} m	L_{linac} m	σ_τ ps	σ_E MeV	P_{peak} MW	Decay %	ε %	$\alpha_{C,max}$ 10^{-3}	σ_x mm	σ_y mm
\multicolumn{14}{c}{0.5 TeV/c Parameters}
4	17	400	5	89	127	91	296	1156	3.7	13	14.88	9.6	1.5
17	70	800	11	367	169	47	579	864	4.8	48	4.04	4.3	1.0
70	190	800	12	996	312	37	720	734	4.0	54	2.63	3.4	0.9
70	190	1600	13	996	231	24	1143	723	4.1	79	1.05	2.2	0.7
190	500	1600	16	2620	430	19	1426	610	4.7	84	0.68	1.7	0.6
190	500	3200	17	2620	323	12	2278	595	4.8	97	0.26	1.1	0.4
\multicolumn{14}{c}{10 TeV/c Parameters}
4	17	400	6	89	107	92	255	797	4.0	11	17.7	9.5	1.3
17	70	800	13	367	144	47	501	575	5.4	45	4.73	4.3	1.0
70	190	1600	15	996	200	24	997	482	4.6	75	1.21	2.1	0.6
190	500	3200	19	2620	288	12	2006	404	5.3	94	0.30	1.1	0.4
500	2200	3200	19	11527	1277	7	3381	503	6.5	96	0.10	0.7	0.4
2200	10,000	3200	19	52396	5475	4	5540	511	6.5	97	0.04	0.5	0.3
\multicolumn{14}{c}{100 TeV/c Parameters}
4	17	200	5	89	179	181	291	186	4.6	1	29.93	17.6	0.7
17	70	400	11	367	238	93	570	153	5.4	6	8.15	8.2	0.5
17	70	800	5	367	364	45	1172	279	3.0	8	1.85	4.2	0.4
70	190	800	13	996	325	47	1126	134	4.4	17	2.11	4.1	0.3
70	190	1600	6	996	490	23	2305	259	2.3	22	0.49	2.0	0.2
190	500	1600	17	2620	455	23	2244	114	5.0	44	0.53	2.1	0.2
190	500	3200	8	2620	676	12	4568	226	2.5	54	0.13	1.0	0.2
500	2200	3200	23	11527	1163	10	5539	112	7.7	79	0.09	0.8	0.2
500	2200	6400	23	11527	920	6	9011	114	7.6	93	0.03	0.5	0.1
2200	10,000	6400	23	52396	3465	4	14352	117	7.6	95	0.01	0.3	0.1
10,000	32,000	6400	23	167668	9314	3	20033	104	6.6	95	0.01	0.1	0.1
32,000	100,000	6400	23	523961	27585	2	28993	103	6.5	95	<0.01	0.2	0.1

Thus, for the 0.5 TeV/c parameters, a lower frequency solution is given when possible, with the number of turns chosen to have about as many decays as in the higher frequency case. Note that the linac tends to get longer; this is probably the primary cost of going to the lower frequency. There is also a slight decrease in efficiency, which is significant but not terribly so. You are trading off the cost of the linac (including its associated power) with the complexity of the arcs by changing the frequency of the linacs.

The first four energy ranges for the 0.5 TeV and the 10 TeV parameters are similar except for a small (less than 15%) change in the longitudinal emittance (the transverse emittance also decreases somewhat). Note that substantial change that can potentially occur in the linac length, energy spread, and peak power requirement as a result. Thus, particularly for the lower energy recirculators, a reduced longitudinal emittance can be of significant advantage in cost savings.

For the 10 TeV parameters, it turns out that an RF frequency of 6.4 GHz is not workable since beam loading becomes too high with 3×10^{12} particles per bunch (Eq. (34) gives a value greater than 1). Thus, for higher energies we continue to use 3.2 GHz RF. In reality, one can use an almost arbitrary number of turns at the higher energies (above 500 GeV), since the relative energy spreads are relatively low (allowing passive switching), and in any case active switching is probably possible since the rings are much longer. 19 turns was chosen arbitrarily, basically to be equal to the number of turns in the 190 GeV/c to 500 GeV/c recirculator. More turns requires more hardware in a multiple arc system. The decay losses start to rise in the higher energy machines due to the increasing length of the linac. The relative small α_C suggests that it may be advantageous to increase the number of sides in the recirculator, allowing a greater synchrotron tune in the recirculator (0.15 per side) and a correspondingly smaller energy spread. The bunch lengths that come out of the calculation are also extremely short, and that is another indicator that it would be helpful to increase the synchrotron tune in this fashion. Many-sided designs will be considered in future work.

For the 100 TeV/c parameters, the longitudinal emittance has increased substantially, and so lower frequencies are often required than are required in the other cases. However, for some of the lower energies, we have provided a higher frequency solution that has a larger energy spread. The relative energy spread is still smaller than it was for the previous stage, so the arcs would be no worse than the arcs in the earlier stages. However, it would be nice to take advantage of the lower energy spreads to construct simpler arcs.

For the 0.5 TeV/c to 2.2 TeV/c recirculator, the 6.4 GHz RF scenario is limited to 23 turns for passive switching, but the maximum number of turns for the 3.2 GHz RF scenario is much higher. Assuming we want to limit the number of turns to limit decays and arc complexity, I have chosen to use a maximum of 23 turns here and for subsequent recirculators. Also, it turns out that going above 6.4 GHz RF leads to beam loading problems like those for the 10 TeV/c case, and thus we will limit ourselves to this frequency.

Arc Design

Clearly one of the greatest challenges lies in the design of the arcs for these recirculating accelerators. The arcs for the low energy systems must accept rather large relative energy spreads, and this is the primary challenge. Arc designs for these large energy acceptances are being considered by several people, including Al Garren, Carol Johnstone, Dejan Trbojevic, Weishi Wan. In addition, these same people are studying single-arc designs where the entire energy range of the recirculator passes through a single arc. It may even end up making sense to have a small number of arcs, where the beam passes through each arc a few times. One of the greatest difficulties in these designs is meeting the requirements on momentum compaction that come from longitudinal considerations, and many (but not all) of these designs have yet to address this issue.

It would be particularly useful to get some kind of rough parameterization of how the dipole packing fraction and other parameters behave with respect to design constraints put on the arcs.

Don Summers has come up with an idea for a different geometry for the recirculating accelerator, which has been called a "dogbone geometry." The idea is to have a single linac through which the particles pass, and have arcs on the end of that linac which return the particles into the same linac. The advantage of this scheme is that lower energy particles can go through a shorter arc than the high energy particles, since the length of the arc is not determined by the distance between linacs. Such a scheme can allow one to reduce decays and/or reduce linac costs in the recirculator. There are issues related to supplying RF power due to the asymmetric way in which the bunches would pass through such a system, and such a system cannot be expanded to many sides for the high energy recirculators, but is certainly an attractive possibility for the low energy recirculators.

At higher energies, a scheme with ramping magnets has been considered for the arcs [2]. While superconducting magnets cannot be ramped fast enough, it is possible that normal conducting magnets could be. So a hybrid scheme is used consisting of interleaved fixed-field superconducting magnets and pulsed normal-conducting magnets. Such a scheme has yet to be examined carefully, in particular the nature of the orbits has yet to be considered. It will have similar difficulties to FFAG systems due to large orbit swings, but in principle it should be better since there is an extra degree of control in the ability to ramp some of the magnets.

Arc designs will not be discussed in much more detail here. Their design is progressing, and the status and other issues with their design will be reported on in the future.

Other Issues

It is possible to use isochronous designs for the recirculators instead of these designs with a finite synchrotron tune. The isochronous designs lack many of the advantages of the non-isochronous designs: energy spreads will be larger, collective instabilities are more difficult to control, and there may be difficulties preserving polarization. In addition, an isochronous design will necessarily increase the longitudinal beam emittance, which is already problematically large (this effect is particularly significant at the lower energies). Isochronous designs do have the advantage that they eliminate one of the primary difficulties associated with a single-arc recirculator, the issue of making the beam arrive at the correct phase of the RF. Isochronous design principles will not be discussed here, but will be treated in a subsequent paper.

Wakefields and their effects have yet to be computed for these kinds of systems, but it is clear that they will be a significant effect, due to the large beam current. At the higher energies, it is important to come up with a scheme which has longer bunch lengths, since the extremely short bunch lengths given in Tab. 3 will create substantial wakefield effects. Going to many-sided designs should be considered for these high energies, not only because of the short bunch lengths but also because the synchrotron tune will be larger and therefore will more readily be able to correct the strong wakefield effects that will occur.

CONCLUSION

We have described a method for designing the acceleration systems for a muon collider. The design method is based primarily on considering the longitudinal phase space dynamics of the bunch. The muons are initially accelerated in a linac which most likely contain several frequencies of RF, and then are accelerated by a series of recirculating accelerators. We have laid out what the parameters might look like for various future high-energy muon colliders, and discussed some of the tradeoffs involved.

The relatively large longitudinal emittances in a muon collider create significant difficulties at lower energies. It requires low-frequency RF in the linac that initially accelerates the muons after the cooling stage, and also requires large energy acceptance arcs in the early stages of recirculation.

Much work remains to be done. The studies of the arcs in the recirculators need to be continued and expanded. The effect of wakefields needs to be considered. Linac designs need to be looked at to determine achievable parameters as well as wakefields. Multi-sided recirculator designs should be looked at, particularly for high energies. Finally, much more work needs to be put into optimization of these designs for cost and performance.

We have laid out a set of parameters for a high-energy muon collider which does not appear to be unrealistic. With these parameters as a starting point, it is possible to attempt to estimate the cost of the acceleration for a high-energy muon collider, and from that determine what direction one should go to reduce those costs.

ACKNOWLEDGEMENTS

This research has been supported in part by the U.S. Department of Energy under contract no. DE-AC02-98CH10886.

REFERENCES

[1] Bruce King, parameters supplied for this conference.
[2] Charles M. Ankenbrandt et al. (Muon Collider Collaboration), *Phys. Rev. ST Accel. Beams* **2**, 081001 (1999).
[3] Carol Johnstone, private communication.
[4] Ronald D. Ruth, *IEEE Trans. Nucl. Sci.* **NS-30**, 2669–2771 (1983).
[5] Alain Blondel, talk given at the Neutrino Factory and Muon Collider Collaboration Meeting, Berkeley, 13–15 December 1999.

Muon Collider Physics at Very High Energies

M. S. Berger

Physics Department, Indiana University, Bloomington, IN 47405

Abstract. Muon colliders might greatly extend the energy frontier of collider physics. One can contemplate circular colliders with center-of-mass energies in excess of 10 TeV. Some physics issues that might be relevant at such a machine are discussed.

INTRODUCTION

The large mass of the muon compared to that of the electron results in a large suppression of bremstrahlung radiation. Consequently it is possible to consider building circular colliders with energies in the multi-TeV regime [1]. Muon colliders have been proposed as Higgs factories and more recently as neutrino factories, but the long-term goal of muon colliders should be to extend the energy frontier. It is not clear at the present time whether advances in accelerator technology will result in electron-positron machines achieving energies of several TeV. In this workshop first attempts were made to explore the feasibility of muon colliders with energies of at least 10 TeV.

It is hard to know what kind of physics might present itself in the 10-100 TeV mass range. After all, physicists have been arguing for a long time about the physics that will manifest itself at the Large Hadron Collider (LHC). The LHC, linear electron-positron colliders, and perhaps muon colliders should give us some clue as to what to expect at the following generation of machines. It is easy to imagine scenarios where a new collider might be necessary, but it is impossible to motivate a specific energy at this time. We can only speculate as to what physics might appear at the LHC or future linear colliders.

LUMINOSITY REQUIREMENTS

The figure of merit for physics searches at a muon collider is the QED cross section $\mu^+\mu^- \to e^+e^-$, which has the value

$$\sigma_{QED} = \frac{100 \text{ fb}}{s \text{ (TeV}^2)} \tag{1}$$

To arrive at a simple estimate of the integrated luminosity needed to study new physics, we assume

$$\left(\int \mathcal{L} dt\right) \sigma_{QED} \gtrsim 1000 \text{ events} \qquad (2)$$

Then the luminosity requirement for this number of events to be accumulated in one year's running is

$$\mathcal{L} \gtrsim 10^{33} \cdot s \text{ (cm)}^{-2} \text{(sec)}^{-1}$$

For the colliders with the center-of-mass energies considered at this meeting:

- $\sqrt{s} \simeq 10$ TeV, requiring

$$\int \mathcal{L} dt \gtrsim 1 \text{ (fb)}^{-1}, \quad \mathcal{L} \gtrsim 10^{35} \text{ (cm)}^{-2} \text{(sec)}^{-1}$$

- $\sqrt{s} \simeq 100$ TeV, requiring

$$\int \mathcal{L} dt \gtrsim 100 \text{ (fb)}^{-1}, \quad \mathcal{L} \gtrsim 10^{37} \text{ (cm)}^{-2} \text{(sec)}^{-1}$$

These luminosities are extremely high, of course, and it is not clear if experiments can be performed in such an environment.

ELECTROWEAK SYMMETRY BREAKING

A 10 TeV muon collider might be very useful for exploring the physics responsible for electroweak symmetry breaking. If Higgs bosons with $m_H < \mathcal{O}(800)$ GeV do not exist then interactions of longitudinally polarized weak bosons (W_L, Z_L) become strong and can be probed by studying vector boson scattering as shown in the figure. Therefore, new physics must be present at the TeV energy scale. While one can study strong $W_L W_L$ scattering at the LHC, linear colliders, or $\mu^+\mu^-$ colliders with a few TeV center-of-mass energy, it might become necessary to go to higher energies to fully explore the multitude of resonances. Indeed we are still studying the analogous spectrum of QCD today.

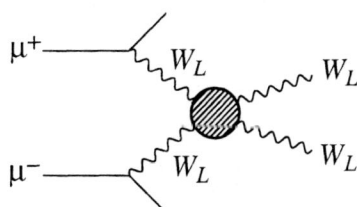

FERMION MASS GENERATION

The mechanism responsible for fermion masses and the mechanism breaking the electroweak symmetry are the same in the Standard Model. A Higgs scalar acquires a vacuum expectation value giving rise to massive gauge bosons and (through Yukawa couplings) masses for the fermions. However, it need not be the case that these mechanisms are the same, and technicolor models were the most prominent examples of theories where the fermion masses arise from a different sector from that responsible for the electroweak symmetry breaking. Hence one should keep an open mind about the origin of fermion masses. Very general constraints one can place on the physics of fermion mass generation are unitarity bounds. The relevant bound for fermions scattering into longitudinally polarized vector boson V_L,

$$f\bar{f} \to V_L V_L \, , \tag{3}$$

is the Appelquist-Chanowitz bound [2] which states that unitarity is violated at the scale

$$\Lambda_f < \frac{8\pi v^2}{\sqrt{3N_c} m_f} \, , \tag{4}$$

where $v = (\sqrt{2} G_F)^{-1/2}$ is the electroweak vev and N_c is the number of colors of the fermion. In the Standard Model this unitarity violation is cured by the inclusion of the s-channel Higgs exchange diagram. The strongest bound comes for the heaviest fermion the top quark for which $\Lambda_t \approx 3$ TeV, indicating that some new physics must occur below this scale.

For a muon one gets $\Lambda_\mu \approx 8,000$ TeV. So if the physics responsible for the muon mass saturates this bound, it is beyond the reach even of a 10-100 TeV muon collider. But one does not really expect that the bound is saturated, but rather that the fermion masses are all generated at a common scale with some masses suppressed by some approximate flavor symmetries. In light of the lower value of Λ_t, one might expect a 10 TeV collider to provide important insight into fermion mass generation if Nature is not so kind to provide a elementary scalar particle. In the typical case one expects the resonances to be broad. In some scenarios [3], one can have strongly interacting Higgs sectors with narrow resonances for which a small energy spread might be helpful.

One can also study the unitarity violation in the subprocess $V_L V_L \to t\bar{t}$, analogous to the case discussed in the previous section for electroweak symmetry breaking. This process could also be sensitive to new physics responsible for the fermion masses, and one would measure the cross sections for $\mu^+ \mu^- \to \nu \bar{\nu} t\bar{t}$ and $\mu^+ \mu^- \to \mu^+ \mu^- t\bar{t}$, and in scenarios where the unitarity is saturated, one might need the energy reach of a very high energy muon collider to probe these strong interactions.

GAUGE BOSONS

A favorite target for new physics is the possibility of new gauge bosons beyond those found in the Standard Model. One might first reveal the existence of these particles via radiative return [4] whereby a vector boson with mass less than the center-of-mass energy is produced in association with an energetic photon. Alternatively one could pinpoint the mass of the vector boson by doing precision measurements of the couplings and asymmetries at energies below the vector boson mass. In either case, one would ultimately want to build a collider with an energy equal to the mass of the vector boson and take advantage of the resonance cross section. An important consideration then is the beam energy spread of the muon collider. The width of the vector boson should scale linearly with its mass. The expectations for a 10 TeV collider is that the energy spread σ_E/E should be something like $10^{-4} - 10^{-3}$ [5], so the spread should be much smaller than the resonance peak in the typical case.

SUPERSYMMETRY

It is possible that the LHC and linear colliders will uncover only part of the supersymmetric (SUSY) spectrum. In fact the lightest two generations of squarks and sleptons might appear at the multi-TeV scale. The absence of certain supersymmetric partners being produced below the TeV energy scale would certainly compel us to go to higher energies.

Beyond the discovery of all the superpartners to the Standard Model particles, another possible role for a very high energy muon collider would be to uncover an entirely new sector responsible for the dynamical breaking of supersymmetry. In gravitationally mediated SUSY breaking, the dynamical sector is hidden and couples only via gravitational couplings to the supersymmetric Standard Model particles. However other scenarios of SUSY breaking are possible, and these can be directly probed with sufficiently energetic collisions. In gauge mediated SUSY breaking scenarios, for example, there is just such another sector (known as the messenger sector) occurs at a scale beyond that which can be probed at the LHC. This messenger sector might perhaps be accessible at a very high energy muon collider. The LHC might indirectly provide clues about the source of SUSY breaking by measuring the spectrum of superpartners and perhaps seeing radiative decays in the case of gauge mediated SUSY breaking. In fact by measuring the location of displaced vertices (relative to the interaction point) from the radiative decay of the next-lightest supersymmetric particle one can put a constraint on the scale of the gauge mediation sector as first suggested in a Very Large Hadron Collider study [6].

COMPTON BACKSCATTERING

It seems at first peculiar to consider backscattering photons off of a muon beam. After all, the reason to employ muon beams rather than the electron beams is to decrease electromagnetic radiation. Eventually however, even for muons, bremsstrahlung radiation would again become a problem at sufficiently high energies in a circular collider. At the energies contemplated here, one can reconsidering employing Compton backscattering to produce photon beams of comparable energies. Kinematics dictates that the highest energy of a backscattered photon that can be obtained is given by

$$\omega_{\max} = \frac{x}{1+x} E_{\text{beam}}, \tag{5}$$

where

$$x = \frac{4 E_{\text{beam}} \omega_{\text{laser}}}{m_\mu^2}. \tag{6}$$

Assuming an incident laser with energy 1.17 eV[1], one obtains maximum backscattered photon energies (shown in the figure) which are still much smaller than the incident muon beam energy. A more energetic photon source would be needed to fully realize the backscattered photon option even at the extremely high muon energies considered here.

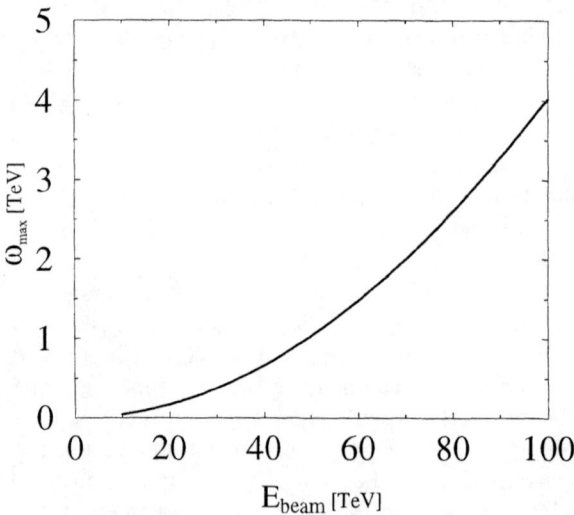

---

[1] For definiteness we take a neodinium glass laser with $\omega_{\text{laser}} = 1.17$ eV which is often considered for Compton scattering at a linear e^+e^- collider. In any case, one expects the laser energy to be in the few eV range.

CONCLUSIONS

It is difficult to motivate a very high energy muon collider without information that will be gleaned after years of operation of the LHC and linear colliders. However, if the past history of particle physics has taught us anything it is that the most important progress has occurred by going to higher and higher energies. It will be interesting in the coming years to learn whether multi-TeV muon colliders are realistic and economical.

ACKNOWLEDGEMENT

Work supported in part by the U.S. Department of Energy under Grant No. DE-FG02-95ER40661.

REFERENCES

1. C. M. Ankenbrandt, et al., *Phys. Rev. ST Accel. Beams* **2**, 081001 (1999), physics/9901022; CERN Report 99-02, eds. B. Autin, A. Blondel and J. Ellis.
2. T. Appelquist and M. Chanowitz, *Phys. Rev. Lett.* **59**, 2405 (1987).
3. P. C. Bhat and E. Eichten, hep-ph/9803468.
4. V. Barger, M. S. Berger, J. F. Gunion and T. Han, *Nucl. Phys. B, Proc. Suppl.* **51A** 13, (1996), hep-ph/9604334; hep-ph/9704290.
5. B. King, these proceedings.
6. G. Anderson, et al., FERMILAB-CONF-97-318-T, hep-ph/9710254.

Fringe Field Effects in Muon Rings

Martin Berz, Kyoko Makino and Béla Erdélyi

Abstract. Because of the predominance of large emittances, muon storage rings have a tendency to being rather sensitive to nonlinear effects. In this paper we study the effects of the nonlinearities due to the lattice elements' fringe fields, which have a fundamentally different behavior from normal multipole terms. It is found that for given scenarios for lattices and emittances, the fringe field effects have dramatic influences on the dynamics and stability of particles and hence require careful study of correction options.

FRINGE FIELD EFFECTS IN BEAM DYNAMICS

The nonlinearities due to fringe fields are a well-known phenomenon in the field of high resolution particle spectrographs [1,2]. In large hadron storage rings, their effects frequently are negligible, but they have a tendency to become noticeable in rings of smaller radius, particularly at larger emittances. In order to understand the fringe effects, we expand the r and ϕ dependencies of the scalar potential in Taylor and Fourier series and have

$$V = \sum_{k=0}^{\infty} \sum_{l=0}^{\infty} M_{k,l}(s) \cos\left(l\phi + \theta_{k,l}\right) r^k. \tag{1}$$

In cylindrical coordinates, the Laplace equation has the form

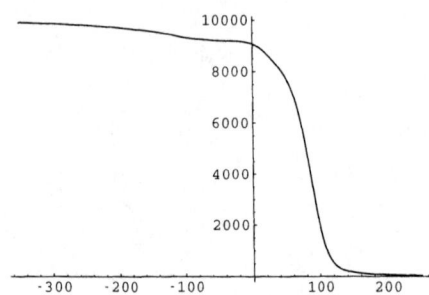

FIGURE 1. The fringe field of an LHC quadrupole.

$$\Delta V = \frac{1}{r}\frac{\partial}{\partial r}\left(\frac{r\partial V}{\partial r}\right) + \frac{1}{r^2}\frac{\partial^2 V}{\partial \phi^2} + \frac{\partial^2 V}{\partial s^2} = 0; \qquad (2)$$

inserting the Fourier–Taylor expansion of the potential, we obtain [3]

$$\Delta V = \sum_{k,l=0}^{\infty}\left\{M_{k,l}(s)\cos(l\phi+\theta_{k,l})\left(k^2-l^2\right) + M''_{k-2,l}(s)\cos(l\phi+\theta_{k-2,l})\right\}r^{k-2}, \quad (3)$$

where the convention has been used that all coefficients $M_{k,l}$ vanish for negative indices. In case $M_{k,l}$ is constant, we obtain that $k = l$, and hence the radial and angular dependencies are coupled in the well-known way. However, in case $M_{k,l}$

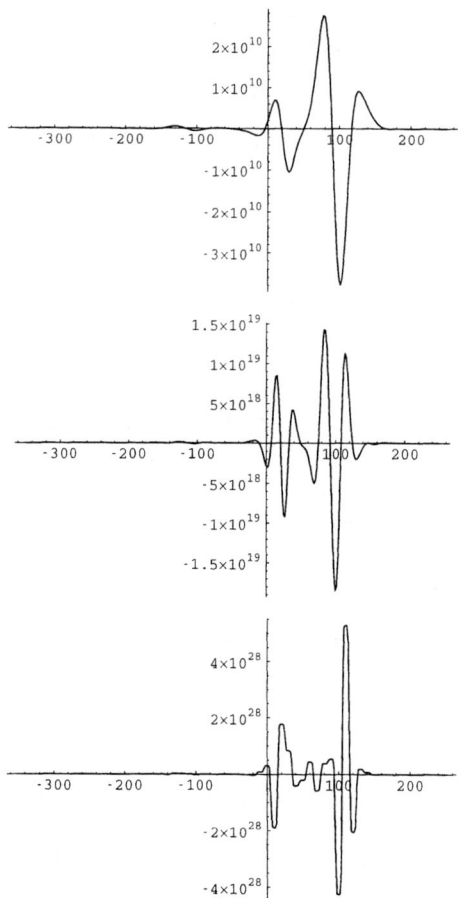

FIGURE 2. The s-derivatives of the quadrupole strength of orders 4, 8, and 12 at the fringing region of an LHC quadrupole.

is not constant, besides the terms $M_{l,l}$, there are also the terms $M_{l+2,l}$, $M_{l+4,l}$, ... which introduce higher-order radial dependencies to a given angular dependence. Specifically, we have

$$M_{l+2n,l}(s) = \frac{M_{l,l}^{(2n)}(s)}{\prod_{\nu=1}^{n}\left((l)^2 - (l+2\nu)^2\right)}, \quad \text{and} \quad \theta_{l,l} = \theta_{l+2,l}, \theta_{l+4,l}, \ldots \qquad (4)$$

where $M_{l,l}^{(2n)}$ is the $2n$th derivative of $M_{l,l}$. In practice this entails that fringe field effects become more and more relevant the more the particles are away from the axis of the element, which of course is directly connected to the emittance of the beam.

The additional non-multipole nonlinearities of the fringe fields couple to higher derivatives of the multipole strength. These derivatives can assume rather striking forms; as an example we show the value of the quadrupole strength as well its derivatives of orders 4, 8,and 12 for an LHC quadrupole from the final focus section in Figure 1 and Figure 2.

The fringe field effects can be particularly easily studied in the map picture using differential algebraic methods [3]. Their complete treatment to any order is possible in the code COSY INFINITY [4–6].

FRINGE FIELD EFFECTS IN MUON STORAGE RINGS

In the following we present observations related to fringe field effects. Similar to the situation in the study of kinematic correction [7], we limit ourselves to the mere observation of the effects, without attempting to devise strategies for their correction through nonlinear elements, which of course should also include the influence of all other relevant nonlinear effects.

The consequences of the fringe field effects influence all orders of the motion, beginning from the linear behavior. The linear effects even affect the tune, and it turned out that for the lattices in question, the effects are significant. However, since the comparison of nonlinear motion is most relevant for the situation where the tunes are identical, in practice a re-fitting of the linear layout back to the original tune is appropriate. To simplify this procedure, COSY has a mode that automatically adds linear correction elements such that the linear map of the system with fringe fields is the same as without fringe fields. This mode was used throughout for the subsequent studies.

Furthermore, the detailed shape of the fringe field fall off influences the details of the nonlinear behavior, yet information of the fall off is not known until the actual mechanical design of the elements. Thus we assumed a default fall off that describes the situation in many multipoles with benign end field design [6].

As a first example, we study the effects on a 30 GeV neutrino factory ring and a 30 GeV Higgs factory ring kindly supplied by Carol Johnstone [8]. The neutrino

TABLE 1. The beginning part of the Taylor transfer map of a 30 GeV neutrino factory ring without fringe field effects.

Expansion coefficients of x,a,y,b depending on the exponents of xayb				
(x,	(a,	(y,	(b,	xayb
-0.1936744	0.1416905	0	0	1000
-6.792904	-0.1936744	0	0	0100
0	0	0.1961760	-0.2456274E-01	0010
0	0	39.14526	0.1961760	0001
0.9450713E-01	0.3330788E-02	0	0	2000
0.1377180	-0.5612056E-01	0	0	1100
-0.5801038	-0.6286416	0	0	0200
0	0	0.1229122	-0.3076403E-03	1010
0	0	0.6994628	0.2591721E-02	0110
0	0	-0.4902813	-0.1229122	1001
0	0	4.130383	-0.6994626	0101
-0.8336143E-02	-0.1539706E-02	0	0	0020
0.2386714	0.1206977E-01	0	0	0011
13.28519	2.453805	0	0	0002
11.55775	-0.1950424	0	0	3000
-81.95874	4.704898	0	0	2100
392.7585	-40.95482	0	0	1200
-1667.934	610.1026	0	0	0300
0	0	-0.7719322	-0.4832862E-01	2010
0	0	14.24336	0.4605147	1110
0	0	-417.3649	-2.944554	0210
0	0	-6.277132	-8.007910	2001
0	0	354.6553	34.10640	1101
0	0	-782.7162	-69.64528	0201
0.2344129	-0.2585728E-01	0	0	1020
-3.593587	1.490467	0	0	0120
26.50950	-1.496885	0	0	1011
-116.8393	6.467636	0	0	0111
2245.991	-29.01786	0	0	1002
-4789.016	244.2371	0	0	0102
0	0	-1.130680	-0.7756062E-02	0030
0	0	-6.403961	-0.4518258	0021
0	0	-430.3112	-29.87579	0012
0	0	-1116.550	-1705.361	0003
-1.322453	-0.9516536E-01	0	0	4000
....	....	....	....	....

TABLE 2. The beginning part of the Taylor transfer map of a 30 GeV neutrino factory ring with fringe field effects.

Expansion coefficients of x,a,y,b depending on the exponents of xayb				
(x,	(a,	(y,	(b,	xayb
-0.1936744	0.1416905	0	0	1000
-6.792904	-0.1936744	0	0	0100
0	0	0.1961760	-0.2456274E-01	0010
0	0	39.14526	0.1961760	0001
0.9450668E-01	0.3330792E-02	0	0	2000
0.1377195	-0.5612046E-01	0	0	1100
-0.5800990	-0.6286441	0	0	0200
0	0	0.9402828	0.2251288E-01	1010
0	0	-11.70894	-0.1896604	0110
0	0	16.57293	1.476914	1001
0	0	-139.6185	-8.654864	0101
-0.1248983	0.2034706E-01	0	0	0020
-5.726400	0.5915340	0	0	0011
-215.4763	16.77799	0	0	0002
478.5009	-5.828819	0	0	3000
-2963.906	119.4893	0	0	2100
11370.25	-1164.119	0	0	1200
-57663.58	24785.65	0	0	0300
0	0	-44.02904	-7.722130	2010
0	0	1084.027	32.25581	1110
0	0	-61208.70	-414.5974	0210
0	0	-2410.660	-1200.200	2001
0	0	9632.604	4243.860	1101
0	0	-128922.6	-5760.175	0201
20.86897	-1.837551	0	0	1020
-418.5652	221.3400	0	0	0120
4203.500	-36.84456	0	0	1011
-8816.418	1001.507	0	0	0111
331876.4	-2632.727	0	0	1002
-596393.2	12424.07	0	0	0102
0	0	-94.33547	-0.6567247	0030
0	0	-681.4684	-25.23468	0021
0	0	-13103.20	-2736.609	0012
0	0	-292594.6	-141126.2	0003
-56.81891	-3.180838	0	0	4000
....	....	....	....	....

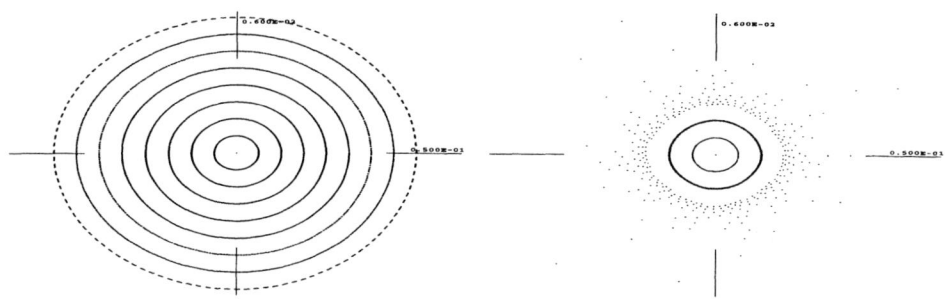

FIGURE 3. Tracking pictures for 1000 turns in a 30 GeV neutrino factory ring without (left) and with (right) fringe field effects in the same scale of 50mm×6mrad. In the absence of fringe field effects, particles seem to survive up to 100mm×15mrad with kinematic correction. With fringe field effects, only those up to 10mm×1.5mrad survive.

factory ring consists of bending elements and quadrupoles, and the Higgs factory ring furthermore has sextupoles. The full gap size of the magnets is assumed to be 10cm for the both rings.

The kinematic correction is included by default [7] in the following results. Tables 1 and 2 show the beginning part of high order Taylor transfer maps of the neutrino factory ring. Table 1 shows the case without fringe field effects, and Table 2 the case with fringe field effects. Because of the re-fitting of the linear behavior, effects can be observed from second order; overall, the nonlinearities are significantly larger than before.

The size of the dynamic aperture was estimated by tracking particles for 1000 turns with transfer maps up to 7th order. In case of no fringe field effects, particles

TABLE 3. Amplitude dependent tune shifts.

Fringe field effects		off	on	Order	Exponents x	y
30 GeV neutrino factory ring	x motion	0.718979	0.718979	0	0	0
		12.0363	469.397	2	2	0
		5.32077	741.941	2	0	2
	y motion	0.218573	0.218573	0	0	0
		5.32077	741.941	2	2	0
		5.92760	472.150	2	0	2
30 GeV Higgs factory ring	x motion	0.864288	0.864288	0	0	0
		35.8432	855.939	2	2	0
		42.0333	2226.02	2	0	2
	y motion	0.665356	0.665356	0	0	0
		42.0333	2226.02	2	2	0
		422.116	4180.00	2	0	2

with x up to 1000mm and a up to 150mrad are stable without kinematic correction and those with x up to 100mm and a up to 15mrad are stable with kinematic correction in the neutrino factory ring [7]. However when the fringe field effects are included, the particles with x up to 10mm and a up to 1.5mrad survive. Figure 3 illustrates the stability of particles in case the fringe fields are on (right side) as compared to off (left side).

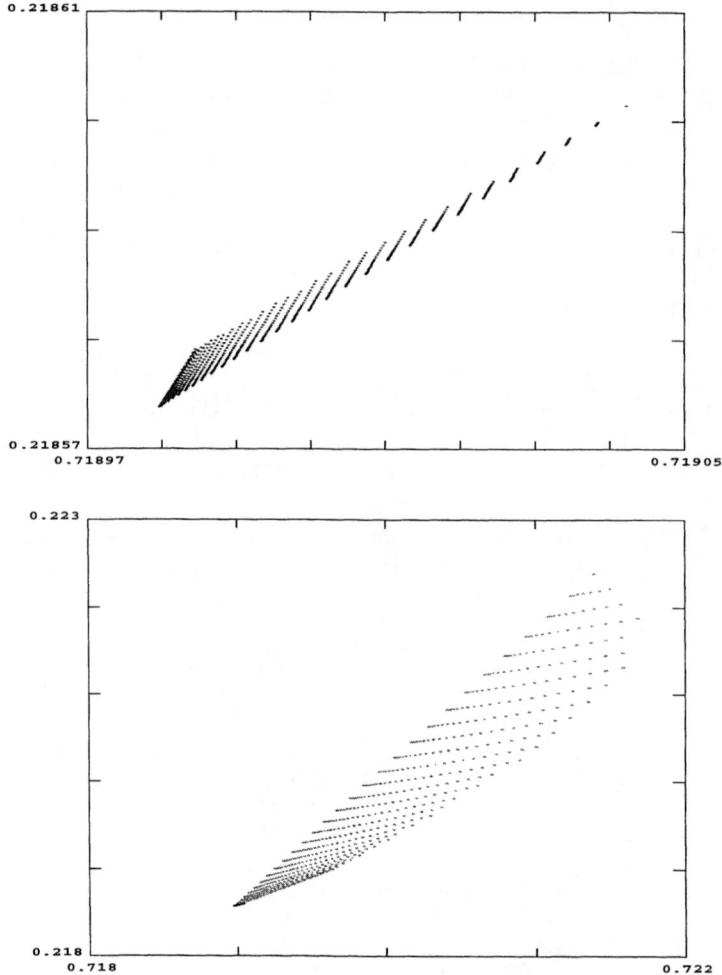

FIGURE 4. Tune footprints of a 30 GeV neutrino factory ring without (upper) and with fringe field effects. The lower picture shows both in the same scale, where the one without fringe field effects is the tiny bar at the lower left.

Based on normal form methods, amplitude dependent tune shifts were computed using COSY INFINITY [9,10]. The results with and without fringe field effects are summarized in Table 3. Figure 4 shows tune footprints of the neutrino factory ring for particles up to 6mm in radius in both x and y directions. The horizontal axis shows the x tune and the vertical shows the y tune. The upper picture shows the tune footprint without fringe field effects, and the lower shows the situation with fringe field effects. The one without fringe field effects occupies a rather small region, and it is shown as a tiny bar at the lower left in the lower picture for comparison.

Altogether, for the design of the neutrino factory ring used in the study, the dynamic aperture decreased by a factor of around 100 in x-a, and there is a large increase by a factor of around 10000 in the tune footprint area.

The 30 GeV Higgs factory ring was studied in the similar way. The tracking pictures in Figure 5 can be used to estimate the size of the dynamic aperture, and a decrease by a factor of 100 in x-a is observed due to fringe field effects. The amplitude dependent tune shifts are listed in Table 3. Figure 6 shows the tune footprints for particles up to 0.5mm in radius in both x and y directions. The upper left picture shows the situation without fringe field effects, and the upper right picture contains fringe field effects. To illustrate the difference of the size more clearly, the lower picture shows the two footprints in the same scale; the small triangle at the lower left is the tune footprint without fringe field effects. An increase of the tune footprint area by a factor of about 400 is observed.

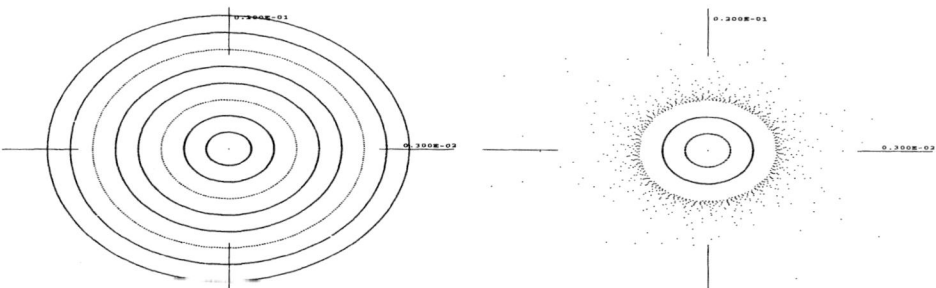

FIGURE 5. Tracking pictures for 1000 turns in a 30 GeV Higgs factory ring without (left) and with (right) fringe field effects in the same scale of 3mm×20mrad. In the absence of fringe field effects, particles seem to survive up to 6mm×50mrad with kinematic correction. With fringe field effects, only those up to 0.6mm×5mrad survive.

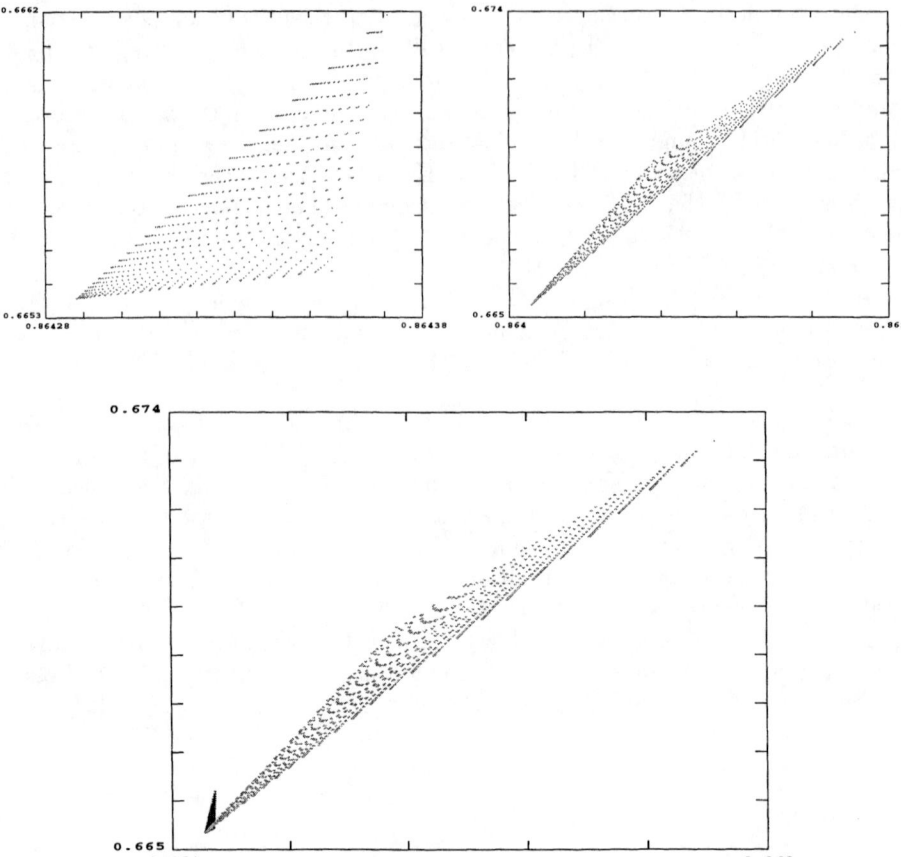

FIGURE 6. Tune footprints of a 30 GeV Higgs factory ring without (upper left) and with (upper right) fringe field effects. The lower picture shows both in the same scale, where the one without fringe field effects is the small triangle at the lower left.

ACKNOWLEDGMENTS

We thank C. Johnstone and W. Wan for providing the storage ring designs for the study. The work was supported by the US Department of Energy and the Alfred P. Sloan Foundation.

REFERENCES

1. N. Anantaraman and B. Sherrill, Editors. Proceedings of the international conference on heavy ion research with magnetic spectrographs. Technical Report MSUCL-685, National Superconducting Cyclotron Laboratory, 1989.
2. M. Berz, K. Joh, J. A. Nolen, B. M. Sherrill, and A. F. Zeller. Reconstructive correction of aberrations in nuclear particle spectrographs. *Physical Review C*, 47,2:537, 1993.
3. M. Berz. *Modern Map Methods in Particle Beam Physics*, volume 108 of *Advances in Imaging and Electron Physics*. Academic Press, San Diego, CA, 1999.
4. M. Berz et al. The COSY INFINITY web page. http://cosy.nscl.msu.edu.
5. K. Makino and M. Berz. COSY INFINITY version 8. *Nuclear Instruments and Methods*, A427:338, 1999.
6. M. Berz. COSY INFINITY Version 8 reference manual. Technical Report MSUCL-1088, National Superconducting Cyclotron Laboratory, Michigan State University, East Lansing, MI 48824, 1997.
7. K. Makino and M. Berz. Effects of kinematic correction on the dynamics in muon rings. In these proceedings.
8. Carol Johnstone. Private communication.
9. M. Berz. *High-Order Computation and Normal Form Analysis of Repetitive Systems, in: M. Month (Ed), Physics of Particle Accelerators*, volume AIP 249, page 456. American Institute of Physics, 1991.
10. M. Berz. Differential algebraic formulation of normal form theory. In *M. Berz, S. Martin and K. Ziegler (Eds.), Proc. Nonlinear Effects in Accelerators*, page 77. IOP Publishing, 1992.

Physics in 2006

S. Dawson

*Physics Department[1]
Brookhaven National Laboratory
Upton, N.Y. 11973*

Abstract. Any consideration of future physics facilities must be made in the context of the Tevatron and the LHC. I discuss some examples of physics results which could emerge from these machines and the resulting questions which would remain for a high energy e^+e^- collider. Particular attention is paid to the electroweak symmetry breaking sector. If a light Higgs boson exists, it will be observed at the LHC and the role of any later accelerator will be to map out the Higg's boson mass and couplings and then determine the space of possible models. If there is no light Higgs boson then some effects of a strongly interacting electroweak symmetry breaking sector will be observed at the LHC and I discuss the role of a high energy linear collider in exploring this scenario.

INTRODUCTION

Physicists are presently faced with a quandry. Plans and designs for the next generation of accelerators need to be formulated in the near future, since the construction of these machines spans many years. The "best" machine to build for the post-LHC era will only become clear, however, when we see what surprises the LHC holds. Faced with our imperfect knowledge, we must examine possible scenarios for LHC physics and determine the machine most likely to answer the physics questions remaining in the years following completion of the LHC and the fulfillment of its physics promise.

Here, I will consider the physics of electroweak symmetry breaking at a high energy e^+e^- collider.[2] This discussion must be made in the context of potential discoveries at the Tevatron and the LHC. I begin by reviewing the current experimental status of electroweak symmetry breaking, both from direct Higgs boson searches at LEP2 and from precision electroweak measurements. Theoretical

---

[1] Supported by the U.S. Department of Energy under Contract No. DE-AC02-76CH00016.
[2] For values of the Higgs boson mass near and slightly above 100 GeV, a $\mu^+\mu^-$ collider operating at the Higgs resonance can make extremely precise measurements of the Higgs mass and couplings. I will not discuss the physics potential of a muon collider here. [1]

expectations for the Higgs boson mass are then reviewed, with emphasis on the implications for physics at higher mass scales.

Next, I review the discovery prospects for the Higgs boson at both the Tevatron and the LHC. The working hypothesis is that a weakly interacting Higgs boson will be discovered, if it exists, at the Tevatron or the LHC, and so the role of the next generation of accelerators will be to study the properties of a Higgs boson. Precision measurements of the mass, decay widths, and production rates will all be necessary in order to verify that a particle is the Higgs boson of the Standard Model.

Aside from the couplings to fermions and gauge bosons, we would also like to know that the Higgs boson self-interactions result from the spontaneously broken scalar potential of the Standard Model. In order to do this, the three- and four-point self-couplings of the Higgs boson must be measured. These couplings can only be probed by multi-Higgs production, which has extremely small rates, both at the LHC and at a high energy e^+e^- collider.

The focus in this note is on verifying the properties of the Standard Model Higgs boson. In order to do this, it is helpful to compare with the predictions of a supersymmetric model since these predictions may be quite different from those of the Standard Model. Distinguishing between the Standard Model and a supersymmetric model is an important test of our understanding of the electroweak sector.

If a Higgs boson is not found at the Tevatron or the LHC, the electroweak symmetry breaking sector must be strongly interacting. I end with a brief discussion of strong electroweak symmetry breaking and a view towards the future.

INFERENCES FROM THE STANDARD MODEL

The Standard Model of electroweak interactions has been verified to the .1% level through precision measurements at LEP and SLD. [2] In fact, the mechanism of electroweak symmetry breaking remains the only unconfirmed area of the Standard Model. The Standard Model predicts the existence of a physical scalar particle, termed the Higgs boson. The search for this particle is therefore a fundamental goal of all current and future accelerators since its discovery is needed to complete our knowledge of the electroweak sector. The mass is a free parameter of the theory and so the Higgs boson must be systematically sought in all mass regions.

The couplings of the scalar Higgs boson, however, are completely specified in terms of the Higgs vacuum expectation value, $v = 246\ GeV$. Hence branching ratios and production rates can be computed unambiguously in terms of the mass. Measurements of ratios of branching rates can then be used to test the validity of the model.

Since the Higgs boson contributes to electroweak radiative corrections at one loop, precision measurements from LEP and SLD can be used to infer a prefered value for the Higgs mass. The contribution of the Higgs boson to electroweak observables is logarithmic and so the limit on the Higgs mass is not nearly as precise as the indirect limit on the top quark mass from precision measurements. The current 95% confidence level limit is, [2]

$$M_h < 230\ GeV, \quad \text{Precision Measurements.} \tag{1}$$

It is important to understand that this limit assumes the validity of the Standard Model. Quantum loops containing new particles can change this limit, as can new operators beyond those of the Standard Model. If there is new physics at the TeV scale, the limit of Eq. 1 can be evaded. [3]

The Higgs boson mass is the only free parameter of the electroweak theory. Although we cannot compute its mass, there are certain theoretical restrictions following from the consistency of the theory. The scalar potential for an $SU(2)$ scalar doublet Φ is,

$$V = -\mu^2 \mid \Phi \mid^2 + \lambda (\mid \Phi \mid^2)^2 \quad . \tag{2}$$

After the electroweak symmetry breaking has occured, there remains the physical scalar Higgs boson h. The quartic coupling, λ, is related to the Higgs boson mass,

$$\lambda = \frac{M_h^2}{2v^2} \quad . \tag{3}$$

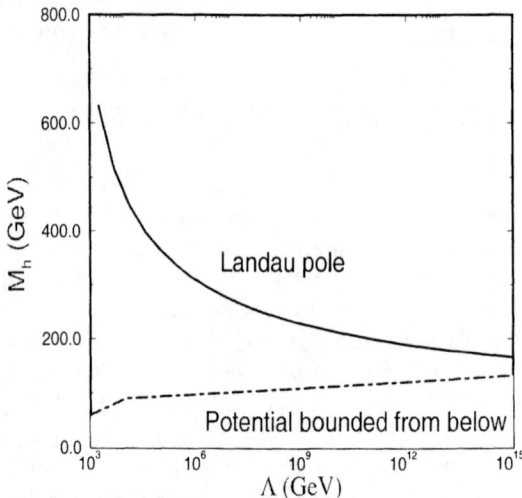

FIGURE 1. Theoretical expectations for a Standard Model Higgs boson, as a function of the scale Λ above which the Standard Model is no longer valid. The region above the solid curve has $\lambda(\Lambda) \to \infty$, while the region below the dotted line has $\lambda(\Lambda) < 0$. The allowed region is the region between the curves.

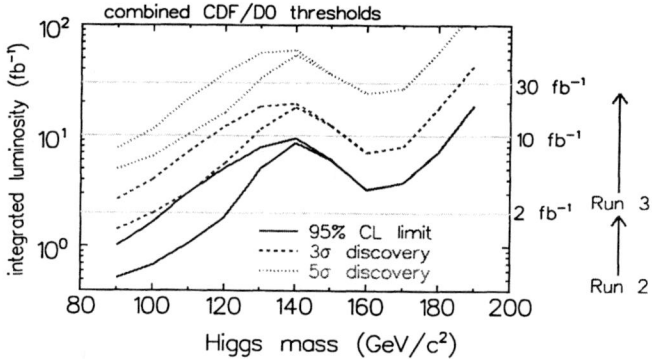

FIGURE 2. Discovery potential for a Standard Model Higgs boson at the Tevatron. The vertical axis shows the required luminosity to observe or exclude a given Higgs boson mass. At the lower values of the Higgs mass two curves are shown; the lower in each set is a neural net analysis, the upper a standard analysis with cuts on the signal and background. From Ref. 11.

Now λ is not a fixed parameter, but scales with the relevent energy, Q, and so Eq. 2 is the potential at the electroweak scale. If λ is large, (corresponding to a heavy Higgs boson), then at a scale Q, [4]

$$Q\frac{d\lambda}{dQ} = \frac{3}{4\pi^2}\lambda, \qquad (4)$$

which can be solved to obtain,

$$\frac{1}{\lambda(\Lambda)} = \frac{1}{\lambda(M_h)} - \frac{3}{4\pi^2}\log(\frac{\Lambda^2}{M_h^2}) \quad . \qquad (5)$$

A sensible theory will have $\lambda(\Lambda)$ finite at all scales, ($\lambda(\Lambda) \to \infty$ is termed the Landau Pole), or correspondingly $\frac{1}{\lambda(\Lambda)} > 0$. This yields an upper bound on λ and hence on M_h^2,

$$M_h^2 < \frac{8\pi^2 v^2}{3\log(\Lambda^2/M_h^2)} \quad . \qquad (6)$$

If the Standard Model is valid to the GUT scale, $\Lambda \sim 10^{16}\ GeV$, then we have an approximate upper bound on the Higgs mass, [4,5]

$$M_h < 170 \text{ GeV} \tag{7}$$

For any given value of Λ, there is a corresponding upper bound on M_h. Λ is often termed the "scale of new physics" since above this scale, the Standard Model interactions are not valid. This bound is the upper curve on Figure 1.

There is also a theoretical lower bound on M_h. If λ is small (light M_h), then

$$Q\frac{d\lambda}{dQ} \sim \frac{1}{16\pi^2}(B - 12g_t^4), \tag{8}$$

where g_t is the Higgs-top quark Yukawa coupling, $g_t = -M_t/v$, and B is a function of the gauge coupling constants, $B = \frac{3}{16}(2g^4 + (g^2 + g'^2)^2)$. We see that the large top quark mass tends to drive λ negative. In order for electroweak symmetry breaking to occur, the potential must remain bounded from below, and λ positive. Solving Eq. 8,

$$\lambda(\Lambda) = \lambda(M_h) + \frac{B - 12g_t^4}{16\pi^2}\log\left(\frac{\Lambda}{M_h}\right) . \tag{9}$$

Requiring $\lambda(\Lambda) > 0$ gives the lower bound on M_h,

$$\frac{M_h^2}{2v^2} > \frac{B - 12g_t^4}{16\pi^2}\log\left(\frac{\Lambda}{M_h}\right) . \tag{10}$$

For large M_t, this relation changes sign and the two-loop renormalization group corrections are important to obtain the numerical bound. Requiring that the Standard Model be valid to the GUT scale, $\Lambda = 10^{16}$ GeV, gives the restriction on the Higgs boson mass [6]

$$M_h > 130 \text{ GeV} \tag{11}$$

This is shown as the lower curve in Figure 1. There have been many theoretical improvements to the naive bounds presented above, but the bottom line is the same: If the Standard Model is valid to the GUT scale, then

$$130 \text{ GeV} < M_h < 180 \text{ GeV}, \qquad \Lambda \sim 10^{16} \text{ GeV}. \tag{12}$$

A Higgs boson outside this mass region would be a signal for new physics at the corresponding scale Λ. The mass region of Eq. 12 is particularly interesting since it could potentially be probed at the Tevatron with an upgraded luminosity.

There are also absolute bounds on the Higgs boson mass which are independent of the scale, Λ. Unitarity of the WW elastic scattering amplitudes requires $M_h < 800$ GeV, while lattice calculations obtain a similar bound, $M_h < 700$ GeV. [7] All of the theoretical bounds of this section predict a Higgs boson comfortably within the discovery range of the LHC and so if the Standard Model is correct, a Higgs boson discovery should be just around the corner.

PROSPECTS FOR DISCOVERY

The current 95% confidence level limit on the Higgs boson mass from direct searches at LEP2 using data from $\sqrt{s} = 189 - 202~GeV$ is [8]

$$M_h > 106~GeV, \quad LEP2. \tag{13}$$

This limit is not expected to improve substantially with further running at LEP2.

The minimal supersymmetric model has two neutral Higgs bosons, h^{SUSY} and H^{SUSY}, a charged Higgs, $H^{\pm}$, and a pseudoscalar, A. The structure of the supersymmetric potential dictates that at lowest order all the couplings can be expressed in terms of two parameters, which are typically taken to be the pseudoscalar mass, M_A, and the ratio of Higgs vacuum expectation values, $\tan\beta$. All masses can then be expressed in terms of these two parameters. [9]

The experimental limit on the Higgs boson mass in a supersymmetric theory typically depends on $\tan\beta$. If we require that the limit be valid for all $\tan\beta$, there is a slightly lower 95% confidence level limit than for the Standard Model Higgs boson, [8]

$$M_h^{SUSY} > 90~GeV \quad LEP2. \tag{14}$$

The minimal supersymmetric theory has the remarkable feature that there is an upper bound on the lightest Higgs boson resulting from the structure of the scalar potential. This bound is roughly

$$M_h^{SUSY} < 110 - 130~GeV, \tag{15}$$

where the exact value depends on assumptions about the parameters of the theory. [10] This is tantalizingly close to the experimental limit of Eq. 14. We see that there is no overlap between the expected mass of the lightest Higgs boson of a supersymmetric model and the Standard Model Higgs boson when $\Lambda \sim M_{GUT}$. Hence an observation of the Higgs boson with even an imprecise value for its mass will help to distinguish between the Standard Model and its minimal supersymmetric extention.

A Standard Model Higgs boson should be discovered at the Tevatron or the LHC. Due to the small rate, the Higgs boson will be extraordinarily difficult to observe at the Tevatron. The signal with the best signature is associated production with a $W^{\pm}$. For $M_h \sim 120~GeV$, the cross section at $\sqrt{s} = 2~TeV$ is $\sigma(p\bar{p} \to W^{\pm}h) \sim .3~pb$. Even with $10~fb^{-1}$, the 5σ discovery level is only $M_h \sim 100~GeV$, below the current LEP2 limit. This underscores the need for the highest possible luminosity.

Figure 2 illustrates the discovery potential for a Standard Model Higgs boson at the Tevatron. [11] For $M_h < 140~GeV$, the dominant signal results from $p\bar{p} \to Wh, h \to b\bar{b}$, while at higher Higgs masses, the decay $h \to WW^*$ becomes the most important. The discovery reach plot combines small signals from many different channels. In fact, the maximum $S/\sqrt{B}$ in any channel is .9 for $\mathcal{L} = 1~fb^{-1}$. A

TABLE 1. Indirect Measurements of M_h

Collider	ΔM_W	ΔM_t	$\frac{\delta M_H}{M_H}$
LEPII, TeV	30 MeV	4 GeV	57 %
LHC	15 MeV	2 GeV	26 %
500 GeV e^+e^-	15 MeV	200 MeV	17 %

Standard Model Higgs discovery at the Tevatron will almost certainly require the full $25 - 30$ fb^{-1} of upgraded luminosity.

The LHC, on the other hand, should discover a Standard Model Higgs boson in any mass region below 1 TeV, as illustrated in Figure 3, even with only 30 fb^{-1}. [12] From $M_h \sim 120$ GeV all the way up to $M_h \sim 700$ GeV, the Higgs boson can be observed through the decay $h \to ZZ \to 4l$. The discovery reach can be extended up to $M_h \sim 1$ TeV through the channels $h \to ZZ \to l^+l^-\nu\bar{\nu}$ and $h \to W^+W^- \to l\nu$ jet jet. With the full luminosity of 100 fb^{-1}, the LHC will see a Higgs signal in multiple channels for all possible masses. The observation in multiple channels will allow preliminary measurements of the Higgs coupling constants, as discussed in the next section.

The Standard Model points to a Higgs boson in the $100 - 200$ GeV mass range, while its minimal supersymmetric extention suggests that the lightest Higgs boson is just above the current experimental limit. In either case, such a light Higgs boson would be kinematically accessible through the process $e^+e^- \to hZ$ at an e^+e^- collider with $\sqrt{s} \sim 350 - 500$ GeV. The rates for Higgs production at an e^+e^- collider are shown in Fig. 4. For an e^+e^- collider with $\sqrt{s} \sim 500$ GeV, the dominant production mechanism is $e^+e^- \to Zh$ for $M_h \sim 200$ GeV. At higher energy, say $\sqrt{s} \sim 1$ TeV, the largest rate is from $e^+e^- \to \nu\bar{\nu}h$ In the next section, we examine the capabilities and the required luminosities for linear colliders to measure the Higgs properties and contrast these potential future measurements with what we will know from the LHC.

PRECISION MEASUREMENTS OF MASS, COUPLINGS, AND BRANCHING RATIOS

Higgs Mass Measurements

There are two complementary approaches to measuring the Higgs boson mass. The first is through the direct observation of the Higgs boson. For most values of M_h, with an integrated luminosity of $\int \mathcal{L} = 300$ fb^{-1}, the LHC will measure $\frac{\delta M_h}{M_h} \sim 10^{-3}$, as shown in Fig. 5. Even at $M_h \sim 800$ GeV, the expected precision is $\frac{\delta M_h}{M_h} \sim 10^{-2}$.

At a high energy e^+e^- collider, the cross section for $e^+e^- \to Zh$ is a sensitive function of the Higgs boson mass and we could hope to obtain an extremely precise measurement of the mass. By measuring the rate as a function of $\sqrt{s}$, a measurement of order [13]

$$\delta M_h \sim 60 \ MeV \sqrt{\frac{\mathcal{L}}{100 \ fb^{-1}}} \qquad (16)$$

could be obtained for a Higgs boson in the $100 \ GeV$ region. An alternate method is to measure the recoil spectrum in the process $e^+e^- \to Zh \to he^+e^-, h\mu^+\mu^-$. This would yield a precision of,

$$\delta M_h \sim 300 \ MeV \sqrt{\frac{\mathcal{L}}{100 \ fb^{-1}}} \ , \qquad (17)$$

again for a Higgs boson in the $100 \ GeV$ region. With $1000 \ fb^{-1}$ the precision on δM_h for a light Higgs boson at an e^+e^- collider could be considerably better than at the LHC, using either the excitation spectrum or the recoil spectrum of the $e^+e^- \to Zh$ process.

Precise measurements of M_W and M_t at future colliders will allow a value of M_h to be inferred [13], as shown in Table 1. (The e^+e^- numbers in this table assume $\int \mathcal{L} = 1000 \ fb^{-1}$.) Since the Higgs boson contributes only logarithmically to electroweak observables, the precision is significantly less than the direct measurement. Consistency between the direct and the indirect measurements will provide an important check of the theory at the quantum level, however.

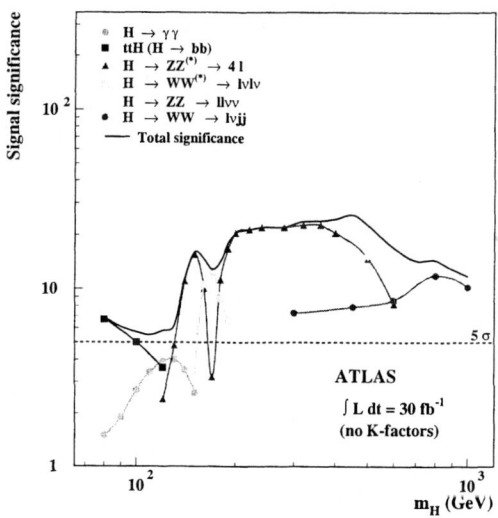

FIGURE 3. Discovery potential for a Standard Model Higgs boson at the LHC, using the ATLAS detector, with $30 \ fb^{-1}$. From Ref. 12.

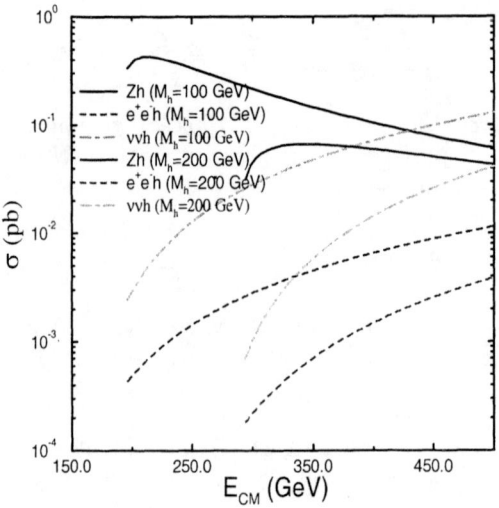

FIGURE 4. Higgs boson production at an e^+e^- collider.

Measurements of Higgs Couplings

The measurement of the Higgs boson couplings is important to differentiate between the Standard Model and other possibilities. In a supersymmetric model, the Higgs couplings to both fermions and gauge bosons can be quite different from those of the Standard Model, as illustrated in Fig. 6 for an arbitrary choice of input parameters. The total decay width can differ by more than an order of magnitude between the Standard Model and a supersymmetric model.

The total Higgs boson width can be measured from the reconstructed Higgs peak at the LHC. This direct measurement is only possible for $M_h > 200 \, GeV$. Below this mass, the width of the resonance is narrower than the experimental resolution. For $M_h > 200 \, GeV$, the Higgs can be observed through the decay $h \to ZZ \to 4l$ and the resulting measurement of the total width is shown in Fig. 7. With $\mathcal{L} = 300 \, fb^{-1}$, the LHC can measure $\Delta\Gamma_h/\Gamma_h < 10^{-1}$ for $300 \, GeV < M_h < 800 \, GeV$.

Measurements of specific branching ratios are probably the most useful quantities for distinguishing between the Standard Model and other models. As an example, I discuss the coupling of the Higgs boson to the top quark. In the Standard Model the Yukawa coupling is given by,

$$g_t = -\frac{M_t}{v}, \tag{18}$$

while in the minimal supersymmetric model the coupling is modified by the factor C_{tth},

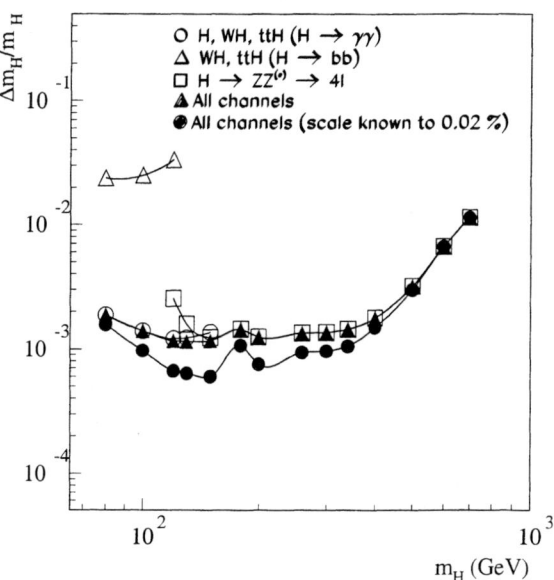

FIGURE 5. Precision measurements of the Higgs boson mass at the LHC with $\int \mathcal{L} = 300\ fb^{-1}$, using the ATLAS detector. From Ref. 12.

$$g_t = -C_{tth}\frac{M_t}{v} \qquad (19)$$

For some values of $\tan\beta$ and M_A, C_{tth} can be quite different from 1, as shown in Fig. 8. Fig. 8 also shows the coupling of the heavier neutral Higgs boson of a supersymmetric theory, H^{SUSY}, to the top quark. Again, the coupling can be far from the Standard Model coupling. Note that for $M_A \to \infty$, $C_{tth} \to 1$, $C_{ttH} \to 0$ and the Standard Model coupling is recovered.

At the LHC, the $t\bar{t}h$ coupling can be measured to roughly 20% through the process $pp \to t\bar{t}h$ in the mass region $M_h \sim 120\ GeV$. [12] (For higher Higgs masses, the cross section becomes quite small.) A similar mass region can be probed at an e^+e^- collider. The signal decays predominantly to $W^+W^-b\bar{b}b\bar{b}$ and so will be spectacular. A study of the signal and background showed that the signal could be extracted from the background using both the semi-leptonic and the hadronic decays of the W's and a measurement of g_t obtained. [14,15] Table 2 shows the expected precision for the measurement of g_t at $\sqrt{s} = 500\ GeV$ and $1\ TeV$. [15] The message is clear. A precision measurement of C_{tth} requires high energy and high luminosity ($L = 1000\ fb^{-1}$) in order to improve on the LHC's measurement.

The total rate for Higgs production in the process $e^+e^- \to Zh$ can be found by measuring the recoil mass of the lepton pair, M_{ll}, from the decay $Z \to l^+l^-$. This measurement is independent of the Higgs boson decay mode. Once the total rate is known, the Higgs branching ratios can be measured by flavor tagging of the Higgs

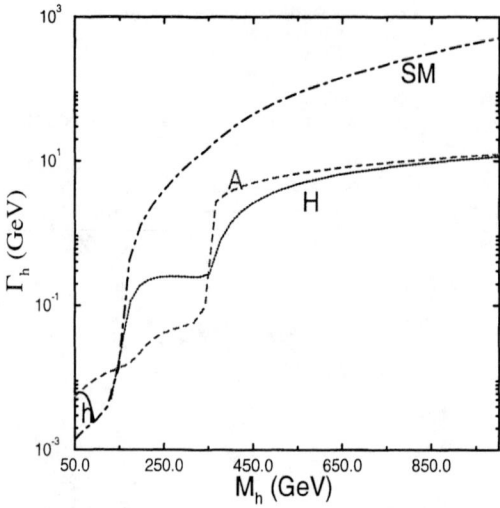

FIGURE 6. Total decay width for Standard Model Higgs boson and the Higgs bosons of a supersymmetric model with $\tan\beta = 2$, $A_t = M_{SUSY} = 1\ TeV$, and $\mu = 100\ GeV$.

TABLE 2. $\frac{\delta g_t}{g_t}$ in e^+e^- interactions With 1000 fb^{-1}. From Ref. 15.

$M_h\ (GeV)$	$\sqrt{s} = 500\ GeV$	$\sqrt{s} = 1\ TeV$
100	.08	.06
110	.12	.06
120	.21	.07
130	.44	.08

decay final states.

The measurements of the Higgs couplings to the lighter quarks can be done with a precision of 5 – 10 % with 500 fb^{-1} at an e^+e^- collider. Ref. [16] found roughly equivalent results for $\sqrt{s} = 350\ GeV$ and $\sqrt{s} = 500\ GeV$. The error on the measurements of the Higgs Yukawa couplings of Ref. [16] is dominated by theoretical uncertainty due to the measured input values of α_s, m_c, and m_b, not by systematic or statistical errors.

Armed with measurements of the Higgs boson branching ratios, we can ask over what region of parameter space the minimal supersymmetric model can be distinguished from the Standard Model. The answer is shown in Figure 9, taken from Ref. [16]. First, the 95% confidence level value of the branching ratio for the Standard Model was computed. Ref. [16] then scans over the parameter space of the minimal supersymmetric model, taking $\tan\beta < 60$ and the mass parameters to be

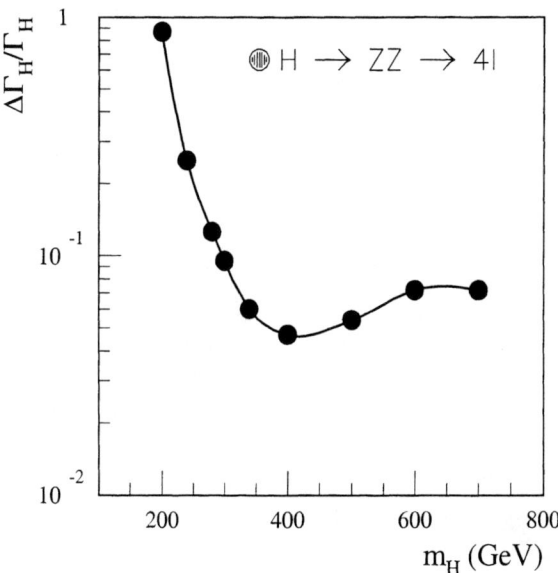

FIGURE 7. Measurement of the total Higgs boson width at the LHC with 300 fb^{-1}, using the ATLAS detector. From Ref. 12.

less than $1-1.5\ TeV$. For a given set of parameters, the Higgs branching ratio was then computed. In Fig. 9, the region to the right of the curves (going from left to right on the figure) has more than 68, 90 or 95% of the supersymmetric model solutions outside of the Standard Model 95% confidence level region. With 500 fb^{-1}, an e^+e^- collider can distinguish between the Standard Model and the minimal supersymmetric model up to $M_A \sim 550\ GeV$, while with 1000 fb^{-1}, the sensitivity is increased to $M_A \sim 730\ GeV$. [16] This is remarkable given the decoupling of the Higgs sector of the minimal supersymmetric model for large M_A.

At the LHC, measurements of the Higgs couplings are less clearcut than at an e^+e^- collider. At the LHC, measurements involving the Higgs boson typically involve combinations of Higgs couplings. For example, a measurement of the ratio of the $h \to \gamma\gamma$ and $h \to ZZ \to 4l$ rates would give the ratio of the $h \to \gamma\gamma$ and $h \to ZZ$ branching ratios, but not the absolute couplings. A study of the combinations of Higgs couplings which can be measured at the LHC is given in Ref. [21].

VERIFYING THE STRUCTURE OF THE HIGGS POTENTIAL

Once a Higgs particle is found, it will be necessary to investigate its self-couplings in order to reconstruct the Higgs potential and to verify that the observed particle is indeed the Standard Model Higgs boson which results from spontaneous symmetry breaking. A first step in this direction is the measurement of the trilinear self-couplings of the Higgs boson which are uniquely specified by the scalar potential of Eq. 2.

After the symmetry breaking, the self-couplings of the Higgs boson are uniquely determined by M_h,

$$V = \frac{M_h^2}{2}h^2 + \frac{M_h^2}{2v}h^3 + \frac{M_h^2}{8v^2}h^4 \quad . \qquad (20)$$

In extensions of the Standard Model, such as models with an extended scalar sector, with composite particles or with supersymmetric partners, the self-couplings of the Higgs boson may be significantly different from the Standard Model predictions.

In order to probe the three- and four- point Higgs couplings, it is necessary to measure multi-Higgs production. Higgs boson pairs can be produced by several mechanisms at hadron colliders:

- Higgs-strahlung $W^*/Z^* \to hhW/Z$,
- vector-boson fusion $WW, ZZ \to hh$,
- Higgs radiation off top and bottom quarks $gg, q\bar{q} \to Q\bar{Q}hh$,
- gluon-gluon collisions $gg \to hh$.

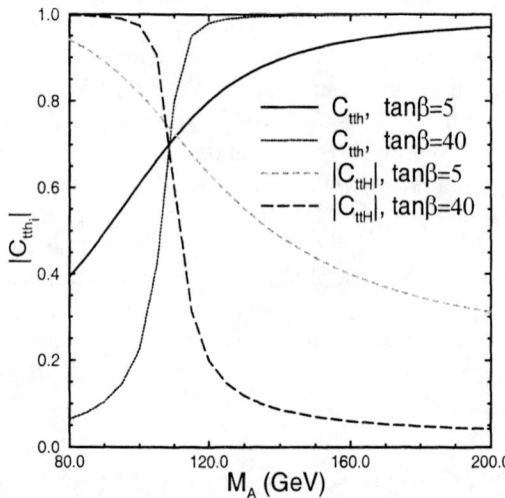

FIGURE 8. Couplings of the neutral Higgs bosons, h^{SUSY} and H^{SUSY}, of a supersymmetric model to the top quark, in units of the Standard Model Higgs-top quark Yukawa coupling.

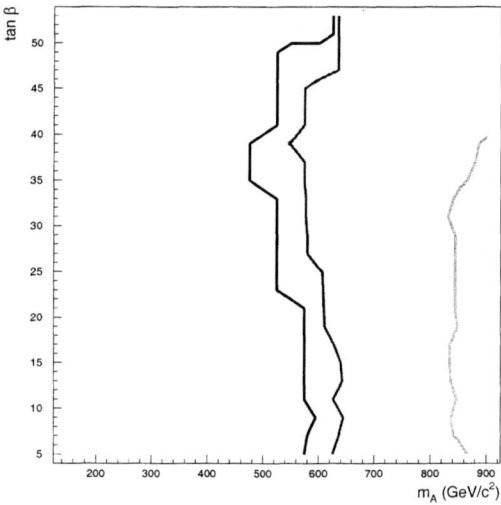

FIGURE 9. Regions of parameter space where the Standard Model and the minimal supersymmetric model can be distinguished. The regions to the right of the curves (moving from left to right) have more than 65%, 90%, or 95% of the minimal supersymmetric model solutions outside of the Standard Model 95% confidence level region. From Ref. 16.

At the LHC, gluon fusion is the dominant source of Higgs-boson pairs in the Standard Model and arises from quark loops, with the dominant contribution coming from top quark loops. The rate, even at the LHC, is quite small as can be seen in Fig. 10. Although the rate is sensitive to the tri-linear coupling, the variation is probably too small to be observed. [17] A detailed study of the signal and background gives the results shown in Fig. 11. [18] This study computed the minimum rate necessary for a 5σ discovery of hh production. It is clear that in the Standard Model, this physics will have to wait for the next generation of accelerators.

In a supersymmetric model, the b- quark contribution to hh production will be enhanced for large $\tan\beta$. Even so, in the absence of large squark loop contributions, with 25 fb^{-1} the Tevatron can only exclude a small region of parameter space with $M_A < 150\ GeV$ and $\tan\beta > 80$. The LHC will be able to exclude an even larger region of M_A and $\tan\beta$ space. [18] However, the situation changes dramatically for light squarks and with the parameters chosen to maximize the squark tri-linear couplings. In this case, it is possible to obtain a significant enhancement of the rate, largely due to resonance effects. This is shown in Fig. 12 for the Tevatron. [18] In this very special situation, even the Tevatron will be extremely sensitive to double Higgs production.

At a high energy e^+e^- collider, Higgs pairs are produced through similar mech-

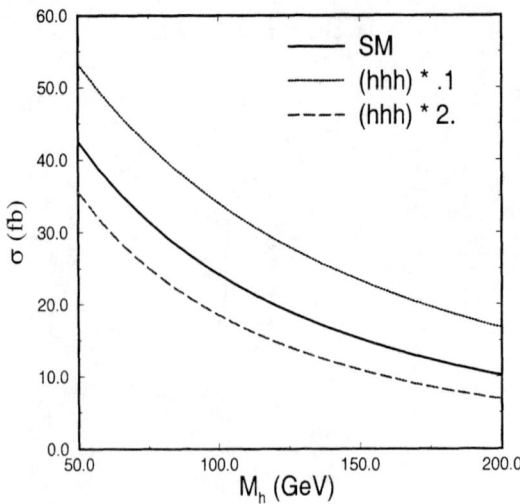

FIGURE 10. Double Higgs production, $pp \to hh$, at the LHC, $\sqrt{s} = 14\ TeV$. The solid line is the Standard Model rate, while the dotted and dashed lines have the tri-linear Higgs couplings modified.

anisms as in hadronic collisions. At intermediate energies, $\sqrt{s} \sim 500\ GeV$, the dominant mechanism is $e^+e^- \to Zhh$, while at TeV scale energies, the process $e^+e^- \to \nu\bar{\nu}hh$ is dominant. Just above the kinematic threshold, the sensitivity to the trilinear coupling is maximal in the $e^+e^- \to Zhh$ process. With $2000\ fb^{-1}$, the tri-linear coupling can be measured to $\sim 15\%$ [19]. The cross sections for all sources of double Higgs production in an e^+e^- collider are small, on the order of a few femptobarn or less for $M_h < 200\ GeV$. [19] This is clearly a measurement which requires the highest possible luminosity in order to isolate the signal from the background and make a measurement of the tri-linear Higgs coupling.

At present, it does not appear possible to measure the Higgs boson four-point coupling. In principle, it could be measured in triple Higgs production, but the rate is miniscule.

STRONGLY INTERACTING SYMMETRY BREAKING

If a Higgs boson is not found at the LHC, then the electroweak symmetry breaking is strongly interacting. Without the addition of some new type of physics, WW scattering will violate unitarity at an energy scale somewhere below $3\ TeV$. There are two classes of effects which could potentially be observed in this scenario.

The first possibility is that whatever new physics unitarizes the WW scattering

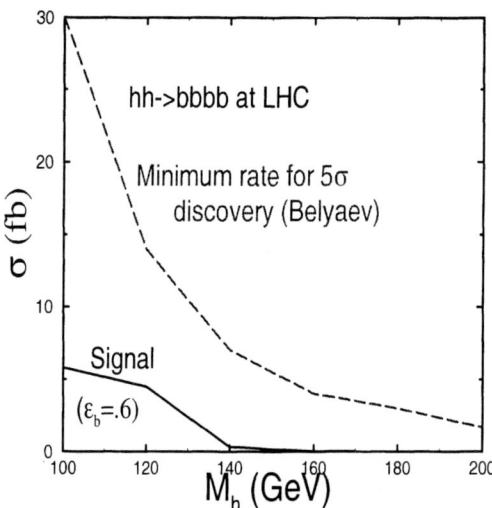

FIGURE 11. Minimum rate for a 5 σ discovery of $pp \to hhX$ at the LHC (dotted line) and the signal using a b tagging efficiency of $\epsilon_b = .6$.

is at too high an energy scale to be observed at either the LHC or an e^+e^- collider with $\sqrt{s} \sim 500\ GeV - 1\ TeV$. In this case the only effects which can be observed are small deviations in absolute rates. The Lagrangian can be written as

$$\mathcal{L} = \mathcal{L}_{SM} + \sum_i \frac{f_i}{\Lambda^2} \mathcal{O}_i, \tag{21}$$

where $\mathcal{L}_{SM}$ is the Lagrangian of the Standard Model with the Higgs boson removed. Without the Higgs boson, the Lagrangian can be written in terms of an expansion in powers of $\frac{s}{\Lambda^2}$, where Λ is the scale of new physics. The f_i are dimensionless coefficients of the new operators, $\mathcal{O}_i$. A complete set of operators at order s/Λ^2 can be found in Ref. [20]. The goal of the LHC or a high energy e^+e^- collider in this scenario would be to measure the f_i and attempt to distinguish between models. At the LHC, there will be a very small number of events [12] and it is doubtful if it will be possible to tell the difference between the various possible models. An e^+e^- collider with $\sqrt{s} \sim 1.5\ TeV$ could measure some of the f_i to $\mathcal{O}(10^{-3})$, but a complete set of measurements will take still higher energy.

In the second case, the new physics which unitarizes the WW scattering amplitudes produces resonances which can be observed. Numerous studies have found that an e^+e^- collider with $\sqrt{s} \sim 1.5\ TeV$ has roughly the same sensitivity to TeV scale resonances as does the LHC. [21] Both machines will be sensitive to resonances on the order of $1.5\ TeV$.

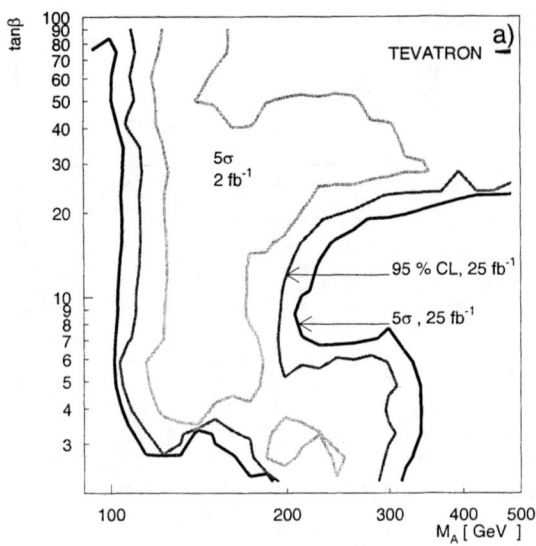

FIGURE 12. Double Higgs production at the Tevatron in a supersymmetric model with enhanced squark contributions. From Ref. 18.

CONCLUSION

Even after the LHC has successfully run for a few years, there will still be unanswered physics questions. If a weakly interacting Higgs boson exists, either from a supersymmetric model or the Standard Model, it will be observed at the LHC. The LHC will make preliminary measurements of the Higgs boson mass and couplings, but a high energy e^+e^- collider with high luminosity will significantly improve on the precision. Precise measurements of the Higgs width are particularly important for differentiating between models. Measurements of double Higgs production and strong symmetry breaking in particular will require the highest possible energy and luminosity.

This note has considered only electroweak symmetry breaking. There will of course be many exciting questions to be answered in other areas of particle physics such as supersymmetry, QCD, CP violation, etc. Interesting times await us!

REFERENCES

1. R. Casalbuoni *et. al.*, JHEP **9908** (1999) 011.
2. M. Swartz, results presented at the 1999 Lepton-Photon Symposium, Stanford, CA, Aug. 9-14, 1999.

3. S. Alam, S. Dawson, and R. Szalapski, *Phys. Rev.* **D57** (1998) 1577; J. Bagger, A. Falk, and M. Schwartz, hep-ph/9908327.
4. J. Casas, J. Espinosa, M. Quiros, and A. Riotto, *Nucl Phys.* **B436** (1995) 3; L. Maiani, G. Parisi, and R. Petronzio, *Nucl. Phys.* **B136** (1978) 115.
5. R. Chivukula, in Proceedings of *NATO Advanced Study Institute on Quantum Field Theory Perspective and Prospective*, Les Houches, France, June 16-26, 1998, hep-ph/9803219.
6. M. Sher, *Phys. Rep.* **179** (1989) 273; G. Isidori, *Phys. Lett.* **B337** (1994) 141; J. Espinosa and M. Quiros, *Phys. Lett.* **B353** (1995) 257.
7. U. Heller *et. al.*, *Nucl. Phys.* **B405** (1993) 555.
8. A. Blondel, report to the LEP Council, Nov. 1999, http://alephwww.cern.ch/ bdl/lepc/lepc.ppt.
9. A review of the phenomenology of the minimal supersymmetric model can be found in J. Gunion *et. al.*, *The Higgs Hunter's Guide* (Addison-Wesley, Redwood City, CA, 1990).
10. S. Heinemeyer *et. al.*, Proceedings of *The International Workshop on Linear Colliders LCWS99*, Sitges, April 28- May 5, 1999, hep-ph/9910285; M. Carena *et al* CERN Yellow Report, CERN-96-01, hep-ph/9602250.
11. J. Hobbs, Proceedings of *The 1999 DPF Meeting*, Los Angeles, Jan. 5-9, 1999, hep-ph/9903494; Report of the Tevatron Higgs and Supersymmetry Working Group, http://fnth37.fnal.gov/higgs.html.
12. ATLAS Detector and Physics Performance Technical Design Report, http://www.usatlas.bnl.gov/physics/phystdr.html.
13. E. Accomando *et. al.*, *Phys. Rep.* **299** (1998)1.
14. S. Moretti, *Phys. Lett.* **B452** (1999) 338.
15. H. Baer, S. Dawson, and L. Reina, *Phys. Rev.* **D61** (2000) 013002.
16. M. Battaglia, Proceedings of *The International Workshop on Linear Colliders LCWS99*, Sitges, April 28- May 5, 1999, hep-ph/9910271.
17. S. Dawson, M. Dittmaier, and M. Spira *Phys. Rev.* **D58** (1998) 115012.
18. A. Belyaev, M. Drees, and J. Mizukoshi, hep-ph/9909386.
19. A. Djouadi, W. Kilian, M. Muhlleitner, and P. Zerwas, *Eur. Phys. Jour.* **C10** (1999) 45; D. Miller and S. Moretti, hep-ph/9906395.
20. T. Appelquist and C. Bernard, *Phys. Rev.* **D22** (1980) 200; A. Longhitano, *Nucl. Phys.* **B188** (1981) 118.
21. T. Barklow *et. al.*, in *New Directions for High Energy Physics: Proceedings of Snowmass 96*, ed. D. Cassel, L. Gennari, and R. Siemann (SLAC, 1997).

A Scaling Radial-Sector FFAG Lattice for a Muon Accelerator

Al Garren

UCLA and LBNL
email: AAGarren@lbl.gov

A lattice example has been designed for a machine intended to be part of the acceleration system for a muon collider or neutrino factory muon storage ring. This machine is a Fixed Field Alternating Gradient (FFAG) of the radial-sector type invented at MURA.[*] The motivation for studying this type of accelerator is the possibility of accelerating particles over a large energy range in a radially compact magnetic structure.

The ring is composed of sixteen arcs separated by straight-section insertions. The arcs are composed of fourteen FODO cells. Each cell is somewhat similar to the cell of a combined-function synchrotron, which has two bending magnets, one radially focusing, and the other defocusing as a result of their gradients being respectively positive and negative. However, whereas the synchrotron magnets all bend the particles in the same direction, the magnets of the machine considered here successively bend the particles inward and outward (relative to the ring center) while the gradients are all outwards. This combination of bending and gradient lead to the same sort of strong focusing as in a synchrotron.

The magnitudes of the bending and gradient fields in the F and D magnets are the same. To produce net inward bending, the F magnets are longer than the D magnets by a factor of about 3/2. As a result, the circumference factor (CF) of the ring R/ρ is about 5. This unfortunate result of the design is forced firstly by the alternate bending, which causes the closed orbits to be nested in a strictly parallel fashion, and secondly by the need to keep the D magnets long enough to preserve vertical stability.

Another difference between the cells of a synchrotron and of this FFAG machine is that the magnets of the former are characterized by a radially constant gradient, and those of the latter by a radially constant field index $n = -r/B \, dB/dr$. The F and D magnets have alternately negative and positive values of *n* (note that r is defined relative to the center of curvature of the magnet in question). The larger the value of *n*, the more closely the orbits are nested. The dense nesting reduces magnet cost, while the large CF increases it. An important result

---

[*] 'Fixed_Field Alternating-Gradient Particle Accelerators'
K. R. Symon, D. W. Kerst, L. W. Jones, L. J. Laslett, and K. M. Terwilliger
Physical Review Vol. 103, No. 6 (1956)

of the FFAG design is that the orbit properties, especially the tunes, are the same for all energies.

The parallel nesting of the closed orbits makes it easy to introduce straight-sections. Short drift straight sections are placed between the F and D magnets of the arc cells. Since these magnets are sector shaped, the orbits remain parallel across the drift spaces. Long straight-section insertions are made with the aid of matching magnets, which have the same values of curvature and field index n as those in the arc cells, but their lengths and positions are adjusted to produce beta-function matching.

The lattice discussed here has 16 long straight sections, equally distributed around the ring, giving it a nearly circular shape. They are to be used for accelerating rf cavities, injection and extraction systems for muons of both signs, diagnostic and correction apparatus, and wiggler magnets for orbit-length correction. In order to reduce the radial spread in the straight sections, pairs of radially uniform bending magnets are placed at the ends of each long straight section. One long straight section is placed along with 14 arc cells in each sector. The sectors are grouped into four superperiods, in each of which one straight section contains wigglers, while the other three are magnet free.

I am grateful for the very helpful discussions and collaboration of Weishi Wan, Carol Johnstone and Frederick Mills.

PHYSICS OPPORTUNITIES WITH A HIGH-ENERGY COLLIDER OF SAME-SIGN MUONS

Clemens A. Heusch

Institute for Particle Physics
University of California, Santa Cruz CA 94010

Abstract. We choose a number of physics processes where collisions of equal-charge muons show particular promise when the center-of-mass energy can be raised well above the few-TeV level, and illustrate the gains that can be realized by clever experimentation. We stress, in this context, the need for more work on higher degrees of muon polarization.

I INTRODUCTION: WHY SAME-SIGN COLLISIONS?

Colliding point-like particles has been a bountiful source of new insight into the spectrum, the symmetries, the interactions of elementary particles over the past decades. Mostly, such gains have been lavished on us by high-energy electron-positron collisions, and the push toward muon collisions to take up some related tasks in the near future has been spawned by the prospect of fewer radiative limitations and a concomitant promise of reaching cleanly defined higher center-of-mass energies.

There has been increasing discussion of providing, in parallel with the e^+e^- collisions that will have to be realized with linear colliders at TeV energies, an e^-e^- version of the same colliders: the chance of experimenting in very clean "exotic" (i.e., non-annihilation) channels with doubly-charged incoming states opens complementary avenues of investigation [1,2]: the ease with which we reach high degrees of polarization for electrons but not for positrons, adds to the attraction of these plans.

In the following, we will show to which extent some of the open questions which a high-energy (say, multi-TeV) version of the muon collider can usefully tackle, can be enriched by the availability of a same-sign version of this machine. For the lower energy range of muon collisions, similar ideas have been advanced in earlier studies [3].

The clearest motivation for our consideration comes from the unique set of simply additive quantum numbers which, in principle, can be used for a clean definition

TABLE 1. For both the ++ and -- versions of a same-sign muon collider, we give the simply additive quantum numbers for incoming beams of given handedness. The fourth and the eighth line show the mixed content of unpolarized incoming muons.

$\lvert\text{in}\rangle$	Q_{el}	s_z	L	L_e	L_μ	I_3^W	Y^W
$\mu_R^+\mu_{+R}$	+2	1	−2	0	−2	1	2
$\mu_R^+\mu_{+L}$	+2	0	−2	0	−2	1/2	3
$\mu_L^+\mu_{+L}$	+2	−1	−2	0	−2	0	4
$\mu^+\mu^+$	+2	1, 0, −1	−2	0	−2	1, 1/2, 0	2, 3, 4
$\mu_L^-\mu_L^-$	−2	−1	2	0	2	−1	−2
$\mu_R^-\mu_L^-$	−2	0	2	0	2	−1/2	−3
$\mu_R^-\mu_R^-$	−2	1	2	0	2	0	−4
$\mu^-\mu^-$	−2	1, 0, −1	2	0	2	0, −1/2, −1	−2, −3, −4

of interactions well beyond the Standard Model, as we show in Table 1: two units of electric and leptonic charges, in conjunction with clean spin projections (if clean polarization states can be attained) permit the definition of significant conserved quantities such as weak isospin and weak hypercharge; operating in the second "generation" of elementary point particles adds a further very relevant twist.

In the following, we will discuss a few significant physics processes well established on the "ten most wanted list" of particle physicists that will benefit from the potential realization of the High-Energy Muon Collider (HEMC) in a version where both muons carry the same charge. We will also address a few technical issues that will define the "overhead" for enriching the muon collider project in this fashion.

II ELECTROWEAK SYMMETRY BREAKING: THE HIGGS SECTOR

It is worthwhile to recall that the detailed study of the Higgs sector may well need energies beyond the 1.5 TeV/4 TeV marks presently set for the "Next" electron/muon collider projects. Central production of any neutral Higgs state has been considered by Minkowski [4]: it shows an essentially saturating cross section, assuring us of a good data rate up to energies well above threshold, as seen in Fig. 1 (in contrast to the "discovery graph" via Higgsstrahlung off Z production in annihilation reactions). In this context, the same-sign version has the advantage of very small backgrounds of scattered $\mu^+\mu^+$ or $\mu^-\mu^-$ when compared with the opposite-sign version.

More significantly, there is ample opportunity to have an extended Higgs sector reach to considerably higher energies, and therefore to be able to profit from

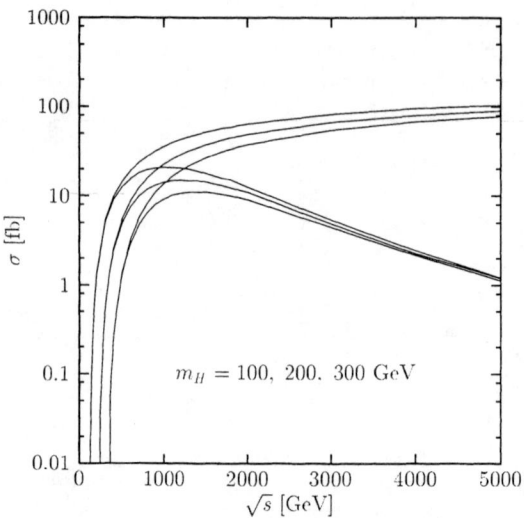

FIGURE 1. Total cross section for central production of a neutral Higgs boson; a sharp rise leads to saturated data ratios (upper curves) with only limited losses to background – depressing angular cuts. From Ref. 4.

the HEMC parameters: Gunion [5] and Rizzo [6] have addressed the usefulness of same-sign electron collisions for a study of "exotic" Higgs sectors such as might emanate from Left-Right Symmetric or higher symmetry models. Singly and equally charged Higgs bosons can easily be pair-produced via WW fusion (see Fig. 2), where backgrounds can be identified by a switch in incoming muon helicity.

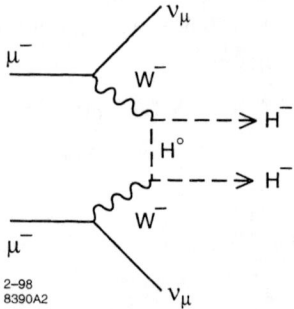

FIGURE 2. Pair production graph for charged Higgs bosons, by WW fusion.

In representations containing doubly charged Higgs bosons, there is an obvious double bonus in high-energy muon collider searches: for the W^-W^- fusion graph (Fig. 3a), which should occur with a full unit of R, there is not only the spectacular s-channel resonance structure with typical Higgs boson widths and decays into W^-W^-, H^-H^-, etc., with its vital dependency on the helicity of incoming muons:

we also have the possibility of a direct H^{--} bilepton coupling (Fig. 3b); and this latter quantity, expected to be minuscule in the e^-e^- case, will greatly increase with the muon mass. We have stressed elsewhere that more exotic bilepton structures can be eliminated from a possible confusion by a change in polarization of the incoming muons. Clearly, the Muon Collider in its like-sign version has a chance to add pivotal information in ascertaining the nature of an extended Higgs sector. Both the energy reach and the muon vs. electron rest mass ratio are vital.

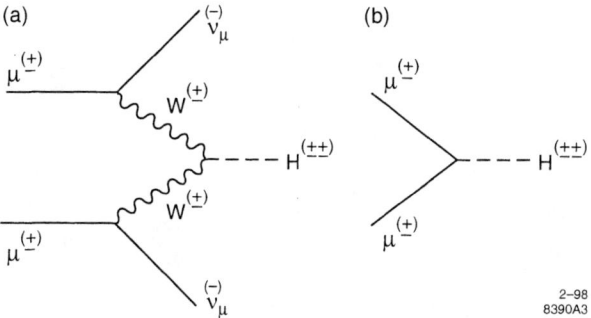

FIGURE 3. a) Production graph for doubly charged Higgs boson from appropriately handed muons, by WW fusion; b) Direct coupling, of unknown strength, of incident muon pair to doubly charged Higgs boson.

III ELECTROWEAK SYMMETRY BREAKING: SUPERSYMMETRY

While we still expect the present or presently approved machines to bring us first actual evidence on possible supersymmetric particles/phenomena, we have no positive evidence on the mass spectrum of SUSY partners for our known particles. The determination of SUSY masses, couplings, phases, and mixing parameters will be a gigantic task.

It has been shown by various authors [7–10] that like-sign lepton collisions offer unique precision in the case of selectron pair production from electron pairs via Majorana neutralino exchange, as shown in Fig. 4: while the great precision of threshold production, and its importance in determining selectron and neutralino masses, are clearly seen from Fig. 4b, this figure lacks radiative correction input. If we now take the same figure and look for like-sign muon pairs producing smuon pairs, we needn't worry about radiative effects in the same manner; better still, we have access to a complete new set of SUSY parameters in the second generation to deal with.

The greater precision in the potential determination of smuon vs. selectron mass parameters then also translates into a greater sensitivity to vital elements of the neutralino mass matrix, as H.C. Cheng illustrates in Figs. 5, 6. We stress that

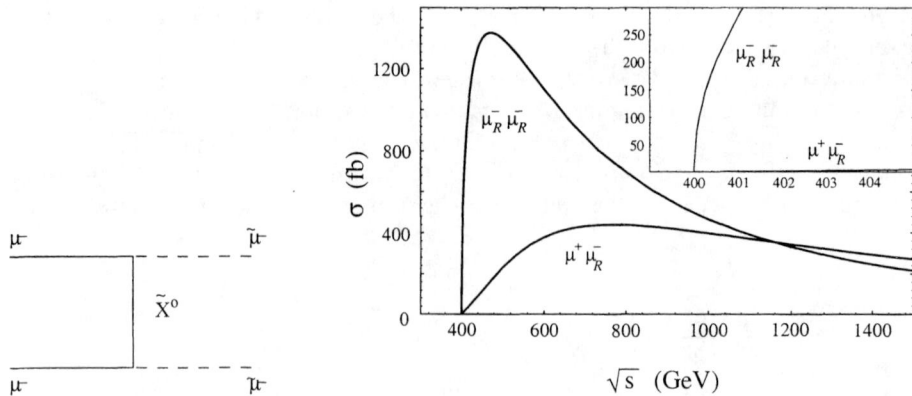

FIGURE 4. a) The contribution to $e^-e^- \to \tilde{e}^-\tilde{e}^-$ from t-channel Majorana neutralino exchange; b) Cross sections $\sigma(e_R^- e_R^- \to \tilde{e}_R^- \tilde{e}_R^-)$ and $\sigma(e_R^+ e_R^- \to \tilde{e}_R^+ \tilde{e}_R^-)$ near threshold. Effects of initial state radiation, beamstrahlung, and the selectron width are not included. From Ref. 8. Exactly the same arguments can be made for the $\mu\mu \to \tilde{\mu}\tilde{\mu}$ case shown here, but radiative effects are much less disturbing, strengthening the importance of the process.

there will be very little, if any, chance elsewhere to measure large gaugino masses. Again, the like-sign muon collider is likely to "own" a fair fraction of the huge Supersymmetry parameter space.

G. Anderson points out [11] that the steep cross section rise at threshold that is at the basis of slepton mass determinations (see above) is equally applicable to the production of two messenger scalars in gauge-mediated Supersymmetry. The relative precision between pair production of messenger scalar particles in $\mu^-\mu^-$ and $\mu^+\mu^-$ collisions is again illustrated, qualitatively, by Fig. 4; but the availability of higher-energy incoming leptons, plus the absence of relevant radiative tails, may well make the vital difference.

IV STRONG SYMMETRY BREAKING

It has been stressed again and again that, in the absence of EWSB in the TeV region, symmetry breaking will necessarily occur via strong phenomena: there will be strong WW scattering in the longitudinal W sector. Hence, there may be as fundamental a set of scattering phenomena to be explored as we did in old times, and at much lower energies, for the scattering of charged pions off each other, in all available J and I channels. That implies a possible resonance structure for $W_L W_L$ scattering at high energies – and that is exactly what the HEMC may find a particularly congenial task.

T. Han [12] pointed out that the exchange of a heavy neutral Higgs boson strongly enhances the cross section of the relevant signature process

$$e^-e^- \to \nu\nu W^- W^- \tag{1}$$

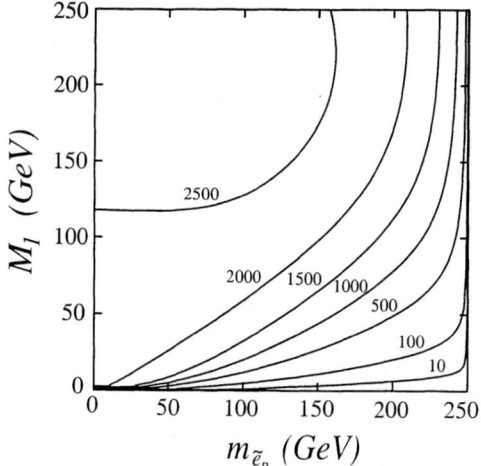

FIGURE 5. Contours of constant $\sigma_R = \sigma(e_R^- e_R^- \to \tilde{e}_R^- \tilde{e}_R^-)$ in fb in the $(m_{\tilde{e}_R}, M_1)$ plane for $\sqrt{s} = 500$ GeV. From Ref. 10. Similarly, measured cross sections of $\mu_R^- \mu_R^- \to \tilde{\mu}_R^- \tilde{\mu}_R^-$ will permit us to read off the relevant M_1 value.

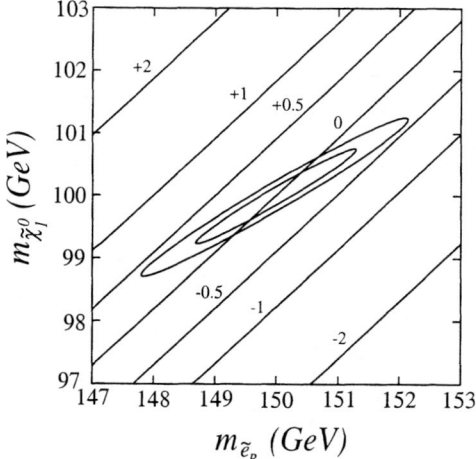

FIGURE 6. The allowed regions, "uncertainty ellipses," of the $(m_{\tilde{e}_R}, m_{\tilde{\chi}_1^0})$ plane, determined by measurements of the end points of final state electron energy distributions with uncertainties $\Delta E = 0.3$ GeV and 0.5 GeV. The underlying central values are $(m_{\tilde{e}_R}, m_{\tilde{\chi}_1^0}) = (150 \text{ GeV}, 100 \text{ GeV})$, and $\sqrt{s} = 500$ GeV. We also superimpose contours (in percent) of the fractional variation of σ_R with respect to its value at the underlying parameters. From Ref. 10. Applying this technique to the smuon pair production process takes advantage of a better definition of muon energies, therefore smaller ΔE ellipses.

w.r.t. the expected SM light-Higgs decay, so that higher energies available at the HEMC will make a sizeable difference. More importantly, he stresses that high polarizations for the incoming leptons are vital for background suppression – which, unfortunately, is more easily exploited in the e^-e^- case than at the muon collider. Still, this is a potentially novel dynamics sector to explore, and its $I = 2$ component is the bailiwick of the machine under consideration here.

V COMPOSITENESS

The question of the ultimate elementarity of our known basic fermions pops up naturally whenever we embark into a new energy regime. It has been stressed repeatedly that Møller scattering of highly polarized leptons is of the greatest sensitivity to any possible substructure [13,14]. Fig. 7 reminds us of the fact that two highly polarized like-sign leptons do indeed display an extraordinary response to high compositeness scales. It is due to crossing from s to u channel in the contact interaction terms mediated at energy Λ, i.e., at compositeness scale $1/\Lambda$. A numerical calculation neglecting Z exchange gives the $\mu^+\mu^+/\mu^-\mu^-$ version an advantage of a factor of $\Lambda^{++/--}/\Lambda^{+-} = 2.38$ for Møller scattering at 90°.

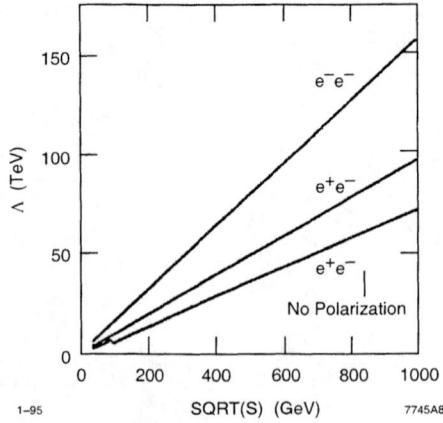

FIGURE 7. The 95% confidence level limits that can be obtained for the compositeness scale Λ_{LL}^+ as a function of the e^-e^- or e^+e^- center-of-mass energy. The luminosity is given, in this example, by $\mathcal{L} = 680\text{pb}^{-1} \cdot s/M_Z^2$. The polarization of the electron beam(s) in the upper two curves is 90%. From Ref. 13. For $\mu^-\mu^-$ or $\mu^+\mu^+$ Møller scattering, we can easily scale this plot such that, say, $\sqrt{s} = 10$ TeV yields a reach to $\Lambda = 1.6$ PeV.

Møller scattering of muons not only has the highest energy potential of stable charged fermions at our disposal – it also gives us access to the second SM generation; and should pointlikeness eventually break down, we would expect the effects to become visible for the relatively massive muons well before an effect might show up for electrons.

VI HEAVY MAJORANA NEUTRINOS

As our last example, we will show the special usefulness of the HEMC in the search for heavy Majorana singlet states advocated as a possible key element for an understanding of the clearly insufficient SM version of the massless single-helicity nature of the known neutrinos.

Heusch and Minkowski [15,16] pointed out that such states, so far undetected/undetectable, might have very attractive features: they might motivate the great mass differences inside higher-symmetry supermultiplets; they could definitively establish the Majorana character of neutrinos (undetectable for zero-mass states); if present, they will furnish definitive experimental signals well above all calculable backgrounds; finally, they might utilize the electron or muon collider for a key experiment on neutrino mass.

Clearly, the symmetry structure of such heavy singlet states N could easily couple them differently to electrons or muons, giving the muon collider a distinctive part in their discovery. The relevant cross section for the diagram shown in Fig. 8 is given by

$$\frac{d\sigma}{d\cos\theta}\left(\mu_L^-\mu_L^- \to W_{\text{long}}W_{\text{long}}\right) \sim \frac{G_F^2}{8\pi}\left|\sum_N \frac{U_{\mu N}^2}{m_N}\right|^2 s^2 \quad \text{for } m_W \gg \sqrt{s} \gg m_N. \quad (2)$$

It has the distinction that it shows explicit dependencies, independently verifiable and measurable, on the center-of-mass energy, on the heavy neutrino masses, and on the mixings/couplings of these states to e, μ, τ. In addition, it operates only for left-handed electrons/muons. The latter feature makes for very easy identification of background processes, i.e., for clean signals.

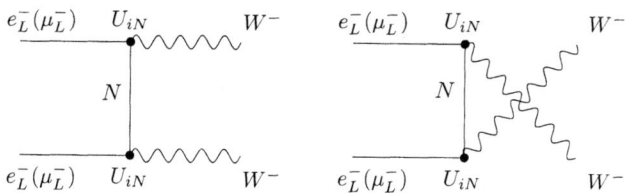

FIGURE 8. Feynman diagrams for the process $e_L^- e_L^-$ (or $\mu_L^- \mu_L^-$) $\to W^- W^-$. The couplings U_{iN} may be different for $i = e, \mu$.

The strong energy dependence $\sim s^2$ (which is valid in the kinematical region of $\sqrt{s}$ between $2m_W$ and m_N) may well make the energy range of the HEMC of vital importance. We do have to stress, however, that the lack of full control over muon polarization is detrimental to an exhaustive search along the lines we describe.

The calculations at the basis of eq. 2 assume the existence of at least 2 such heavy neutrino states. This feature is of key importance for the other valuable feature a muon collider project might exhibit for this heavy neutrino search: These states

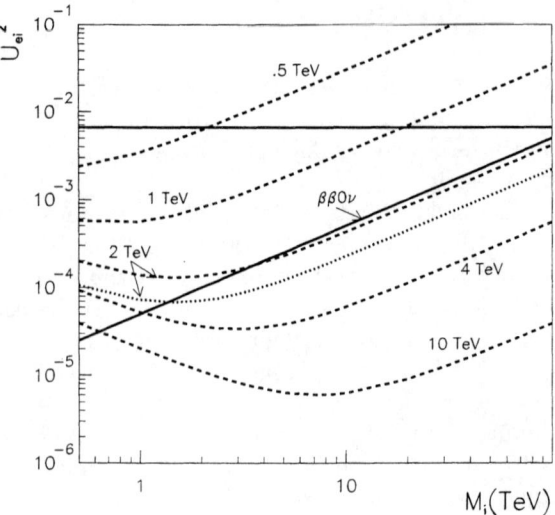

FIGURE 9. Discovery limits for $e^-e^- \to W^-W^-$ at equal-sign lepton colliders with various center of mass energies(dashed lines). M_i is the heavy neutrino mass while U_{iN}^2 is the square of the relevant mixing angle. Unpolarized beams are assumed together with a luminosity of $80(s/1\text{ TeV}^2)\text{fb}^{-1}$. The sensitivity is increased by a factor of 4 for polarized incoming leptons, so that the dotted curve in the 2 TeV case corresponds to polarized beams. The region being probed is above the curves in all cases and 10 events are required for discovery. The diagonal solid line is the incorrect $\beta\beta_{0\nu}$ constraint of Bélanger et al. [17], while the horizontal solid line arises from universality. Here, the allowed region is below that line. From Ref. 18.

would naturally exhibit different couplings to electrons vs. muons. Fig. 9 shows an exclusion plot for the observability of heavy Majorana neutrinos of given mass between between 0.5 and 100 TeV, for given couplings and center-of-mass energies. While the exclusion due to non-observation of neutrinoless double beta decay, also shown in Fig. 9 [18] is erroneous [19] and should be neglected, the value of the high energies open to searches at the HEMC becomes quite obvious.

VII TECHNICAL OVERHEAD FOR THE LIKE-SIGN MUON COLLIDER

Whereas, in the electron collider case, there is essentially no overhead in providing a like-sign version in addition to the opposite-sign one, the general scheme of the muon collider does not make the implementation of our variant quite as easy. Still, the only major addition is a two-vertically-superimposed-ring structure for the final collider ring (as shown in Fig. 10): this added feature, while not trivial, does not constitute a major design challenge.

The muon sources and cooling schemes are clearly compatible. Here, the main

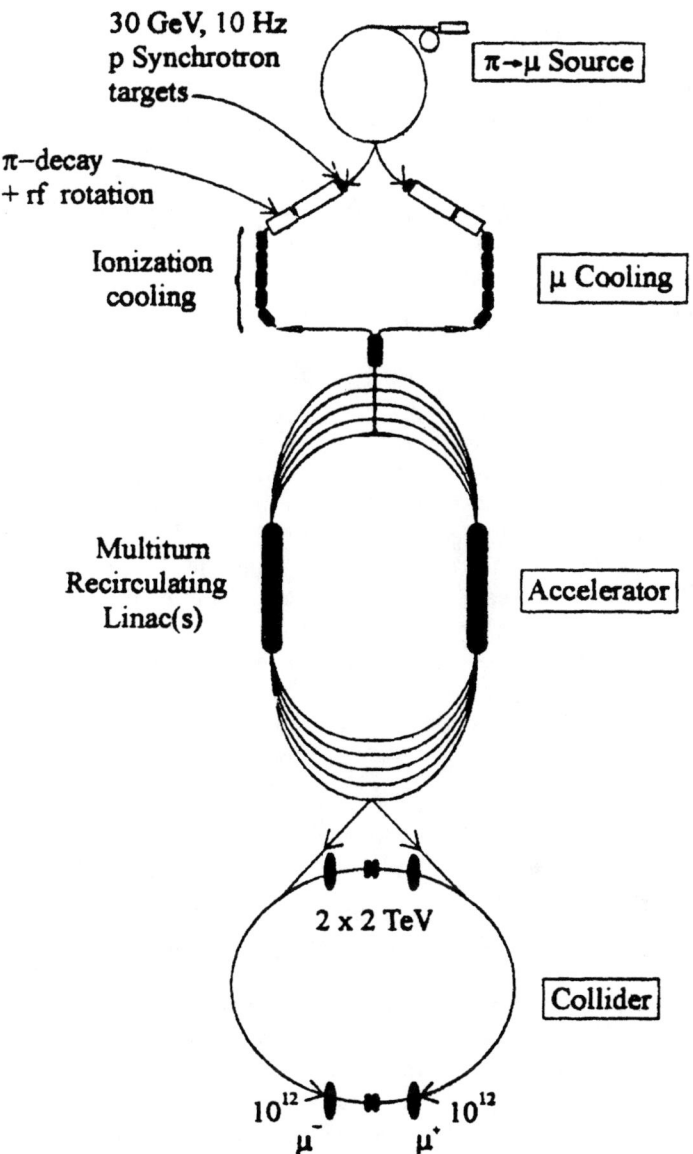

FIGURE 10. Schematic layout of a high-energy muon collider. The final collider has to be realized with two rings of superimposed magnets for the $\mu^+\mu^+$ or $\mu^-\mu^-$ version. From Ref. 20.

challenge lies in the problem of providing the highest degrees of polarization possible. This is clearly problematic, and deserves an intensive study for better solutions than what has been proposed up to now [21]. We have stressed time and again in the physics descriptions above, the great need for the best polarization values attainable – preferably close to those of the electron colliders. It is clear, however, that the prevalent cooling schemes mix up muons that were forward- and backward-emitted during pion decay, hence are differently polarized. Better separation entails losses in beam intensity, hence in luminosity.

Moreover, one regrettable implication of the (otherwise most welcome) large rest mass of the muon when compared to the electron is the difficulty involved in reversing the polarization: while this is a trivial point in the electron gun, there is no nanosecond-range scheme in which the muon sources could even attempt to equal that performance. This has the consequence that we will not be able to do simple signal-vs.-background checks by quick polarization reversals.

This disadvantage for muons vs.electrons is, of course, caused by the gamma factor in the spin precession condition, which does not allow a "spin rotator" to be activated on a pulse-by-pulse basis in the muon case. It was pointed out, however, [22] that the muon energies in the collider rings may be chosen such that $\gamma = (2n+1)/(g-2)$; in this case, and if the momentum definition is sharp enough to enforce this definition (with n an integer), the helicity of the muons can in fact be inverted from turn to turn in the collider; and since we have two superimposed rings there, we can do so independently for both incoming beams. We cannot, however, choose longer runs with chosen helicities, at the same energies, at will - as is easily done in the electron case.

VIII CONCLUSIONS

We summarize this brief attempt at making a case for providing a like-sign option as an integral part of the High-Energy Muon Collider project with the simple reiteration that a machine project as distinctive as that of the HEMC should be given every chance to serve all possible important searches for novel physics results within and beyond the Standard Model: in a number of relevant cases, this will not be feasible unless there is a like-sign version that is compatible, and interchangeable, with the opposite-sign one. The physics payoff, by no means exhaustively covered here, is rich and unique. We urge the inclusion of this option in further planning; we also reiterate the need for an optimization of muon source schemes for higher degrees of polarization.

ACKNOWLEDGMENT

It is a pleasure to thank Bruce King and the staff of the HEMC Workshop for a well-prepared and well-orchestrated set of discussions on many different physics topics that may well be served by the Muon Collider project in this high-energy regime.

REFERENCES

1. e^-e^- 1995, Proceedings of the Electron-Electron Linear Collider Workshop, C.A. Heusch, editor, *Int. J. Mod. Phys.* **A11**, 1 (1996).
2. e^-e^- 1997, Proceedings of the 2nd Electron-Electron Linear Collider Workshop, C.A. Heusch, editor, *Int. J. Mod. Phys.* **A13**, 1 (1998).
3. C.A. Heusch and F. Cuypers, *Proceedings of the 2nd Workshop on Physics Potential and Development of $\mu^+\mu^-$ Colliders*, Sausalito, CA, 17-18 November 1994, D.B. Cline, editor (AIP Conference Proceedings 352, 1995) p. 219; C.A. Heusch, *Proceedings, 4th International Conference on Physics Potential and Development of $\mu^+\mu^-$ Colliders*, 10–12 December 1997, San Francisco, CA, D. Cline, ed., (AIP Proceedings, 1998) p. 116.
4. P. Minkowski, *Int. J. Mod. Phys.* **A13**, 2255 (1998).
5. J.F. Gunion, ibidem, p. 2277.
6. T.G. Rizzo, *Int. J. Mod. Phys.* **A11**, 1563 (1996).
7. M.E. Peskin, *Int. J. Mod. Phys.* **A13**, 2299 (1998).
8. J.L. Feng, ibidem, p. 2319.
9. S. Thomas, ibidem, p. 2307.
10. H.C. Cheng, ibidem, p. 2329.
11. G. Andersen, these proceedings.
12. T. Han, *Int. J. Mod. Phys.* **A13**, 2337 (1998).
13. T.L. Barklow, *Int. J. Mod. Phys.* **A11**, 1579 (1996).
14. A. Czacnecki and W.J. Marciano, *Int. J. Mod. Phys.* **A13**, 2235 (1998).
15. C.A. Heusch and P. Minkowski, *Nucl. Phys.* **B416**, 3 (1994); *Phys. Lett.* **B374**, 116 (1996).
16. C. Greub and P. Minkowski, *Int. J. Mod. Phys.* **A13**, 2363 (1998).
17. G. Bélanger, F. Boudjema, D. London and H. Nadeau, *Phys. Rev.* **D52**, 6292 (1996).
18. T. Rizzo, *Int. J. Mod. Phys.* **A11**, 1613 (1996).
19. C.A. Heusch, *Int. J. Mod. Phys.* **A13**, 2383 (1998).
20. R. Palmer, D. Neuffer and J. Gallardo, *Proceedings, 2nd Workshop on Physics Potential and Development of Muon Colliders*, 17-19 November 1994, Sausalito, CA, D.B. Cline, editor (AIP Conference Proceedings 352, 1995) p. 209.
21. B. Norum and R. Rossmanith, *Nucl. Phys. Proc. Suppl.*, **51A** (1996).
22. B. King, private communication.

Collective Effects in High-Energy Muon Colliders

Eberhard Keil

CERN, Geneva, Switzerland

Abstract. Collective single-beam effects, driven by impedances and wake fields in the vacuum chamber of high-energy muon colliders HEMC, are evaluated: (i) using techniques that have been applied to proton-proton colliders such as the LHC and extrapolations from it, (ii) using a new technique more applicable to nearly-isochronous HEMC, adding longitudinal and transverse kicks from longitudinal and transverse loss factors. Results from both techniques are presented for HEMC at 10 and 100 TeV centre-of-mass energies.

INTRODUCTION

In this contribution, I evaluate collective effects in high energy muon colliders HEMC in two different styles:

1. In the Section entitled "An-isochronous HEMC" I use a style similar to that used for large proton storage rings, e.g. LHC and VLHC. I present tables comparing collective effects in LHC and a high energy muon collider at 10 GeV centre-of-mass energy, and a VLHC and a high energy muon collider at 100 GeV centre-of-mass energy, respectively.

2. In the Section entitled "Isochronous HEMC" I assume that the HEMC is isochronous, and that longitudinal motion is absent. I derive criteria for the longitudinal and transverse loss factors.

AN-ISOCHRONOUS HEMC

An-isochronous HEMC are similar to the LHC [1] and the VLHC [2–4]. The arcs consist of FODO cells, the momentum compaction does not vanish, and an RF system keeps the muon beams bunched. The resistive wall instability, coherent synchrotron tune shift, longitudinal microwave instability, and transverse mode coupling instability TMCI are important for LHC [5] and for the VLHC [6], and might be important for the HEMC. I first list the formulae that I shall use.

The growth rate of the resistive wall instability τ_w^{-1} is [5]:

$$\tau_w^{-1} = \frac{r_c \bar{\beta} I F_w}{e Z_0 \gamma b^3} \sqrt{\frac{\mu_0 \rho_w C c}{\pi(n-Q)}} \qquad (1)$$

Here, r_c is the classical muon radius, $\bar{\beta} \approx R/Q$ is the average β-function, I is the total beam current, Z_0 is the impedance of free space, γ is the usual relativistic factor, b is the radius of the beam screen, ρ_w is the resistivity of the beam screen, C is the circumference, and $(n-Q) = 0.25$ is the tune of the n-th mode. The wall penetration factor F_w describes the effects of a beam screen similar to that in the LHC [1], consisting of a thin inner Cu layer and a thicker layer of stainless steel [7].

To preserve longitudinal Landau damping, the synchrotron tune shift must remain smaller than the synchrotron tune spread. This leads to an upper limit for the imaginary part of the effective longitudinal impedance [6]:

$$\Im\left(\frac{Z_L}{n}\right)_{\text{eff}} \leq \frac{6}{\pi^3} \frac{h_{\text{RF}}^3 V_{\text{RF}}}{I_b} \left(\frac{2\pi\sigma_s}{C}\right)^5 \qquad (2)$$

Here, h_{RF} and V_{RF} are the harmonic number and peak voltage of the RF system, I_b is the bunch current, and σ_s is the bunch length. All effective impedances Z_{eff} are the weighted sums of $Z(\omega)$ times the bunch power spectrum [5].

The bunches are stable against the longitudinal microwave instability if the following inequality holds [5,6]:

$$\left|\left(\frac{Z_L}{n}\right)_{\text{eff}}\right| \leq \frac{12}{\pi^3} \frac{h_{\text{RF}} V_{\text{RF}}}{I_b} \left(\frac{2\pi\sigma_s}{C}\right)^3 \qquad (3)$$

The transverse mode coupling instability is caused by the shift of the $m=0$ head-tail mode towards the $m=-1$ mode due to the broad band transverse impedance. The stability condition is [5,6]:

$$\Im(Z_T)_{\text{eff}} \leq \frac{8 Q_s}{\bar{\beta} I_b} \frac{2\pi\sigma_s}{C} \frac{E}{e} \qquad (4)$$

Here, Q_s is the synchrotron tune, and E is the energy of the circulating muon beam.

Comparison between LHC and HEMC10

Tab. 1 shows the LHC and HEMC10 parameters needed for evaluating (1) to (4). I generated the LHC parameters at collision energy in a *Mathematica* notebook [8] that I have used for testing packages for the design of storage rings [9]. They are close to the official LHC parameters [1]. Similarly, I generate many of the HEMC10 parameters in a *Mathematica* notebook [10], and replace them by parameters in B. King's Parameter Table [11] before I evaluate (1) to (4).

Tab. 2 shows a comparison of the results for growth rates and thresholds for the LHC and HEMC10. In the LHC at injection energy, the resistive wall growth rate is

TABLE 1. LHC and HEMC10 Parameters

Parameter	LHC	HEMC10
CoM energy E/TeV	14	10
Circumference C/m	26658	14916
Average β-function $\bar{\beta}$/m	68.1	31.8
Momentum compaction η	$3.1 \cdot 10^{-4}$	$4 \cdot 10^{-5}$
Vacuum chamber radius b/mm	19	19
Vacuum chamber material	Cu/Fe	W
Vacuum chamber temperature/K	20	300
Vacuum chamber resistivity ρ_w/nΩm	0.55	55
Harmonic number h_{RF}	35560	24871
Peak RF voltage V_{RF}/MV	16	100
Synchrotron tune Q_s	0.00202	0.0018
Bunch current I_b/mA	0.188	9.8
Beam current I/mA	669	68.4
Bunch length σ_s/mm	75	2.2
RMS relative energy spread $\sigma_e/10^{-3}$	0.11	0.6

larger, and the threshold impedance of the transverse mode-coupling instability is lower than the values at collision energy. The resistive-wall growth rate in HEMC10 is smaller than in the LHC. The three impedance thresholds are all worse to much worse in the HEMC10 than in the LHC. The reason for this is the fact that the bunch length in the HEMC10 has to be much smaller than in the LHC, in order to permit the very small value of the β-function at the interaction point. Note also that the number of synchrotron oscillations in a muon life time is of order unity.

TABLE 2. Comparison of growth rates and threshold impedances between LHC and HEMC10. The machine and beam parameters are shown in Tab. 1

	LHC	HEMC10
Resistive wall growth rate τ_w^{-1}/s^{-1}	6.6	2.6
Coh. synchrotron tune shift $\Im(Z_L/n)_{\text{eff}}/\Omega$	1.3	$15 \cdot 10^{-6}$
Long. μ-wave instability $|(Z_L/n)_{\text{eff}}|/\Omega$	6.45	$4 \cdot 10^{-3}$
TMCI threshold $\Im(Z_T)_{\text{eff}}$/MΩm^{-1}	160	1.9

In the LHC, the wall resistivity is approximately doubled by the 10 % of the circumference with a Cu vacuum chamber at room temperature [5]. I ignore this factor in the calculation of the growth rates.

Comparison between VLHC and HEMC100

Tab. 3 shows the VLHC and HEMC100 parameters needed for evaluating (1) to (4). I generate also the VLHC and HEMC100 parameters in *Mathematica* notebooks [12,13]. Again, I replace several parameters by parameters in B. King's Parameter Table [11] before I evaluate (1) to (4). In HEMC100, the synchrotron radiation damping time is of the order of the muon life time. Compensating the synchrotron radiation loss needs a substantial RF voltage. It remains to be checked whether avoiding the quantum excitation of the transverse emittance requires an arc lattice such that the equilibrium emittance is in the neighbourhood of the assumed one [4].

TABLE 3. VLHC and HEMC100 Parameters

Parameter	VLHC	HEMC100
CoM energy E/TeV	100	100
Circumference C/km	120	105
Average β-function $\bar{\beta}$/m	542	255
Momentum compaction η	$8.9 \cdot 10^{-4}$	$2.4 \cdot 10^{-4}$
Vacuum chamber radius b/mm	30	19
Vacuum chamber material	Cu/Fe	W
Vacuum chamber temperature/K	20	300
Vacuum chamber resistivity ρ_w/nΩm	0.55	55
Harmonic number h_{RF}	159960	104825
Peak RF voltage V_{RF}/MV	7	25000
Synchrotron tune Q_s	0.00165	0.026
Bunch current I_b/μA	7.2	277
Beam current I/mA	231	6.9
Bunch length σ_s/mm	47.3	2.5
RMS relative energy spread $\sigma_e/10^{-3}$	0.0046	0.113

Tab. 4 shows a comparison of the results for growth rates and thresholds for the VLHC and HEMC100. The resistive-wall growth rate in HEMC100 is smaller than in the VLHC. The three impedance thresholds are all worse to much worse in the HEMC10 than in the LHC. The reason for this is again the fact that the bunch length in the HEMC100 has to be much smaller than in the VLHC. Contrary to HEMC10, the number of synchrotron oscillations in a muon life time is much larger than unity.

ISOCHRONOUS HEMC

In an isochronous HEMC, synchrotron oscillations are absent, and the positions of the muons along the bunch are frozen, whatever RF accelerating fields and longitudinal and transverse wake fields they see. The longitudinal wake field changes the energy of the muons.

TABLE 4. Comparison of growth rates and threshold impedances between VLHC and HEMC100. The machine and beam parameters are shown in Tab. 3

	VLHC	HEMC100
Resistive wall growth rate τ_w^{-1}/s^{-1}	3	0.14
Coh. synchrotron tune shift $\Im(Z_L/n)_{\text{eff}}/m\Omega$	28	0.003
Long. μ-wave instability $\lvert(Z_L/n)_{\text{eff}}\rvert/\Omega$	0.63	0.017
TMCI threshold $\Im(Z_T)_{\text{eff}}/M\Omega m^{-1}$	420	28

In the longitudinal direction, I derive the following inequality for the longitudinal loss factor $k_\parallel$ by imposing the condition that the energy change accumulated over the relativistic muon lifetime T_μ, expressed as the number of turns n_{turn}, is not larger than the rms beam energy spread $E\sigma_e/2$:

$$k_\parallel \leq \frac{(E/e)\sigma_e}{2Nen_{\text{turn}}} = \frac{(E/e)\sigma_e}{2I_b T_\mu} \qquad (5)$$

Here, E is the centre-of-mass energy of the muons, i.e. twice the muon beam energy, e is the muon charge, σ_e is the relative rms energy in the beam, N is the bunch population, and I_b the bunch current. Strictly speaking, $k_\parallel$ should be interpreted as the variation of the loss factor along the bunch. That part of $k_\parallel$ which is constant along the bunch simply changes the energy of all muons, but does not contribute to the energy spread.

In the transverse direction, I derive the following inequality for the transverse loss factor $k_\perp$ by imposing the condition that the transverse kicks accumulated over the relativistic muon lifetime, expressed as the number of turns n_{turn}, are not larger than the rms beam divergence at the interaction point σ':

$$k_\perp \leq \frac{(E/e)}{2Nen_{\text{turn}}\bar{\beta}} = \frac{(E/e)}{2I_b T_\mu \bar{\beta}} \qquad (6)$$

Here, $\bar{\beta}$ is the average value of the β-function in the components that drive $k_\perp$. There are no coherent transverse kicks when the beam is perfectly centred in the vacuum chamber. I therefore assumed that the typical offset of the beam around the circumference is of the order of the rms beam radius there in order to derive (6). The transverse kicks have two effects on the beams. They displace the beam centres, thus eventually causing mis-crossings at the interaction points, while their variation along the bunch causes an growth of the emittance.

Tab. 5 shows the maximum tolerable loss factors $k_\parallel$ and $k_\perp$, found by applying (5) and (6) to HEMC10 and HEMC100, respectively. I leave to the reader a comparison between these values, and the values of the loss factors that are likely to occur in HEMC10 and HEMC100.

TABLE 5. Maximum tolerable loss factors $k_\parallel$ and $k_\perp$ in HEMC10 and HEMC100

	HEMC10	HEMC100
CoM energy E/TeV	100	100
RMS relative energy spread $\sigma_e/10^{-3}$	0.0046	0.113
Average β-function $\bar{\beta}$/m	542	255
Muon life time n_{turn}	2090	2975
Longitudinal loss factor $k_\parallel/\text{VpCb}^{-1}$	2.94	19.7
Transverse loss factor $k_\perp/\text{V(pCbm)}^{-1}$	$4.91 \cdot 10^3$	$1.74 \cdot 10^5$

REFERENCES

1. The LHC Study Group, CERN/AC/95-05 (LHC) (1995).
2. G. Dugan et al., Proc. 1996 DPF/DPB Summer Study on New Directions in High-Energy Physics (1997) 251, also Cornell University CBN 96-18 (1996).
3. E. Keil, CERN/SL/97-13 (AP) (1997).
4. E. Keil, Proc. Particle Accelerator Conference PAC'97 (Vancouver, 12-16 May 1997) 104; also CERN-LHC-PROJECT-REPORT-102.
5. F. Ruggiero, Particle Accelerators **50** (1995) 83.
6. J.T. Rogers, Proc. 1996 DPF/DPB Summer Study on New Directions in High-Energy Physics (1997) 337, also Cornell University CBN 96-14 (1996).
7. E. Keil and B. Zotter, Proc. 6th European Particle Accelerator Conference EPAC'98 (Stockholm, 22-26 June 1998) 963; also CERN-SL-98-021-AP.
8. E. Keil, http://wwwslap.cern.ch/~keil/Math/lhc.nb
9. E. Keil, CERN-SL-99-053-AP (1999).
10. E. Keil, http://wwwslap.cern.ch/~keil/Math/hemc10TeV.nb
11. B. King, http://pubweb.bnl.gov/people/bking/heshop/hemc_para.html
12. E. Keil, http://wwwslap.cern.ch/~keil/Math/vlhc50GeV.nb
13. E. Keil, http://wwwslap.cern.ch/~keil/Math/hemc100TeV.nb

Prospects for Colliders and Collider Physics to the 1 PeV Energy Scale [1]

Bruce J. King

Brookhaven National Laboratory
email: bking@bnl.gov
web page: http://pubweb.bnl.gov/people/bking

Abstract.

A review is given of the prospects for future colliders and collider physics at the energy frontier. A proof-of-plausibility scenario is presented for maximizing our progress in elementary particle physics by extending the energy reach of hadron and lepton colliders as quickly and economically as might be technically and financially feasible. The scenario comprises 5 colliders beyond the LHC – one each of e^+e^- and hadron colliders and three $\mu^+\mu^-$ colliders – and is able to hold to the historical rate of progress in the log-energy reach of hadron and lepton colliders, reaching the 1 PeV constituent mass scale by the early 2040's. The technical and fiscal requirements for the feasibility of the scenario are assessed and relevant long-term R&D projects are identified. Considerations of both cost and logistics seem to strongly favor housing most or all of the colliders in the scenario in a new world high energy physics laboratory.

I INTRODUCTION

No clear-cut consensus currently exists on the best long-term strategy for experimentation in high energy physics (HEP) over the next 50 years. This paper puts the case for continuing to aggressively raise the frontier energy reach of both hadron and lepton colliders. It is argued that a continuation at the historical rate of progress in the log-energy reach of colliders is plausible and would provide us with outstanding prospects for deepening our understanding of the elementary entities and organizing principles of our physical Universe.

In order to demonstrate the possible feasibility of such a push to higher collider energies, a proof-of-plausibility scenario is presented for future colliders that would continue at the historical pace in log-energy reach and would, by about the year 2040, attain a constituent energy reach of 1 PeV (i.e. 1000 TeV).

---

[1] This work was performed under the auspices of the U.S. Department of Energy under contract no. DE-AC02-98CH10886.

The proof-of-plausibility scenario is only one choice from a parameter space of plausible scenarios that might advance the energy reach of colliders at the historical pace and, even if viable, no claim is made that it is in any sense optimal. Instead, it is intended as a spur to constructive criticism and future research that will lead to its refinement and to alternative scenarios. Any such discussions will help us to assess the future prospects of HEP and to identify long-term R&D needs. In turn, this will enable the field to make more informed planning decisions towards our long-term future.

The paper begins with motivational background on the essential role of past and future colliders for our understanding of elementary particle physics. It then reviews the technical challenges of energy frontier colliders and presents, and then evaluates, the aforementioned proof-of-plausibility scenario.

In more detail, the paper is organized as follows. Section II provides a brief historical review of the impressive historical gains in physical understanding from past accelerator experiments and then turns to an outline of the heady physics goals for future colliding accelerators. Section III gives a wish-list, motivated by physics considerations, for the technical specifications and capabilities of future colliders. These include increased energy (mainly), specification of the physics requirements for luminosity, and the physics advantages in being able to study more than one type of projectile collisions. Section IV reviews the rate of historical progress in the energy reach of colliders as characterized by the famous Livingston plot. It then introduces the proof-of-plausibility scenario as an extension to the Livingston plot. The technical challenges and potential energy reaches of future e^+e^-, hadron and $\mu^+\mu^-$ colliders are briefly assessed in sections V through VII, respectively, and justifications for the specific collider parameter choices of the proof-of-plausibility scenario are embedded in the more general discussions of these sections. The scenario as a whole is then assessed in section VIII, before concluding with a summary of the issues highlighted by the paper.

II COLLIDERS AT THE ENERGY FRONTIER ARE INDISPENSABLE

The continuing motivation for colliding accelerators (colliders) is to explore the most fundamental mysteries of the natural Universe: what are its fundamental entities and organizing principles ? What is the nature of space and time ? How did the Universe originate and evolve ?

Accelerators provide us with experimental insights on our Universe that could not plausibly be obtained in other ways and, to quote Harvard theorist Sidney R. Coleman (1), "Experiment is the source of imagination. All the philosophers in the world thinking for thousands of years couldn't come up with quantum mechanics". The properties and spectrum of elementary particles have been no less hidden from theoretical understanding than was quantum mechanics.

This section reviews the central historical importance of accelerators in uncovering what is known today as the Standard Model of elementary particles. It then turns to our future aspirations for a yet deeper understanding of the elemental entities and physical principles of our Universe. In our most optimistic hopes, this might ultimately be described by a complete and logically self-consistent "theory of everything".

A The Historical Importance of Accelerators

The past fifty years of experiments at accelerators have lead to remarkable progress in our understanding of the elementary processes and building blocks of our physical Universe. We discuss, in turn, the new insights gained on photons and on the building blocks of everyday matter, and then briefly summarize our current state of knowledge as encapsulated in the Standard Model of elementary particles.

A Context for Understanding Photons

Surprisingly, accelerators have greatly expanded our understanding of the multiple roles that photons play in the make-up and runnings of our Universe. It is manifestly obvious that these important insights could not have been attained without accelerator experiments – and this despite the fact that the photons themselves are all around us and can be studied in many ways.

In a famous quote (2), Albert Einstein once confessed that "All the fifty years of conscious brooding have brought me no closer to the answer to the question, "what are light quanta?"". While we certainly still can't claim full understanding, discoveries at accelerators have at least moved us towards a context and framework for understanding the photon that even Einstein could not have suspected. Far from being an isolated entity, the photon has massive siblings – known as the Z^0 and $W^\pm$ – that have been produced and studied at colliders, with observed masses of $M_Z = 91.2 \, \text{GeV}/c^2$ and $M_W = 80.4 \, \text{GeV}/c^2$, respectively. Their close relationship to the photon has been well established from their observed properties, and the interactions and relative couplings of the photon, Z and W to other elementary particles are now precisely specified by a theory, known as the electroweak theory, that unites the electromagnetic and weak interactions. (Experiments at upcoming colliders may further expand the scope of this theoretical framework to include the mechanism for generating all mass, as will be presented in the following subsection II B.)

In stark contrast to the photon itself, the large masses and immediate decays of its sibling Z and W bosons clearly preclude their direct study outside collider experiments. However, their effects are still seen in our everyday world since it is their interactions that turn out to be ultimately responsible for many radioactive decays – an everyday phenomenon that has been made a little less mysterious through the understanding provided by the electroweak theory.

Understanding Matter

Besides photons, our understanding of the building blocks of everyday matter has been revolutionized by accelerators. Electrons have passed progressively more stringent experimental tests of whether they are indeed point particles. On the other hand, protons and neutrons have been exposed as being composite entities rather than point particles. Experiments at accelerators have found them to be composed of hitherto unknown elementary particles: up-type, down-type and a few strange-type quarks bound together by gluons.

A Periodic Table of Elementary Particles

Accelerator experiments and, to a lesser extent, cosmic ray experiments have also shown us convincingly that the limitation of everyday matter to electrons, protons and neutrons is merely an "accident" of our low energy environment. Heavier forms of matter exist but cosmic ray interactions are the only naturally occurring process on Earth to supply the energy densities required to produce them. The few such particles that are produced sporadically by cosmic rays then decay almost instantaneously (sometimes in cascades of several stages) down to the familiar everyday particles – photons, electrons, protons and neutrons – plus the ghost-like neutrinos that are also all around us but are almost undetectable. Accelerator experiments are the only place where these particles can be systematically studied.

The list of additions to our everyday matter is impressive. Besides the aforementioned W and Z bosons, the electron has been found to have heavier siblings – the muon and the tau particle, and each of these has a neutral counterpart known as a neutrino that interacts so feebly that it is difficult to detect. Hundreds of quarks-plus-gluons bound states besides the proton and neutron have also been discovered, including bound states containing other, heavier quarks than the up and down quark – the so-called second generation strange and charm quarks and the third generation bottom (or beauty) and top quarks.

The new elementary particles fill out a veritable periodic table, which is shown in figure 1. Although smaller than the more familiar periodic table of the elements, the structural patterns are more complicated. The grouping of the particles in figure 1 reflects the particles' properties and the interactions they participate in. These properties and interactions are all well described by the so-called Standard Model of elementary particles (4), which will be discussed further in the following subsection.

To summarize our past progress, accelerator experiments have already revolutionized our understanding of the elementary building blocks and interactions in the Universe around us. They have led to the standard model of elementary particles and have exposed the intrinsic naivete of any pre-accelerator picture of the Universe that had the proton, neutron, electron and photons as the sole elementary particles. However, they have also highlighted our continuing naivete. Further collider

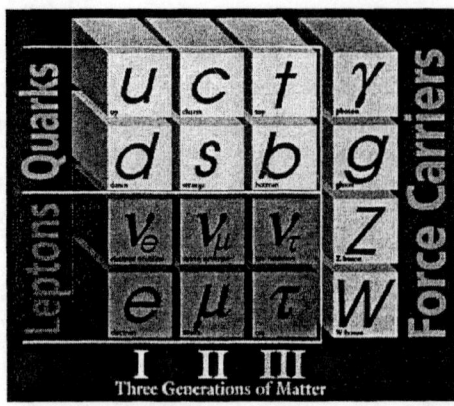

FIGURE 1. The known elementary particles. Illustration reproduced from reference (3).

experiments will be essential to achieve a more satisfying level of understanding, as will now be explained.

B Heady Physics Goals for Future Colliders

The Standard Model is only a Stop-Gap Theory

Even though the crowning achievement of the past 50 years of HEP has been the construction of the Standard Model, our intellectual goals for its future are very much more wide-reaching than simply filling out the "periodic table" of figure 1 and further detailing its properties. For, despite its predictive power for existing HEP experiments, we know the Standard Model to be no more than a stop-gap theory with a limited domain of applicability. It is phenomenological rather than fully predictive, incorporating 19 free parameters that need to be determined by experiment. Even more damning, it becomes logically inconsistent when we try to extrapolate it to experimentally inaccessible energy scales.

Instead, the quest is for a deeper knowledge of where the Standard Model comes from and also for understanding its connection to the existence and bulk properties of the Universe such as its preponderance of matter over antimatter and its gravitationally-curved 3+1 dimensional space-time structure.

Strategies for Advancing beyond the Standard Model

This paper deals largely with the paramount importance of energy frontier colliders as discovery machines to further uncover the secrets of elementary particles and, hence, to learn more about the physical foundations of the Universe. This emphasis on energy reach is further justified in section III A.

Apart from the approach emphasized here, it is worthwhile to briefly mention some other of the well-developed alternative or complementary strategies for examining these questions at the current and next generations of colliders. More detailed expositions can be found elsewhere in these proceedings (5; 6; 7).

Very briefly, the origin of all mass in the Universe is hypothesized, in the Standard Model, to be intimately tied in with an as yet undiscovered particle known as the Higgs boson. This is a hypothesized extension to the electroweak theory that was discussed previously – that is to say, our context for understanding photons could further lead to a context for understanding the generation of all mass! As such, it provides a beautiful example of how experimental advances in elementary particle physics might build up a level of understanding of our Universe that would have been otherwise unattainable. The LHC has been optimized to search for this Higgs particle and proposed TeV-scale linear colliders have been optimized for follow-up precision studies of such a particle, if it exists. (Note, however, that both the LHC and e^+e^- colliders will anyway be at the energy frontier, so this is merely a shading of the approach emphasized in this paper rather than a true alternative strategy.)

As another thread (6), colliders have the potential to reproduce and study the extreme energy densities that existed in the first moments after the Universe formed in the hypothesized "Big Bang". Collisions of heavy ions – at the RHIC collider, for example – will provide the largest volumes under such conditions, even though these volumes are admittedly still miniscule.

Related to this, it has been hypothesized that the observed dominance of matter over antimatter in the observable Universe might have originated from the properties of exotic heavy particles and their anti-particles that could have been routinely produced in the earliest moments of the Universe and whose interactions could have involved large matter-antimatter asymmetries – this is known as CP violation in the obscuring lingo of the field. We may get an experimental handle on such possibilities from experiments at B factory colliders and elsewhere that study the effects of CP violation.

The "Theory of Everything"

Ultimately, the "holy grail" of high energy physics is to advance from the Standard Model to the hypothesized "Theory of Everything" that describes and explains the elementary entities, structure and organizing principles of our Universe and is predictive – at least in principle, even if not calculationally – for any experiment we could conceive of that involves elementary particles.

The nature of the Theory of Everything and any possible intermediate levels of understanding towards the Theory of Everything are the subjects for current speculation and future discoveries at accelerators, coupled to theoretical breakthroughs in interpreting these discoveries. Speculation on the elementary particle physics phenomena we might find at future accelerators is helpful in stimulating theoretical progress towards the Theory of Everything and also for the design of the future colliders and their experiments. Several example scenarios for what we could find at many-TeV muon colliders have been presented in these proceedings (8; 9; 10), along with a very helpful classification scheme for the possibilities (11).

The Cambridge theorist and physics popularizer Stephen Hawking gives the field a 50% chance (12) of attaining the Theory of Everything within the next 20 years. This is the most optimistic assessment I am aware of and had many of the theorists at this workshop shaking their heads. At the opposite extreme, it is even logically possible that a unified understanding of the Universe is simply beyond human intellect or, as Steven Weinberg put it, like trying to explain Newtonian mechanics to your dog!

In any case, we know it will not be *easy* to reach a complete understanding of the physical foundations for our Universe, and collider experiments at the energy frontier will presumably be our main experimental tool in this heady quest.

III A WISH-LIST FOR THE PARAMETERS OF FUTURE COLLIDERS

This section begins by stressing the paramount importance of energy reach in determining the physics potential of future colliders. (It should be acknowledged that the viewpoint expressed here has certainly been influenced by other presentations at the HEMC'99 workshop and contributions to these proceedings, particularly that of Samios (13) and the summary presentation by Willis (6).)

The auxiliary requirements for, and benefits of, adequate luminosity and utilizing a variety of colliding projectiles are then discussed in subsections III B and III C. Beam polarization and clean event reconstruction are other relevant experimental capabilities whose discussion is left to elsewhere in these proceedings, in references (14) and (15; 16), respectively.

A The Paramount Importance of Energy Reach

Energy Reach versus Less Direct Experimental Strategies

Some specific experimental strategies for extending our knowledge beyond the Standard Model were discussed in the preceding section. However, the only way we can *directly* examine an energy scale is by cranking up the energy of our colliders to reach that energy scale. This will then allow us to observe any exotic particles or even more complicated entities (11) that might exist at that energy scale, whether or not they had been previously forecast on theoretical grounds. Hence, a direct frontal assault on the collider energy frontier is intrinsically more powerful and more likely to result in major break-through discoveries than are alternative, more indirect experimental approaches.

Energy Frontier Colliders Can Also Do Lower Energy Physics

In weighing the balance for future frontier machines versus lower energy colliders it should be borne in mind that, besides their primary mission of discovery, frontier machines can also do well at studying lower energy processes – often even better than at dedicated lower-energy facilities.

As an example, the LHC, with $E_{CoM} = 14$ TeV, will be one of the best places to do studies with B ($M \simeq 5\ GeV/c^2$) and charm ($M \simeq 1.7\ GeV/c^2$) particles. Even lepton colliders have the general property that lower mass particles are produced in higher order processes and in the decays of heavier particles.

A currently relevant example with lepton colliders is given by collider parameter sets for the 10 TeV and 100 TeV muon colliders that were studied at this workshop (17). The specified luminosities would correspond to the production of more than 10^7 Standard Model Higgs particles if these existed at the 100 GeV mass scale. This would be orders of magnitude more events than at any of the lower energy electron or muon colliders that have been proposed with the principal goal of studying such a Higgs. (Admittedly, less precise event reconstruction may somewhat dilute the statistical advantage.) Therefore, at least some aspects of any Higgs particle, such as rare decay modes, might be better addressed at frontier colliders than at dedicated Higgs machines operating at the few hundred GeV energy scale.

Besides examples using future colliders, the case can also be made in a historical context, as now follows.

An Alternative History: The Standard Model could have been Reconstructed from Today's Energy Frontier Experiments Alone

We now consider a historical "what if" question that highlights both the paramount importance of energy reach and the ability of energy frontier colliders to perform analyses concerning lower energy scales.

Consider the state of elementary particle physics a half century ago, in 1950. The positron (1933), muon (1937) and pion (1947) had been discovered in cosmic ray experiments, following up on the discovery of the neutron (1932) and the inferred existence of the neutrino from beta decay spectra (1932-3). The historical gedanken experiment is to imagine that, instead of the newly commissioned 184-in synchrocyclotron at Berkeley, which could produce pions, the HEP community of 1950 had been immediately gifted with today's energy-frontier hadron and lepton colliders – the 1.8 TeV Tevatron proton-proton collider and the 90–200 GeV LEP electron-positron collider – along with the technology for their modern-day general purpose collider experiments.

We can then ask the following question: how much of today's current understanding of elementary particles (i.e. the Standard Model) would have been promptly reconstructed from the data and what, if anything, would have been missed ?

It can be argued that the basic structure of the Standard Model would have been quickly recovered – either in its entirety or nearly so – since the Tevatron and LEP see evidence for all of the particles in table 1 (redundantly, in most cases) and provide measurements of their interactions and couplings.

In more detail, the copious production of W's and Z's would quickly arrive at the electroweak theory that was mentioned previously in section II A. Knowledge of the strong interaction and of the point-like quarks and gluons it acts on would also come easily, from observations of the "jettiness" of hadronic events at both colliders and from other evidence, and these event signatures would also show immediately that the Tevatron's proton projectiles were composed of these quarks and gluons.

Probably the last piece of the Standard Model structure to be experimentally established in this scenario would be the complex phase in the CKM quark mixing matrix that accounts for CP violation. The energy frontier collider experiments are poorly optimized for observing the small effects of CP violation in kaon decays where this phenomenon was experimentally discovered, and the alternative of experimental evidence for CP violation in B decays still has only marginal statistical significance at both LEP (18) and the Tevatron (19).

Even though CP violation is the part of the Standard Model least suited for study at the Tevatron and LEP, their data would certainly still provide the CKM matrix as a theoretical construct and would show it to be non-diagonal. From there, theoretical conjecture on a possible complex phase would be natural and this could well lead to a re-optimized detector (e.g. similar to the B-TeV detector that has been proposed for the Tevatron) that could follow up with more definitive measurements of CP violation to complete the picture of the Standard Model.

To summarize, the outcome of the above gedanken experiment reinforces the previous conclusions of this section by demonstrating that today's energy frontier colliders can quickly provide access to all of the elementary particle physics structure that we are aware of from our 50 years of historical progress.

B Desirable Luminosities and their Scaling with Energy

The luminosity of future high energy colliders is the machine parameter that is second in importance only to energy reach.

A rule of thumb for hadron colliders that came into prominence in the 1980's is that the physics gain from a factor of 10 in a hadron collider's luminosity corresponds roughly to factor of 2 in energy reach for hadron colliders. The possibilities for such a trade-off are presumably more limited for the point-like projectiles of lepton colliders, where E_{CoM} gives a more precise measure of the discovery energy reach.

Probably the best way to define the luminosity goals for energy frontier colliders is that the luminosity should be sufficient to gather good statistical samples for the study of any elementary particles existing at the energy scale, $E \leq E_{\text{CoM}}$.

This definition raises a conundrum for discovery colliders at the energy frontier: the number of events is given by the product of the production cross section for the particle, σ, and the time integral of the luminosity, $\mathcal{L}$:

$$\text{no. events} = \sigma \int \mathcal{L}\,dt; \tag{1}$$

yet how can we predict the cross sections for unknown particles?

Fortunately, it is common knowledge that very approximate upper limits for production cross sections as a function of collider energy can be guessed at just from general considerations of relativistic quantum mechanics. We now give a version of the type of hand-waving argument that makes this connection. This argument works for the point-like projectiles of lepton colliders at any chosen E_{CoM}. The very approximate luminosity specifications that result could arguably be extended to hadron colliders by replacing E_{CoM} with the equivalent energy reach, $E_{\text{reach}}^{pp} \simeq E_{\text{CoM}}/6$, for the collisions of the quark and gluon sub-components of protons. (See section IV for further discussion on the parameter E_{reach}^{pp}.)

The venerated Heisenberg uncertainty principle of non-relativistic quantum mechanics,

$$\Delta p \, \Delta x \geq \frac{\hbar}{2}, \tag{2}$$

can, for elementary particles, be recast into the very approximate relativistic form

$$\Delta E \, \Delta x \sim \hbar c, \tag{3}$$

where Δp is the momentum spread of a particle's wave-function, Δx is the position spread, $\hbar$ is the reduced Plank's constant, ΔE can be considered as the energy scale of an interaction and Δx gives the corresponding spatial extent, and the conversion from equation 2 to equation 3 uses the approximate ultra-relativistic relation $E \simeq pc$ that neglects the incoming particle masses. The cross sectional area over which the interaction can occur will be of order the square of the spatial extent of the interaction, roughly $(\Delta x)^2$, so the maximum cross section for a given center-of-mass energy will be roughly:

$$\sigma_{max} \sim (\Delta x)^2 \sim \left(\frac{\hbar c}{E_{\text{CoM}}}\right)^2, \tag{4}$$

or, numerically,

$$\sigma_{max}[pbarn] \sim \frac{400}{(E_{\text{CoM}}[TeV])^2}, \tag{5}$$

where units are given in square brackets in this equation and throughout this paper, and 1 picobarn (pbarn) is 10^{-12} barn or 10^{-36} cm^2.

The crude estimate of equation 5 actually does surprisingly well at predicting the largest cross section at today's 100-GeV-scale e^+e^- colliders: it predicts $\sigma_{max} \sim 50$ nbarn at the energy of the Z pole, 91.18 GeV, which agrees well with the actual cross section of 38 nbarn.

Apart from the Z resonance, however, most cross sections for point-like interactions have been observed to fall several orders of magnitude below the value of equation 5. This can be explained away, in the hand-waving spirit that the equation was derived, by saying that any coupling suppressions arising from the detailed physics process will generally reduce the probability of an interaction occurring for even the closest encounters.

As an acknowledgment that large coupling suppressions are the norm rather than the exception, the luminosities of lepton colliders are commonly bench-marked to a process other than resonant Z production, namely, to lepton-antilepton annihilations to fermions through photon exchange, e.g.,

$$e^+e^- \xrightarrow{\gamma} \mu^+\mu^-. \tag{6}$$

This has a cross section of:

$$\sigma_R[pbarn] = \frac{0.087}{(E_{CoM}[TeV])^2}. \tag{7}$$

(To be precise, this only gives accurate predictions for the cross section at energies well below the Z resonance, at 91 GeV. At energies above this, the cross section is substantially modified by interference with the corresponding process involving Z exchange. Instead, equation 7 is intended as the *definition* of a benchmark cross section that can be used at all energies, as we now explain.)

The inverse of the characteristic cross section of equation 7 defines a unit of integrated luminosity known as a "unit of R" such that a collider that collects one unit of R of integrated luminosity will produce, on average, one event that has the cross section of equation 7.

It was the guidance (20) of SLAC theorist Michael Peskin that the luminosity for this workshop's straw-man muon collider parameter sets should, if possible, allow an accumulated inverse luminosity of 10 000 units of R. To convert this to an average luminosity it can be noted that obtaining this integrated luminosity over 5×10^7 seconds of running (five "Snowmass accelerator years") requires an average luminosity of:

$$\mathcal{L}^{desired}[cm^2.s^{-1}] = 2.3 \times 10^{33} \times (E_{CoM}[TeV])^2. \tag{8}$$

This can be contrasted with the much more modest luminosity that would be needed to acquire 10 000 events produced at the approximate maximum cross section specified by equation 5:

$$\mathcal{L}^{borderline}[cm^2.s^{-1}] = 5 \times 10^{29} \times (E_{CoM}[TeV])^2. \tag{9}$$

The straw-man parameters for both the 10 TeV and "100 TeV ultra-cold beam" examples met or even exceeded Peskin's request, each with 8700 units of R per detector in a single year. On the other hand, the "100 TeV evolutionary extrapolation" parameter set specified only 87 units of R per detector per year. This reflects the escalating luminosity demands with E_{CoM} due to the $1/E_{CoM}^2$ cross section scaling of equations 5 and 7.

C The Complementarity of Different Projectile Types

The different experimental conditions and, particularly, the different interacting projectiles of hadron and lepton colliders will generally lead to different sensitivities for specific processes at the energy scale under consideration, so the two types of colliders are also complementary to a certain extent and there are advantages to operating both types of machines. This complementarity also applies to the two types of lepton colliders – e^+e^- and $\mu^+\mu^-$ colliders – but to a lesser extent.

There are also many other possibilities for the colliding projectiles that will not be discussed further in this paper: gamma-gamma collisions, heavy ion colliders, like-sign lepton colliders (14) (e^-e^- and $\mu^-\mu^-$) and any one of the several options that collide dissimilar projectiles. These options all have some potential for complementary physics studies and should be looked at further. However, it should be noted that several of them are understood to be less suitable for exploring the energy frontier for various reasons, a few of which are discussed elsewhere in these proceedings (21).

In the past, energy frontier hadron colliders have been regarded more as discovery machines while lepton colliders, following later but with cleaner experimental conditions, have been considered mainly as follow-up machines for precision studies. The following section reviews the history of collider facilities that led to this assignment, as well as introducing a speculated scenario for future colliders.

IV THE LIVINGSTON PLOT FOR PROGRESS IN THE ENERGY REACH OF COLLIDERS – PAST, PRESENT AND FUTURE

A Presentation and Interpretation of the Plot

Figure 2 is the famous Livingston plot showing the historical exponential growth with time in the energy reach of both lepton and hadron colliders. The data for past and present lepton and hadron colliders has been taken from reference (22) and is discussed and parameterized in the following subsection.

The logarithmic energy scale in figure 2 is physically appropriate under the reasonable assumption that the underlying physical importance of the mass spectrum will lie in the *ratios* of particle masses as opposed to mass *differences*. As some

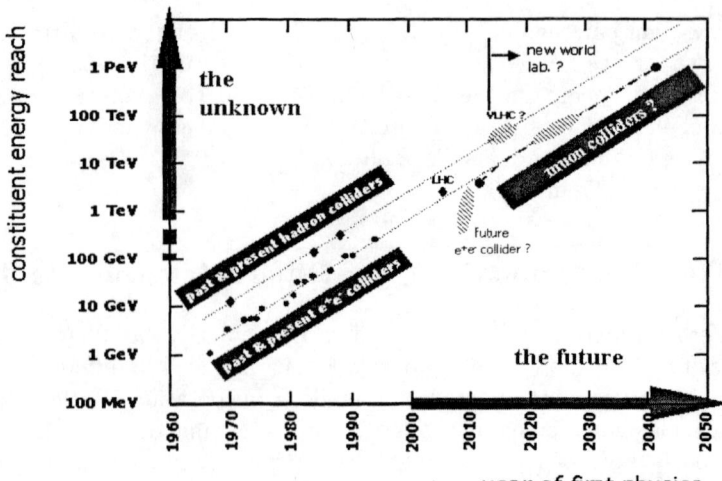

FIGURE 2. The Livingston plot showing the historical progress in the constituent energy reach of lepton and hadron colliders. Each point on the curve represents a collider. A possible scenario for future colliders to continue the exponential progress in hadron and lepton energy reach has been added. The definitions and data points in the plot are discussed further in the text, as is the scenario for future colliders.

confirmation of this assumption, the masses of the known elementary particles do indeed fall relatively evenly along a log-energy scale rather than being bunched at the low energy end. To rephrase this in a way that might sound depressing to accelerator builders, the past exponential progress in the energy reach of colliders can be considered to have corresponded to merely a steady (linear) rate of advance in their physics capabilities since the logarithm of the energy rather than the energy itself is the appropriate metric for assessing the discovery reach of colliders.

Some speculated future colliders beyond the LHC have been added in to figure 2. In sum, they are intended to comprise a straw-man proof-of-plausibility scenario to show that a sufficiently motivated and adequately funded HEP community may be able to continue constructing accelerators that lie on or near the lepton and hadron Livingston curves and that extend up to the PeV constituent energy scale (where 1 PeV = 1000 TeV). Discussion on this scenario occupies the final subsection in this section, subsection IV C, as well as much of the remainder of this paper.

B Parameterizations for the Historical Progress in the Energy Reach of Hadron and Lepton Colliders

The constituent energy reach for lepton colliders in figure 2 has been defined simply to be their center-of-mass energy,

$$E_{reach}^{lepton} \equiv E_{CoM}. \tag{10}$$

The fact that protons are composite particles rather than point-like elementary entities dilutes the constituent energy reach of hadron colliders relative to lepton colliders by a factor that depends on both the physics process and the collider luminosity. The choice of dilution factor for past hadron machines was copied from reference (22) and we have chosen the similar dilution factor of 6 for future hadron colliders:

$$E_{reach}^{pp} \equiv E_{CoM}/6. \tag{11}$$

For any given hadron collider, other estimates for the dilution factor may differ by a factor of two or more from this choice, in either direction. There is also a slight arbitrariness in some of the other data choices, so the reader is warned that the details in Livingston plots may vary from publication to publication.

The two dashed lines drawn through the data points give parameterizations for the constituent energy reach, E_{reach}, versus year of first physics, Y, for lepton and hadron colliders. They have equations:

$$\log_{10}(E_{reach}^{lepton}[TeV]) = (Y - 2002)/13 \tag{12}$$

and

$$\log_{10}(E_{reach}^{pp}[TeV]) = (Y - 1994)/13, \tag{13}$$

respectively and substituting the energy dilution factor from equation 11 into equation 13 gives the required CoM energy reach for future hadron colliders:

$$\log_{10}(E_{CoM}^{pp}[TeV]) = (Y - 1984)/13. \tag{14}$$

Equation 12 informs us that a decade of energy increase in lepton colliders has historically occurred every 13 years. Proton colliders have advanced at the same rate of progress as lepton colliders – an energy decade every 13 years – but have been about 8 years ahead of lepton colliders in attaining a given constituent energy reach.

As well as past and present colliders, figure 2 plots the planned 2005 completion date for the Large Hadron Collider (the LHC, currently under construction) and a region representing roughly the range of predicted or proposed turn-on dates and energies for contemplated electron-positron colliders. Beyond these, later and more speculative collider points at higher energies have also been added to figure 2. A straw-man "target" region for a Very Large Hadron Collider (VLHC) is shown with an energy range of 200–400 TeV corresponding to first physics dates of 2014-8. Three muon collider points/regions are also shown, extending up to the lepton Livingston curve and with CoM energies up to 1 PeV.

It can be noted that the cancelation of the SSC collider has hampered progress in the energy reach of hadron colliders, as is also stressed elsewhere in these proceedings (13). The SSC would have been above the Livingston curve for hadrons if it had already been built at its design energy of $E_{CoM} = 40$ TeV. Instead, it can be seen from figure 2 that the LHC has already fallen below the Livingston curve; to have stayed on the curve, equations 11 and 13 specify that it would have needed to have either turned on last year (1999) or, for its planned 2005 turn-on, had an E_{CoM} of 42 TeV rather than 14 TeV.

C Introducing the Proof-of-Plausibility Scenario for Future Colliders

TABLE 1. A time-line for energy frontier muon colliders that would hold to the historical rate of progress in the highest collision energies attained with lepton colliders.

year	E_{CoM} & type	description
2009-11	0.4-3 TeV e^+e^-	EW-scale/energy frontier linear collider
2012	4 TeV $\mu^+\mu^-$	lab. site-filler with low beam current
2014-18	200-400 TeV pp	400 km circum. is boundary for new lab.
2021-8	30-100 TeV $\mu^+\mu^-$	high luminosity collider in new world lab.
2041	1 PeV $\mu^+\mu^-$	linear collider with electron drive beams

Table 1 summarizes the collider type, energy reach and year of first operation for the colliders in the proof-of-plausibility scenario to extend the Livingston plot beyond the LHC.

The scenario is economical in requiring only 5 colliders to reach all the way up to the 1 PeV constituent energy scale – one each of e^+e^- and hadron colliders and three $\mu^+\mu^-$ colliders. The correspondingly large leaps in energy continue the necessary trend that was first set by the SSC, with $E_{CoM} = 40$ TeV, which improves by more than a factor of 20 from the existing energy frontier at $E_{CoM} = 1.8$ TeV. (The SSC ended up being too expensive, but this should not be interpreted as a fundamental flaw in the concept.) The increasingly large energy jumps are dictated by the rising cost of each successive machine. They are also desirable to give each new machine a big enough window for a good chance to discover new physical phenomena or, at minimum, to rule out a big enough energy range for this to have theoretical significance.

In order to limit the number of colliders in table 1 to five, the energies of the colliders have all been pushed up as far as was considered practical towards, in each case, a natural and rather sharply defined technological bound. For the first collider in table 1 – a TeV-scale e^+e^- collider – the fundamental energy bound comes from the sharp rise with energy in the interactions between TeV-scale electrons and electromagnetic fields, as will be discussed in section V.

Only one extra hadron collider was included in the scenario since the energy scale of the LHC, with $E_{CoM} = 14$ TeV, will already be enough energy to probably warrant jumping in a single step up to the few hundred TeV energy bound that results from synchrotron radiation. The limit and the 200-400 TeV hadron collider will be discussed in section VI.

The 3 muon colliders all butt up against different technological energy limits, as will be further addressed in subsection VII B. The 4 TeV muon collider is about at the size limit for fitting on a laboratory site and, more importantly, is fighting against the energy-cubed rise in the off-site neutrino radiation hazard for populated locations. The middle muon collider pushes the synchrotron radiation energy limit for circular muon colliders, which is probably in the range of 30-100 TeV. Finally, 1 PeV is a limiting energy scale for any linear muon colliders that can't call on exotic acceleration schemes to obtain accelerating gradients that greatly exceed today's technological bounds.

While the scenario looks plausible at first sight, it is important to stress that this is only a first presentation, and the scenario's feasibility or otherwise should be better established with further studies. Also, no claim is made that the scenario is in any way optimal and it is hoped that follow-up studies and technical progress will lead to refinements and also to alternative scenarios. This would give us more options and possibilities to progress with foresight towards a productive future for HEP.

V FUTURE ELECTRON-POSITRON COLLIDERS

A Energy Limitations on e^+e^- Colliders

The fundamental energy limitation for electron-positron colliders is the projectile itself. At only 1/1837 of the proton mass, electrons are relativistic enough at TeV energy scales and beyond to be overly sensitive to electromagnetic fields. They become difficult to bend, focus or collide because of their excessive tendency to throw off photons due to synchrotron radiation in the magnetic fields of bending and focusing magnets and then to beamstrahlung in the electromagnetic fields of the oncoming bunch at collision.

Even to advance to the TeV scale, these difficulties have already required the major technology shift from circular colliders to single pass linear colliders. Synchrotron radiation scales as the fourth power of the beam energy (or as the third power, for some parameters), so further technological innovations would clearly be needed to advance to still higher energies. To solve each of the three classes of problems, we would need all three of:

1. acceleration that doesn't have a prohibitive linear cost coefficient with energy

2. focusing to collision that doesn't induce excessive synchrotron radiation

3. some way of damping out the beamstrahlung at collision.

Research is underway to attempt solutions to each of these three problems, which have arguably been presented in order of increasing difficulty. Potential solutions include electron drive beams for acceleration, plasma focusing or focusing induced by auxiliary beams, and 4-beam schemes that also use additional beams to partially cancel out the electromagnetic fields at collision.

B This Scenario: A Single TeV-Scale Electron-Positron Collider

Planning for proposed TeV-scale e^+e^- colliders is far enough along that the shaded region for e^+e^- in figure 2 simply represents the very approximate range of energies and turn-on dates for proposed machines.

It can be seen that the shaded region extends well below the Livingston curve. This reflects the technological problems associated with energy that were discussed in the preceding subsection and, somewhat related to this, the widespread support for the alternative motive of studying identified physics processes – such as production of top quarks or Higgs particles in the 100-200 GeV mass range, if they exist – rather than a dedicated thrust towards the energy frontier.

The scenario assumes that exactly one TeV-scale e^+e^- collider will be constructed before the mantle of frontier energy lepton collider is passed on to muon colliders. Besides its own physics value, this e^+e^- collider would continue to establish the technology needed for the later muon colliders and, in particular, for the assumed 1 PeV muon collider that completes the scenario. The technological overlap between the two types of lepton colliders is discussed further in section VIII A.

VI FUTURE HADRON COLLIDERS FOR THE ENERGY FRONTIER

This section discusses current research towards future hadron colliders and assesses the constraints on their ultimate potential energy reach. As a particular constraint on their energy reach, it is explained why linear hadron colliders will probably never be viable. Finally, the 200-400 TeV pp collider in the proof-of-plausibility scenario is discussed.

A Current Research Towards Future Hadron Colliders

Beyond the LHC, research is underway for a follow-up hadron collider, usually referred to as the "Very Large Hadron Collider" (VLHC), whose goal would be to reach substantially higher energies than the LHC without being much more expensive. The U.S. research efforts towards a VLHC are conveniently summarized in the annual report of the U.S. Steering Committee for a VLHC (23).

Much of the VLHC research involves magnet design because the bending magnets for the collider storage ring are assumed to be a large, if not dominant, component of the cost for a VLHC. (For comparison, dipole magnet costs for the collider ring of the SSC were budgeted to comprise roughly 25% of the total cost.) The current emphasis on magnet R&D is largely divided between the design of low field (2 Tesla) superferric magnets and very high field (greater than 9 Tesla) magnets.

The low field superferric magnet designs might be very cheap to construct. As one of the challenges for this option, there is concern that the attainable luminosities for the low field superferric option may be limited due to beam instabilities caused by the combined effects of the small magnet apertures, the large collider circumference and the lack of synchrotron radiation damping. The potential lack of stability and tune-ability for such simple magnets is also a concern.

A major motivation given for very high field magnets, as opposed to intermediate field magnets, is their potential to cause a beneficial level of synchrotron radiation damping of the beam at an assumed beam energy of 50 TeV. The radiation damping is higher for high field magnets because the fractional energy loss per turn, $\frac{\Delta E}{E_{CoM}}$ due to synchrotron radiation scales with E_{CoM} and with the average bending magnetic field, B, according to:

$$\frac{\Delta E}{E_{CoM}} \propto E_{CoM}^2 \times B. \qquad (15)$$

Very high field magnets cannot use the niobium-titanium superconductor that has been used in all collider magnets to date since this conductor has an impractically low critical current at magnetic fields above 9 Tesla. Other superconductor materials must be used, such as niobium-tin or the new high-T_C superconductors, and these are – at least at present – considerably more expensive than niobium-titanium and have inferior mechanical properties.

Regardless of progress in superconducting materials, the mechanical stresses in magnets scale as the square of the magnetic field and will always conspire to raise the cost per unit length of high field magnets relative to those at lower fields. On the other hand, a more relevant yardstick than cost per unit length is the cost per Tesla-meter since a collider ring at a given energy will need a fixed number of Tesla-meters to bend the beams in a circle. This favors higher field magnets so long as the cost per meter increases less than proportionally to the field strength. Also, higher fields should reap further cost savings from the consequent reduction in tunnel length. Because of these trade-offs, the field strength for superconducting magnets that would give the optimal cost for a future hadron collider is not at all well established. (However, see the discussion in subsection VI C regarding a study by Willen (24) that uses *today's* superconducting magnet technology.)

More generally, it appears that the cost and technology optimizations used for the VLHC could usefully be extended to include varying the energy of future hadron colliders away from the VLHC's assumed $E_{CoM} = 100$ TeV, as will be addressed in subsection VI C.

B   The Ultimate Energy Reach for Hadron Colliders

Limits from Synchrotron Radiation

For the VLHC studies at $E_{CoM} = 100$ TeV, the beneficial damping effects of synchrotron radiation must already be balanced against the problems it causes. Given the strong power-law rise in synchrotron radiation with energy, it can be surmised that synchrotron radiation should lead to a fairly sharp technical cut-off in the viability of hadron colliders by the few hundred TeV range. More quantitatively, the power radiated due to synchrotron radiation, P_{synch}, is given approximately by:

$$P_{synch}[MW] \simeq 0.6 \times I[A] \times B[T] \times (E_{CoM}[100 \text{ TeV}])^3, \qquad (16)$$

where I is the average current in each proton beam and B is the bending magnetic field which, for this equation, is simplistically assumed to be constant around a circular collider ring.

Constraints from the Experimental Environment

Probably the biggest technical challenges for the SSC and LHC came, not from the colliders themselves, but from the extreme operating conditions anticipated for the collider detectors. The experiments at future energy frontier hadron colliders will have even worse problems coping with luminosities that, as was shown in section III B, should ideally rise as the square of E_{CoM}. The problems arise from the large cross section for soft background interactions:

$$\sigma^{pp}_{TOT} \simeq 200 \text{ millibarns}, \qquad (17)$$

rising only slightly with E_{CoM}. The average number of background events per bunch crossing, n_b, is given by,

$$n_b = \frac{\mathcal{L} \times \sigma^{pp}_{TOT}}{f}, \qquad (18)$$

with f the bunch-crossing frequency, or, numerically,

$$n_b \simeq 2 \times \frac{\mathcal{L}[10^{34} \text{ cm}^{-2}.\text{s}^{-1}]}{f[\text{GHz}]}. \qquad (19)$$

Desirable luminosities in the range 10^{35-36} cm^{-2}.s^{-1} will require major advances in detector technology and event analysis to resolve the background event pile-up predicted from equation 19 and also to cope with event-induced radiation damage to central detectors. Particular attention will need to be paid to the radiation hardness of the central tracker and electromagnetic calorimeter, and to fast timing, triggering and read-out of the events.

Equation 19 shows that the bunch crossing frequency, f, should be made as large as is practical in order to minimize the event pile-up. Ideally, the time between crossings, $1/f$, should be comparable to the resolving time of the detector, which tends to be limited to on the order of nanoseconds. (The LHC is designed for one bunch crossing every ?? nanoseconds.) However, in practice this turns out to be an inefficient way to produce luminosity, requiring large stored beam currents that exacerbate the synchrotron radiation problem of equation 16 and bring on other technical headaches.

The Implausibility of Single Pass Hadron Colliders

An extreme example of the inefficiency of frequent bunch crossings would be provided by any attempted design for a *single pass* hadron collider that would use the linear accelerator technology developed for e^+e^- colliders. The magnitude of the problem appears to prohibit any serious speculation on using linear hadron colliders to extend the energy frontier, as we now show.

To obtain a crude scaling argument, we note that the luminosity as a function of the bunch crossing repetition rate, f, the number of particles per bunch, N_b, and the transverse beam dimensions, σ_x and σ_y, is given roughly by:

$$\mathcal{L} \simeq \frac{f\, N_b^2}{\sigma_x \sigma_y} \propto \frac{1}{f} \cdot \frac{(P_{\text{beam}})^2}{\sigma_x \sigma_y}, \qquad (20)$$

where we haven't bothered to keep track of numerical factors depending on the precise definitions of σ_x and σ_y (several different conventions are in common use!) and have ignored effects of order unity such as the pinch enhancement of luminosity, and the second expression follows from the first because the average beam power, P_{beam}, is proportional to the average beam current, $P_{\text{beam}} \propto f N_b$. Equation 20 shows that the luminosity at linear colliders falls off inversely with the repetition rate – at least to the extent that the pinch enhancement can be neglected and, as a more substantial caveat, provided that the beam power and transverse beam dimensions are fixed. Therefore, very low repetition rates are strongly favored and the nanosecond-scale repetition rates desired for hadron colliders are strongly disfavored.

To set the numerical scale, the first very speculative straw-man parameter set for a 1 PeV linear muon collider (25) assumes a luminosity of $\mathcal{L} = 5.4 \times 10^{35}$ cm^{-2}.s^{-1} and a repetition rate of only $f = 3.2$ Hz. (This corresponds to a very impressive per-collision integrated luminosity of 170 inverse nanobarns per collision!) If this parameter set was translated to a 1 PeV hadron collider rather than a muon collider then equation 19 predicts a manifestly unmanageable 3×10^{10} interactions per collision!

Even if one allows for a re-optimization of parameters from this rather extreme example, it is hard to imagine a single-pass hadron-hadron collider with both an interesting luminosity and viable experimental conditions. The conclusion of this subsection is then that the ultimate center of mass energy for hadron colliders will almost certainly be attained using circular hadron colliders and will probably be limited to a few hundred TeV.

C A 200-400 TeV Hadron Collider for the Proof-of-Plausibility Scenario

Basic Specifications

The preceding subsection established that the collision energy often assumed in VLHC studies, $E_{CoM} = 100$ TeV, is rather close to the ultimate energy scale possible for future hadron colliders, at – we will guess – perhaps 200 to 400 TeV. It is notable that the energy jump from the LHC ($E_{CoM} = 14$ TeV) to this energy would be about the same factor of twenty-or-so as was planned to go from the Tevatron (1.8 TeV) to the SSC (40 TeV). Both rate-of-progress and economy therefore suggest that it makes eminent sense to try to reach this frontier energy in a single step. This motivates the choice for our proof-of-plausibility scenario of a single hadron collider after the LHC, at $E_{CoM} =$ 200-400 TeV. (The possibility has not been excluded for an upgrade from the lower end to the higher end of this energy range.)

For definiteness in the overall scenario, we can also assume a 400 km circumference for the 200-400 TeV hadron collider ring. As will be seen later, this fits in with the size scale for a new world HEP laboratory that also includes a muon collider.

Magnets and Synchtrotron Radiation Damping

A 400 km circumference corresponds to an average bending magnetic field around the collider ring of 5.3 (10.5) tesla for the assumed energy range of $E_{CoM} = 200$ (400) TeV.

Because the 200-400 TeV energy range is higher than the $E_{CoM} = 100$ TeV value usually considered for the VLHC, it is notable that it might give a better match between, on the one hand, the desirable level of synchrotron radiation damping sought by those designing high field VLHC magnets and, on the other hand, the lower magnetic field strengths that might correspond to a cost minimum. In fact, as a very intriguing possibility that is more general than this scenario, it is not ruled out that the global cost optimum for optimal synchrotron damping might be at lower magnetic fields but higher collision energies than are currently considered for the high field 100 TeV VLHC, i.e. the extra energy reach could conceivably come for free! The synchrotron radiation and cost aspects of the 200-400 TeV hadron collider will now be discussed in turn.

The levels of synchrotron radiation damping in this scenario are easily seen to range from just slightly above, to far beyond, those encountered in very high field magnet studies at a 100 TeV VLHC. From equation 15, the 5.3 T average field for the 200 TeV scenario gives the same damping as an average field 4 times larger at $E_{CoM} = 100$ TeV, i.e. 21 tesla, which is slighty above the maximum considered for VLHC studies. In contrast, an unrealistic factor of 16 in magnetic field strength at 100 TeV would be needed to compensate for a four-fold increase in energy, to 400 TeV, so the level of synchrotron radiation damping in this scenario is obviously much larger. It is a subject for further studies to determine whether the level of synchrotron radiation at 400 TeV is desirable in, or even compatible with, any self-consistent set of hadron collider parameters that would presumably utilize lower beam currents, smaller spot sizes and a stronger final focus than any current VLHC parameters.

Cost Considerations

The basis for cautious optimism on the magnet costs for this scenario comes from a careful study, by superconducting magnet expert Erich Willen (24), of the cost optimum in field strength that would be obtained by assuming *today's* superconducting magnet technology.

Willen's cost evaluation is for a $E_{CoM} = 200$ TeV hadron collider. (As a cautionary note, he actually refers to his parameter sets as being for 100 TeV colliders, but close inspection reveals this to be the energy per beam rather than E_{CoM}.) The costing for the dipole magnets is a careful scaling of the costs for the dipole magnets used in the existing RHIC collider. The scaling takes into account some suggested design modifications to increase the magnetic field and reoptimize the magnet length, aperture and superconducting coil layout, all of which are compatible with currently available technology. The cost vs. magnetic field strength characteristic was found to have a rather broad minimum reaching down to 1423 ?? 1990 ? $ U.S. per tesla-meter at a field strength of 5.7 T. In table 7 of reference (24), Willen presents an estimated cost of 6 $B for the dipole magnets in a 200 TeV collider using two rings of 5.7 T dipole magnets with 80% packing, for an average bending field of 4.6 T and a circumference of 460 km. Willen also priced the tunnelling costs for the collider ring at a little over 0.4 $B , costed at $900/m after studies at the 1996 Snowmass Workshop, i.e. a much smaller component of the total colider cost.

To use Willen's study as a benchmark for the 200-400 TeV collider considered here, Willen's cost minimum at a 4.6 T average field is rather close to the required 5.3 T field for the 200 TeV collider with a 400 km circumference, so his 6 $ B cost estimate for today's magnet technology is directly applicable. Improving technology in high field superconducting magnets can then be expected to both lower the cost per tesla-meter and raise the field strength of the cost minimum, i.e., move the magnetic field strength some distance in the direction of the 10.5 T average bending field that has been assumed for the 400 TeV energy.

As a more quantitative statement of the progress that might be demanded in order for 200 TeV or 400 TeV hadron colliders to become economical, it would be helpful reduce the dipole magnet cost to about 3 $B. This would require a factor of about 2 or 4 reduction, respectively, in magnet costs-per-Tesla-meter from Willen's estimate of $1423 ??/Tesla-meter. It is not unrealistic to hope that such savings could come from economies of scale and from a decade of technological advances in magnet components, design and manufacture that builds on the current magnet R&D program for the VLHC.

VII   MUON COLLIDERS

A   Circular and then Linear Muon Colliders to 1 PeV and Beyond ?

Switching from Electrons to Muons

All of the problems with TeV-scale e^+e^- colliders that were discussed in section V are associated with the relative smallness of the electron mass. The proposed technology of muon colliders aims to solve, or at least greatly reduce, these problems by instead colliding muons, which are leptons that are 207 times heavier than electrons.

Replacing the mass-related problems of e^+e^- colliders, the main problems at muon colliders arise because muons are unstable particles, with an average lifetime of approximately 2.2 microseconds in their rest frame. The preparation, acceleration and collision of the muon beams must all be done quickly and the supply of muons must be replenished often. The products of the muon decays also cause problems: the decay electrons deposit energy all along the path of the muon beams and create backgrounds in the detectors and, more surprisingly, the neutrinos can cause a radiation hazard in the surroundings of the collider ring (26; 27).

The technology and status of R&D on muon colliders has been covered in detail in reference (28) and the specific issues involving many-TeV muon colliders were examined at this workshop and form the topic of many of the papers in these proceedings. A focal point for the studies at this workshop was provided by three self-consistent parameter sets for muon collider rings, one set at 10 TeV and two sets at 100 TeV. Reference (17) of these proceedings discusses the parameter sets and their evaluation at the workshop. A very significant development at the workshop was the presentation (25) of parameter sets for *linear* muon colliders at energies ranging from 3 TeV all the way up to 1 PeV. The general assessments on the energy reach for both circular and linear muon colliders will now be briefly reviewed.

Synchrotron Radiation Limits for Circular Muon Colliders

The potential energy reach for *circular* muon colliders appears to hit a fairly hard limit at about $E_{\text{CoM}} = 100$ TeV, where the radiated power from synchrotron radiation has risen to become approximately equal to the beam power. Additional constraints arise from beam heating due to the quantum fluctuations in synchrotron radiation, as was pointed out in the workshop and is discussed elsewhere in these proceedings (29). Like the beam power, the quantum fluctuations also rise as a relatively high power of the beam energy (both rising as energy cubed, for some benchmark parameters – c.f. equation 16 for hadron colliders) – and this further pins the ultimate potential for circular muon colliders down to the 100 TeV energy scale. See reference (17) for further discussion on the limits imposed by synchrotron radiation.

Single Pass Muon Colliders

Following the historical path of e^+e^- colliders, muon collider energies above 100 TeV can be contemplated by switching to the technology of *linear* colliders, as was shown by Zimmermann (25). All of Zimmermann's parameter sets – at 3, 10, 100 and 1000 TeV – require specified "exotic" technologies both for preparing the muon beams and for acceleration, although these are not implausible at first reading by this author, who is admittedly not particularly knowledgable about linear colliders; see reference (17) for further discussion. (Zimmermann's provocative parameter sets clearly need further review by people more expert than this author.)

It is a remarkable feature of the progression in parameters that the final parameter set, at $E_{CoM} = 1$ PeV, appears to be not so much more "exotic" or less technologically plausible than the initial parameters at 3 TeV. This justifies the inclusion of a PeV-scale linear muon collider as the last collider in the proof-of-plausibility scenario, where it can benefit from about three decades of R&D to develop and refine the necessary technologies. Conversely, circular muon colliders have been favored over linear colliders up to the 100 TeV scale, where their technologies seem better established.

Continuation of linear muon colliders to even beyond the PeV scale has not yet been rigorously excluded, although the technological challenges would obviously be formidable. For example, the final focus design is far removed from anything we can seriously contemplate today. Also, the 2 linacs for even a 1 PeV collider are each already 500 km long for an assumed accelerating gradient of 1 GV/m, giving a maximum depth of 5 km below a spherical Earth. (See reference (27) for an illustration of the geometry.)

A 10 PeV collider would need either 10 times the tunnel length or 10 times the accelerating gradient of the 1 PeV example, or some compromise between these parameters. However, the maximum depth below the Earth's surface goes as the square of tunnel length, so trying to lengthen the tunnel quickly gets one into hot lava! The barriers against increasing the gradient to 10 GV/m have more loopholes. Gradients of 1 GV/m are already pushing the material limits for any known surfaces in the accelerating structures, even at the highest frequencies people consider, but the solution to this might plausibly come from exotic acceleration schemes using plasmas driven by lasers. Various experiments to test such acceleration schemes are already underway.

B Muon Colliders for the Proof-of-Plausibility Scenario

This subsection provides discussion that is specific to the 3 muon colliders of table 1, at E_{CoM} = 4 TeV, 30-100 TeV and 1 PeV. In order to spell out the entire progression, however, we begin by discussing a neutrino factory – a simpler accelerator that is not actually a collider but which would likely serve as an important staging point towards the construction of the 4 TeV muon collider that is the first $\mu^+\mu^-$ collider entry in table 1.

Leading-in with a Neutrino Factory Muon Storage Ring

Because no muon collider has yet been built, a convincing demonstration of muon collider technology would be prudent before investing in an energy frontier collider. A neutrino factory would afford this demonstration while providing much useful and complementary physics to the colliders.

A neutrino factory is a simpler, non-colliding muon storage ring that would be optimized for the collimated decay of muons into one or more intense neutrino beams aimed at neutrino physics experiments. As an aside on the choice of beam energy for the neutrino factory, current studies are concentrating mainly on neutrino factories with beam energies of 20 to 50 GeV that are optimized for a particular range of neutrino oscillation studies. However, neutrino factories at higher beam energies of perhaps 100–200 GeV are more optimal for the complementary high-rate experiments that study neutrino interactions and also for some other neutrino oscillation scenarios. This option for neutrino factories at higher energies might also be further considered since it would better test the acceleration technology needed for high energy muon colliders and should also be easier to upgrade to such a collider.

In the straw-man scenario presented here, the design of the neutrino factory demonstration machine would be made compatible with an upgrade to a site-filling 4 TeV muon collider, to occur immediately on completion if not beforehand. The neutrino factory provides an ideal intermediate step in the construction of a 4 TeV collider because it would be a valuable HEP facility in its own right if, for some reason, the continued upgrade to a collider was found not to be technologically feasible. It can therefore be built before a final decision has been made on the collider technology, and its construction can better inform that decision.

As another nice feature of this staging scenario, the experimental investments and the incremental physics from neutrinos is not lost with the upgrade to a collider. Quite the contrary, it turns out (30) that much of the high-rate neutrino physics can anyway be performed even better in parasitic running at a multi-TeV collider than in a dedicated lower-energy neutrino factory. The schedule for the proof-of-plausibility scenario would require a decision by 2007-8 on whether or not to proceed to a site-filling muon collider. By this time, the decision could presumably be guided by preliminary results from 14 TeV CoM proton-proton collisions at the LHC.

A 4 TeV Muon Collider

Four TeV was the muon collider energy chosen for the design study (31) presented at the Snowmass'96 workshop. However, it has since become more widely appreciated that neutrino radiation from the collider ring is a concern at these energies (26), so the muon beam current would likely be limited to more than an order of magnitude below the Snowmass'96 parameters. The reduced current would anyway save on the cost and power of the proton driver, which might have been considerable for the Snowmass'96 specifications.

Reference (32) contains a self-consistent parameter set for a low-current 4 TeV muon collider that is appropriate for the proof-of-plausibility scenario given here. The parameter set has a luminosity of $\mathcal{L} = 6 \times 10^{33}$ cm^{-2}.s^{-1} and an average neutrino radiation dose in the plane of the collider that is below one-thousandth of the 1 mSv/year U.S. federal off-site limit.

The Highest Energy Circular Muon Collider, at 30-100 TeV

The next step up in energy is assumed to carry all the way to the highest feasible energy for a circular machine, which we assume to be in the range E_{CoM} = 30-100 TeV.

In order to consolidate resources and minimize the overall cost of the scenario, it is sensible that the 30-100 TeV $\mu^+\mu^-$ collider should be constructed in the same laboratory whose 400 km boundary is defined by the 100-200 TeV hadron collider of the preceding section. A boundary of such a size will anyway be dictated by the requirements of neutrino radiation, as is discussed in detail elsewhere in these proceedings (27). Briefly, the radiation disk that is emitted in the plane of the collider ring would rise to approximately 300 meters above the Earth's surface for such a boundary and in the approximation of a spherical Earth. This is assured to be far enough above any structures that no practical off-site radiation hazard would remain, as is discussed further in (27).

The high end of the energy range, E_{CoM} = 100 TeV, corresponds to two of the straw-man parameter sets for this workshop – see the parameter table in reference (17) – while parameters for the low end, 30 TeV, can be estimated by interpolating between the 10 TeV and 100 TeV straw-man parameter sets in reference (17). As we learned in the workshop (29), beam heating effects from synchrotron radiation push the parameters in the direction of a lower average bending field than the 10.5 T assumption of the workshop's parameter sets. A sensible value might, e.g., be half of this, i.e. a 5.3 Tesla average for a 200 km circumference at 100 TeV, as was indicated in reference (17). Such effects should be much less important if the maximum collider energy turns out to be limited towards the lower end of the energy range under consideration. A 30 TeV collider would presumably still use as high an average bending field as is practicable in order to maximize its luminosity, e.g., perhaps as high as a 10.5 T average. This would correspond to the much smaller circumference of only 30 km.

As for the hadron collider, bending magnets are likely to be the major cost component of this $\mu^+\mu^-$ collider. The collider ring magnets should be much cheaper for a 30-100 TeV muon collider than for the 200-400 TeV hadron collider as they would require an order of magnitude less tesla-meters of total bending: the beam energy is several times lower and there is a further factor-of-two saving because only one ring of magnets is needed instead of two (counter-rotating beams of opposite charges can share the same magnet ring). Indeed, a presentation at this workshop by Mike Harrison (33) suggested a total magnet cost of only about 400 million dollars for the collider ring magnets in the 10 TeV parameter set. This is very encouraging to the extent that it can be scaled up to higher energy muon colliders (although see Harrison's caveats regarding such a scaling). Instead, the magnets for the muon acceleration are likely to be a much larger cost component than the collider ring magnets, with perhaps several times the total tesla-meters of bending as well as additional costs associated with the need to transport large momentum spreads. (See reference (17) of these proceedings for a more detailed discussion.) The design of the magnets for both the acceleration and collider rings will also need to cope with the energy deposited from decay electrons and sychrotron radiation, and this should also feed down into more expensive magnets than at hadron colliders. The upshot of the discussion in this paragraph is that the cost of the 30-100 TeV muon collider might plausibly be similar to that of the 200-400 TeV hadron collider in the scenario. There would probably also be a substantial overlap in the magnet technologies that drive the costs of the two colliders.

It is noted that, as a follow-up to the workshop's parameter sets (17), new $\mu^+\mu^-$ collider parameter sets are currently being generated (34) at the $E_{CoM} = 30$ TeV and 100 TeV lower and upper limits of the range specified here.

The 1 PeV Linear Muon Collider

The final collider in table 1 is a 1 PeV linear $\mu^+\mu^-$ collider. This item in the proof-of-plausibility scenario simply defers to an expert in linear colliders by assuming the 1 PeV parameters that were presented elsewhere in this workshop by Zimmermann (25) and have already been discussed previously in this paper. (If a change to Zimmermann's scenario were to be guessed at, it might be in the direction of more frequent bunches with smaller emittances but fewer muons per bunch. Such a refinement appears to move towards the anticipated potential capabilities of emittance reduction using the proposed method of optical stochastic cooling (35).)

Zimmermann does not specify a linac length or accelerating gradient, so this scenario makes the additional assumption of an accelerating gradient of 1 GV/m.

This corresponds to a 500 km total length in each of the 2 linac tunnels. This gradient is rather ambitious, as was pointed out in the preceding subsection. The new laboratory site should have provision for a 1000 km long linear tunnel centered in the laboratory site, as is shown in figure 3.

In order for the beam acceleration to have some hope of an acceptable cost scaling with energy, Zimmermann assumes an electron-drive beam technology for the acceleration, such as is now under development for the CLIC linear e^+e^- collider. To be affordable at 1 PeV, this technology will need to reduce the cost-per-unit-length of the linac by more than an order of magnitude over the more conventional klystron-driven technology, to below $ 10 000/meter, corresponding to a total linac cost below $ 10B.

The luminosity of Zimmermann's 1 PeV parameter set is 5.4×10^{35} cm^{-2}.s^{-1}. This corresponds to only about half a unit of R in integrated luminosity per 10^7 second accelerator year (see section III B) and so falls squarely on top of the "borderline" luminosity assignment of equation 9:

$$\mathcal{L}^{\text{borderline}}[\text{E}_{\text{CoM}} = 1\,\text{PeV}] = 5 \times 10^{35}\,\text{cm}^{-2}.\text{s}^{-1}. \tag{21}$$

The 1 PeV linear $\mu^+\mu^-$ collider is the farthest extrapolation in time, energy and technology of all the colliders in the proof-of-plausibility scenario, and its overall feasibility and choice of parameters also have the largest uncertainties. Even so, reference (25) was really only a first look at linear $\mu^+\mu^-$ colliders and there is presumably much that can be done to further assess and develop this possibility with studies that wouldn't be too time-consuming. The interested reader is encouraged to take a further look!

VIII ASSESSMENT OF CHALLENGES AND PROSPECTS FOR THE PROOF-OF-PLAUSIBILITY SCENARIO

The individual challenges for each of the e^+e^-, pp, and $\mu^+\mu^-$ collider technologies in the proof-of-plausibility scenario have been briefly addressed in the preceding sections. The three subsections in this section now discuss the global features: the common technologies, overall costs and the potential physics rewards.

A Technological and Logistical Requirements for the Overall Scenario

Common Technologies

It has been noted in the preceding sections that pp and circular $\mu^+\mu^-$ colliders share the technology of high field bending magnets as their cost drivers. In fact,

TABLE 2. Technology overlaps for future colliders at the energy frontier.

Technology	pp	e^+e^-	$\mu^+\mu^-$
magnets:			
SC conductors	Y	n	Y
cost reduction	Y	n	Y
v. high B dipoles	Y	n	Y
v. high B quads.	Y	n	Y
heat removal	Y	n	Y
FFAGs	n	n	Y
acceleration:			
high grad. SC rf	n	Y	Y
high grad. normal rf	n	Y	Y
drive-beam	n	Y	Y (linacs)
laser-driven plasma	n	?	far future?
other hardware:			
OSC	?	n	Y (linacs)
beam diagnostics & feedback	Y	Y	Y
active magnet movers	n	Y	Y (linacs)
design and simulations:			
hard final focus design	?	Y	Y
beam stability studies	Y	Y	Y

it is true more generally that the technologies for progress in the three accelerator types are very much intertwined, as is itemized in table 2.

The considerable overlap serves to remind us of the common future of the field and to provide further impetus for cooperative technical studies between experts in each of the 3 accelerator types, and the technological health and vigor of each one of these accelerator types will trickle down to affect the others.

All of these technologies will require considerable R&D if a viable rate of progress towards the energy frontier is to be maintained. This implies a concerted and, probably, expanded commitment of resources to accelerator technologies, as was already pointed out several years ago by the Nobel laureate experimental physicist Samuel C. C. Ting (1): "We need revolutionary ideas in accelerator design more than we need theory. Most universities do not have an accelerator course. Without such a course, and an infusion of new ideas, the field will die."

The Desirability of a New World HEP Laboratory

As well as the technological overlap, the proof-of-plausibility scenario assumed that the 200-400 TeV hadron collider and the final circular and linear $\mu^+\mu^-$ colliders would be housed in the same new world HEP laboratory. An "artist's conception" example illustration of the layout for this scenario is shown in figure 3 and its large

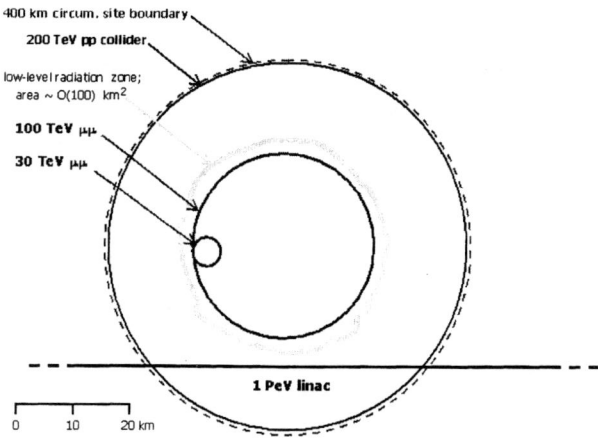

FIGURE 3. The ultimate high energy physics laboratory ?

size scale has been emphasized in figure 4 by placing it somewhat arbitrarily on a map of Australia.

The choice of an appropriate large laboratory site would obviously be a major project involving a considerable amount of research followed by detailed political negotiations. Site selection would involve the optimization of many factors and the satisfaction of several requirements besides isolation (36). For example, the site should be in a politically and economically stable country and should have ready access to an industrial base and resources such as power, transport and cooling water. Besides Australia, other candidate regions for a site would obviously include the U.S.A., Canada and several parts of Northern Europe or Asia. The example of figure 4, showing the site in an unpopulated desert region of Australia, would presumably use closed cooling loops to conserve water. It recalls the discussion of around 1980 on the "desertron", which was the original popular name for a mooted energy frontier hadron collider that then evolved into the SSC project.

The consolidation of resources into a single new HEP laboratory should be more generally beneficial in all scenarios that envisage more than one collider extending the high energy frontier, not to mention that even a single such collider will anyway be too big to fit on existing laboratory sites. This would also help financially by avoiding duplication of laboratory operating costs – an aspect that will be covered in more detail in the following subsection.

B Could the Scenario be Affordable ?

For a plausible funding scenario, note that the straw-man scenario includes 5 energy-frontier colliders beyond the LHC (1 e^+e^-, 1 pp and 3 muon colliders) over

FIGURE 4. An example to illustrate the size of a HEP laboratory with a 400 km site boundary circumference. A circle of this diameter has been drawn in the Great Victoria Desert (just above the "A" in the label "South Australia"), showing that the outline of a laboratory of this size would even be visible on a map of Australia. The 1000 km long strip of land for the 1 PeV linear collider would perhaps not be included inside the original laboratory boundary. The choice of country and positioning of the site are for illustration only.

a time period of about 40 years. The following suggested cost goals appear to be plausible and correspond to an assumed level of worldwide construction funding on these machines of 1 B$/year over these 40 years:

1. 10 $B for the combined cost of the e^+e^- collider, the neutrino factory and the 4 TeV muon collider

2. 8 $B for the 200-400 TeV hadron collider

3. 10 $B for the final circular muon collider, at 30-100 TeV

4. 12 $B for the 1 PeV linear muon collider.

The 10 $B cost combined cost for the first machines in the scenario can be estimated with slightly more confidence than the others and appears difficult but plausible. The cost drivers contributing to the rest of these guessed figures have been at least touched on in the preceding sections but the potential reasonableness or otherwise of such cost goals is clearly a subject requiring much more careful and detailed study. Ideally, a global cost assessment of funding scenarios would benefit from explicit funding algorithms that have been benchmarked to recent large accelerator projects such as the SSC, LEP, HERA, RHIC and the LHC, and then peer-reviewed and continually refined by the HEP community as our technical understanding improves.

Besides the costs of the accelerators themselves, additional costs would include accelerator R&D and the commissioning and operating costs of the new world HEP laboratory. The laboratory would presumably cost several billion dollars to set up, although this might reasonably be partially or fully covered by a one-off contribution from the host country. Operating costs and upkeep for such a large laboratory might amount to several hundred million dollars per year, considering that the world's largest current laboratory, CERN, has an annual budget in the range of half a billion dollars. Electricity for the colliders would be a significant part of the laboratory's operating cost; to set the scale, operating for a 10^7 second accelerator year with a total wall-plug power of 1 Gigawatt would cost 140 million dollars at an assumed rate of 5 cents per kilowatt hour.

All-in-all, the potential viability of the proof-of-plausibility scenario assumes something around 2 $B per year as the world-wide HEP budget devoted to the high energy frontier. This can be compared to the total funding for HEP in the U.S. alone (37), which has bounced around at an average slightly below 1 $B per year since the early 1960's but has risen as high as 1.2 $B in 1970 and 1.4 $B in 1992, at the height of SSC funding. (These figures are in FY 1995 U.S. dollars.) This indicates that the scenario could not easily be funded by one country acting alone but would instead require the combined commitments of all of the U.S., continental Europe, Britain, Japan and other, smaller contributors.

To summarize the cost discussion, the obvious answer to the question in the section title is "not easily". The cost constraints will be extremely challenging in the straw-man collider scenario presented here, or in any other that holds to the historical precedent for progress along the Livingston plot. The costs of each machine will need to be agressively minimized and, even so, the field will need to make a united and convincing case for concerted world-wide funding at a level that is at least comparable to the historical norm for each of the contributing countries.

C Benchmarking the Experimental Potential of Future Colliders

The preceding discussion in this section has addressed the "pain" involved in keeping to the Livingston curves for accelerator progress. We now turn to the "gain" that would make all the effort worthwhile. If the future colliders covered in this proof-of-plausibility scenario turned out to be indeed feasible and were constructed, it would extend our energy reach for elementary particles from the current 300 GeV flagship in energy reach (the 1.8 TeV proton-antiproton Tevatron) all the way up to the 1 PeV energy scale.

The true significance that such an experimental advance would have is unknowable until it happens since it depends on what we find and how successful we are in tying it in to our theoretical understanding of the Universe. All we can do for now is extrapolate from past experience in particle physics. Such an advance of $3\frac{1}{2}$ decades in energy reach can be judged against these benchmarks:

- the past and present colliders on the Livingston plot span from about 1 GeV to 300 GeV. These $2\frac{1}{2}$ energy decades of collider experiments have been sufficient to revolutionize our knowledge of the elementary constituents of our universe. The final span of 6 energy decades would be an increase on this by a factor of 2.4, and it would be pessimistic to not predict that this would again revolutionize our level of understanding.

- those elementary particles in figure 1 that are not massless have a mass spectrum extending from the electron, with $m_e \simeq 5.11 \times 10^{-4}$ GeV, to the top quark, with $m_t \simeq 1.75 \times 10^2$ GeV. (We exclude for the moment the very preliminary and indirect experimental evidence for non-zero neutrino masses that, if confirmed, would be much lower.) Therefore, the additional $3\frac{1}{2}$ energy decades could potentially broaden the known spectrum of elementary particle masses from the current $5\frac{1}{2}$ decades in mass to 9 decades, a substantial increase of more than 60%.

- $3\frac{1}{2}$ decades in energy reach would explore more than 20% of the log-energy gap from our current experimental reach all the way up to the Planck mass scale, at 10^{19} GeV. The Planck scale is defined by where quantum gravitational effects necessarily become important enough that even the framework for our current theories no longer makes sense and some theory that would surely be a close approximation to the Theory of Everything would be required to describe the physical processes. By biting off a significant fraction of this log-energy span we would presumably be giving ourselves a good shot at such an elevated level of understanding of our Universe.

IX   CONCLUSIONS

This paper has reviewed the past and future importance of accelerators for understanding our cosmos and its elementary constituents. The prospects for future e^+e^-, pp and $\mu^+\mu^-$ colliders were also reviewed and a proof-of-plausibility scenario was presented, incorporating the 5 plausible future colliders in table 1, that is able to hold to the historical rate of progress in the log-energy reach of hadron and lepton colliders and to reach the 1 PeV constituent mass scale by the early 2040's.

While the challenges to a further half century of concerted progress on energy frontier colliders are great, the potential rewards are grander. Experimental discoveries at colliders would be expected to feed further advances in theoretical areas such as, e.g., string theory and, to reciprocate, any such theoretical advances would then motivate and inspire continued advances in colliders and collider experiments. The side-by-side progress of experiment and theory have the common and lofty goal of uncovering the long sought after "Theory of Everything" – the sum total of the elementary entities and organizing principles that underly the structure and processes of our physical Universe. Such an immortal pillar of knowledge and understanding, if attained, could justifiably be regarded as the greatest scientific achievement in all of human history, no less!

The high stakes and long-term nature of this scientific endeavor underly the wisdom of devoting some small fraction of our energies towards better understanding the possible options and required technologies for the years and decades to come. Proof-of-plausibility scenarios such as the one presented in this paper can guide this planning. However, they encompass diverse and often speculative areas of expertise and so will necessarily start out poorly informed and in need of refining in the furnace of scientific peer review, to then be augmented by alternative scenarios and either developed further or else disgarded as unrealistic.

Such a spirit of friendly and cooperative problem-solving, constructive criticism and model-building was a hallmark of this workshop. HEMC'99 was restricted to exploring the technologies and collider physics of many-TeV muon colliders but planning is underway for a follow-up study and workshop, in the Summer and Fall of 2001, that will explore the long-term prospects for all types of energy frontier colliders and the physics processes they might illuminate.

REFERENCES

[1] Quote reproduced from Scientific American, February, 1994.
[2] Albert Einstein, letter to M. Besso (1951).
[3] Figure reproduced from http://www.fnal.gov/pub/standardmodel.html. Used with permission.
[4] See any textbook on elementary particle physics. For example, David Griffiths, *Introduction to Elementary Particles*, pub. John Wiley and Sons, QC793.2.G75.
[5] S. Dawson, *Physics in 2006*, these proceedings. Also available from http://pubweb.bnl.gov/people/bking/heshop/hemc_papers.html.
[6] Bill Willis, *Muon Collider Workshop Summary*, ibid.
[7] Z. Parsa, *High Energy Physics Potential at Muon Colliders*, ibid.
[8] M.S. Berger, *Muon Collider Physics at Very High Energies*, ibid.
[9] Kenneth Lane, *Technicolor Signatures at the High Energy Muon Collider*, ibid.
[10] Thomas G. Rizzo, *Kaluza-Klein Physics at Muon Colliders*, ibid.
[11] Joseph D. Lykken, *New Physics and the Post-LHC Era*, ibid.
[12] Response by Stephen Hawking to an email question during a televised question and answer session, 1999.

[13] N.P. Samios, *Might These Machines Be Affordable*, these proceedings. Also available from http://pubweb.bnl.gov/people/bking/heshop/hemc_papers.html.

[14] Clemens A. Heusch, *Physics Opportunities with a High-Energy Collider of Same-Sign Muons*, ibid.

[15] O. Benary, S. Kahn and I. Stumer, *Detector Backgrounds for a High Energy Muon Collider*, ibid.

[16] P. Rehak et al., *Detector Challenges for $\mu^+\mu^-$ Colliders in the 10-100 TeV Range*, ibid.

[17] B.J. King, *Parameter Sets for 10 TeV and 100 TeV Muon Colliders, and their Study at the HEMC'99 Workshop*, ibid.

[18] The OPAL Collaboration, "Investigation of CP violation in $B^0 \to J/\psi K_0^S$ decays at LEP", Eur. Phys. J. C 5, 379-388 (1998).

[19] The CDF Collaboration, "A Measurement of $\sin(2\beta)$ from $B \to J/\psi K_0^S$ with the CDF Detector", Phys. Rev. D 61, 072005 (2000).

[20] Private communication with Michael Peskin.

[21] Valery Telnov, *Problems and Stoppers for $\gamma\gamma$, $\gamma\mu$, μp Colliders Using Very High Energy Muons*, these proceedings. Also available from
http://pubweb.bnl.gov/people/bking/heshop/hemc_papers.html.

[22] The NLC Design Group, *Zeroth-Order Design Report for the Next Linear Collider*, LBNL-PUB-5424, SLAC Report 474, UCRL-ID-124161, May 1996.

[23] Annual report from Steering Committee for a Very Large Hadron Collider, March, 2000, available from http://vlhc.org/.

[24] Erich Willen, *Superconducting Magnets*, Presented at the INFN Eloisatron Project 34th Workshop, "Hadron Colliders at the Highest Energy and Luminosity", Erice, Sicily, November 4-13, 1996.

[25] F. Zimmermann, *Final Focus Challenges for Muon Colliders at Highest Energies*, these proceedings. Also available from
http://pubweb.bnl.gov/people/bking/heshop/hemc_papers.html.

[26] B.J. King, *Assessment of the prospects for muon colliders*, paper submitted in partial fulfillment of requirements for Ph.D., Columbia University, New York (1994), available from LANL preprint archive as *physics/9907026*; B.J. King, *Potential Hazards from Neutrino Radiation at Muon Colliders*, available from LANL preprint archive as *physics/9908017*. An abbreviated version of this paper is available as Proc. PAC'99, New York, 1999, pp. 318-320.

[27] B.J. King, *Neutrino Radiation Challenges and Proposed Solutions for Many-TeV Muon Colliders*, these proceedings. Also available from
http://pubweb.bnl.gov/people/bking/heshop/hemc_papers.html.

[28] The Muon Collider Collaboration, *Status of Muon Collider Research and Development and Future Plans*, Phys. Rev. ST Accel. Beams, 3 August, 1999.

[29] Valery Telnov, *Limit on Horizontal Emittance in High Energy Muon Colliders due to Synchrotron Radiation*, these proceedings. Also available from
http://pubweb.bnl.gov/people/bking/heshop/hemc_papers.html.

[30] nuphysref ... (nufnal97,sf97,jh99,nubook) B.J. King, *Neutrino Physics at a Muon Collider*, Proc. Workshop on Physics at the First Muon Collider and Front End of a Muon Collider, Fermilab, November 6-9, 1997, available from http://xxx.lanl.gov/ as

hep-ex/9907033; B.J. King, *High Rate Physics at Neutrino Factories*, Submitted to Proc. 23rd Johns Hopkins Workshop on Current Problems in Particle Theory, "Neutrinos in the Next Millenium", Johns Hopkins University, Baltimore MD, June 10-12, 1999, available from http://xxx.lanl.gov/ as **hep-ex/9911008**; I.I. Bigi et al., *The potential for High Rate Neutrino Physics at Muon Colliders and Other Muon Storage Rings*, in preparation for publication in Physics Reports.

[31] The Muon Collider Collaboration, $\mu^+\mu^-$ *Collider: A Feasibility Study*, BNL-52503, Fermilab-Conf-96/092, LBNL-38946, July 1996.

[32] B.J. King, *Discussion on Muon Collider Parameters at Center of Mass Energies from 0.1 TeV to 100 TeV*, Proc. EPAC'98, BNL–65716. available from LANL preprint archive as *physics/9908016*.

[33] Mike Harrison, *Magnet Challenges: Technology and Affordability*, oral presentation at this workshop. Transparency copies can be viewed at
http://pubweb.bnl.gov/people/bking/heshop/hemc_papers.html.

[34] B.J. King, parameters and paper in preparation for the EPAC 2000 conference.

[35] A.A. Zholents, *The Potential for an Optical Stochastic Cooling After-Burner*, oral presentation at this workshop. Transparency copies can be viewed at
http://pubweb.bnl.gov/people/bking/heshop/hemc_papers.html.

[36] Colin Johnson, Siting Options for a High Energy Muon Collider, *ibid*.

[37] *HEPAP Subpanel on Vision for the Future of High-Energy Physics, Appendix A*, May, 1994, available at http://www.hep.net/ssc/new/history/appendixa.html.

Parameter Sets for 10 TeV and 100 TeV Muon Colliders, and their Study at the HEMC'99 Workshop [1]

Bruce J. King

Brookhaven National Laboratory
email: bking@bnl.gov
web page: http://pubweb.bnl.gov/people/bking

Abstract. A focal point for the HEMC'99 workshop was the evaluation of straw-man parameter sets for the acceleration and collider rings of muon colliders at center of mass energies of 10 TeV and 100 TeV. These self-consistent parameter sets are presented and discussed. The methods and assumptions used in their generation are described and motivations are given for the specific choices of parameter values. The assessment of the parameter sets during the workshop is then reviewed and the implications for the feasibility of many-TeV muon colliders are evaluated. Finally, a preview is given of plans for iterating on the parameter sets and, more generally, for future feasibility studies on many-TeV muon colliders.

I INTRODUCTION

Self-consistent example parameter sets for the acceleration and collider ring parameters of many-TeV muon colliders were an important focal point for the discussions at the HEMC'99 Workshop – "Studies on Colliders and Collider Physics at the Highest Energies: Muon Colliders at 10 TeV to 100 TeV", held at Montauk, NY from September 27-October 1, 1999. They served as straw-man examples to be criticized, fleshed-out and improved upon by the accelerator experts attending the workshop, and the physics-related parameters helped the experimental and theoretical physicists at the workshop in their evaluations and comments on the physics potential of such colliders.

Three acceleration and collider parameter sets were used at HEMC'99: one at a center-of-mass energy of 10 TeV (set A) and two at 100 TeV (sets B and C). The collider ring and accelerator parameters are presented in tables 1 and 2, respectively. For comparison, table 1 also includes the parameter ranges for the lower energy

---

[1] This work was performed under the auspices of the U.S. Department of Energy under contract no. DE-AC02-98CH10886.

muon colliders that have been studied by the Muon Collider Collaboration (MCC). This paper describes the methods used to generate the parameter sets, details the motivations and assumptions for the specific choices of parameters and summarizes the evaluations, conclusions and suggestions for the parameter sets that were given by the workshop participants.

In more detail, the collider ring parameters are presented first, in section II, since they were considered the more critical of the two for assessing the feasibility of many-TeV muon colliders. They also determine the initial assumptions used for the acceleration parameters, which are then discussed in section III. The level of understanding advanced substantially during the workshop, and section IV goes over the issues and viewpoints raised during the workshop as well as referencing the more detailed studies that are included elsewhere in these proceedings and discussing their impact on our assessment of the parameter sets. Finally, the Outlook and Conclusions section, section V, summarizes the results discussed in the preceding section in the more general context of what they imply for the feasibility of many-TeV muon colliders. This concluding section also discusses the outlook for iterations and refinements on the parameter sets and, more generally, previews some plans for further studies on many-TeV muon colliders.

II STRAW-MAN MUON COLLIDER RING PARAMETER SETS AT 10 TEV AND 100 TEV

A Generation of the Parameter Sets

The parameter sets in table 1 were generated through iterative runs of a stand-alone computer program, as has been described previously (2; 3).

The most important physics parameter for a specified collider energy is the luminosity, $\mathcal{L}$. This is derived in terms of several input parameters according to the formula (2):

$$\mathcal{L}[\text{cm}^{-2}.\text{s}^{-1}] = 2.11 \times 10^{33} \times H_B \times (1 - e^{-2t_D[\gamma\tau_\mu]})$$
$$\times \frac{f_b[\text{s}^{-1}](N_0[10^{12}])^2 (E_{\text{CoM}}[\text{TeV}])^3}{C[\text{km}]}$$
$$\times \left(\frac{\sigma_\theta[\text{mr}].\delta[10^{-3}]}{\epsilon_{6N}[10^{-12}]}\right)^{2/3}, \qquad (1)$$

where the input variables are the CoM energy (E_{CoM}), the collider ring circumference (C), the beams' fractional momentum spread (δ) and 6-dimensional invariant emittance (ϵ_{6N}), the time until the beams are dumped (t_D), the bunch repetition frequency (f_b), the initial number of muons per bunch (N_0), and the beam divergence at the interaction point (σ_θ). Units in equations throughout this paper are given in square brackets. (The time-to-dump, t_D, is given in units of the boosted

TABLE 1. Self-consistent collider ring parameter sets for many-TeV muon colliders. The parameters are as evaluated in the HEMC'99 workshop with the exception of the neutrino radiation parameters, which have been updated to incorporate the improved estimates from reference (1).

parameter set center of mass energy, E_{CoM} additional description	 0.1 to 3 TeV MCC status report	A 10 TeV evol. extrap.	B 100 TeV evol. extrap.	C 100 TeV ultracold beam
collider physics parameters:				
luminosity, $\mathcal{L}$ [10^{35} cm^{-2}.s^{-1}]	$8 \times 10^{-5} \to 0.5$	10	10	1000
$\int \mathcal{L}dt$ [fb^{-1}/year]	$0.08 \to 540$	10 000	10 000	1.0×10^6
No. of $\mu\mu \to ee$ events/det/year	$650 \to 10\,000$	8700	87	8700
No. of 100 GeV SM Higgs/year	$4000 \to 600\,000$	1.4×10^7	2.1×10^7	2.1×10^9
CoM energy spread, σ_E/E [10^{-3}]	$0.02 \to 1.1$	0.42	0.080	0.071
collider ring parameters:				
circumference, C [km]	$0.35 \to 6.0$	15	100	100
ave. bending B field [T]	$3.0 \to 5.2$	7.0	10.5	10.5
beam parameters:				
(μ^- or) μ^+/bunch, N_0[10^{12}]	$2.0 \to 4.0$	3.0	0.80	0.19
(μ^- or) μ^+ bunch rep. rate, f_b [Hz]	$15 \to 30$	27	7.9	65
6-dim. norm. emit., ϵ_{6N} [10^{-12}m^3]	$170 \to 170$	85	10	1.0×10^{-3}
ϵ_{6N} [10^{-4}m^3.MeV/c^3]	$2.0 \to 2.0$	1.0	0.12	1.2×10^{-5}
P.S. density, N_0/ϵ_{6N} [10^{22}m^{-3}]	$1.2 \to 2.4$	3.5	8.0	19 000
x,y emit. (unnorm.) [$\pi.\mu$m.mrad]	$3.5 \to 620$	0.81	0.018	4.4×10^{-4}
x,y normalized emit. [π.mm.mrad]	$50 \to 290$	38	8.7	0.21
long. emittance [10^{-3}eV.s]	$0.81 \to 24$	21	47	8.1
fract. mom. spread, δ [10^{-3}]	$0.030 \to 1.6$	0.60	0.113	0.100
relativistic γ factor, E_μ/m_μ	$473 \to 14\,200$	47 300	473 000	473 000
time to beam dump, $t_D[\gamma\tau_\mu]$	no dump	no dump	1.0	1.0
effective turns/bunch	$450 \to 780$	1040	1350	1350
ave. current [mA]	$17 \to 30$	55	4.0	7.8
beam power [MW]	$1.0 \to 29$	131	100	198
synch. rad. critical E [MeV]	$5 \times 10^{-7} \to 8 \times 10^{-4}$	0.012	1.75	1.75
synch. rad. E loss/turn [GeV]	$7 \times 10^{-9} \to 3 \times 10^{-4}$	0.017	25	25
synch. rad. power [MW]	$1 \times 10^{-7} \to 0.010$	0.91	99	195
beam + synch. power [MW]	$1.0 \to 29$	130	200	390
power density into magnet liner [kW/m]	$1.0 \to 1.7$	4.3	1.2	2.4
interaction point parameters:				
spot size, $\sigma_{x,y}$ [μm]	$3.3 \to 290$	1.3	0.21	0.015
bunch length, σ_z [mm]	$3.0 \to 140$	2.2	2.5	0.49
$\beta^*_{x,y}$ [mm]	$3.0 \to 140$	2.1	2.5	0.49
ang. divergence, σ_θ [mrad]	$1.1 \to 2.1$	0.63	0.086	0.030
ip compensation factor: $N_0/N_{0,\text{eff}}$	1	1	1	10
beam-beam tune disruption, $\Delta\nu$	$0.015 \to 0.051$	0.085	0.100	0.100
pinch enhancement factor, H_B	$1.00 \to 1.01$	1.08	1.11	1.11
beamstrahlung frac. E loss/collision	negligible	6.8×10^{-8}	1.5×10^{-6}	9.0×10^{-7}
final focus lattice parameters:				
max. poletip field of quads., $B_{5\sigma}$ [T]	$6 \to 12$	15	20	20
max. full aper. of quad., $A_{\pm 5\sigma}$ [cm]	$14 \to 24$	22	19	6.6
quad. gradient, $2B_{5\sigma}/A_{\pm 5\sigma}$ [T/m]	$50 \to 90$	140	210	610
β_{max} [km]	$1.5 \to 150$	580	19 000	64 000
ff demag., $M \equiv \sqrt{\beta_{max}/\beta^*}$	$220 \to 7100$	17 000	89 000	360 000
chrom. quality factor, $Q \equiv M \cdot \delta$	$0.007 \to 11$	10	10	45
neutrino radiation parameters:				
collider reference depth, D[m]	$10 \to 300$	100	100	100
ave. rad. dose in plane [mSv/yr]	$2 \times 10^{-5} \to 0.02$	2.3	10	20
str. sec. len. for 10x ave. rad. [m]	$1.3 \to 2.2$	1.1	1.0	4.2
ν beam distance to surface [km]	$11 \to 62$	36	36	36
ν beam radius at surface [m]	$4.4 \to 24$	0.8	0.08	0.08

muon lifetime, $\gamma\tau_\mu$.) This formula uses the standard assumption from the Muon Collider Collaboration that the ratio of transverse to longitudinal emittances can be manipulated freely in the muon cooling channel to maximize the luminosity for a given ϵ_{6N}. The pinch enhancement factor, H_B, is very close to unity (see table 1), and the numerical coefficient in equation 1 includes a geometric correction factor of 0.76 for the non-zero bunch length, $\sigma_z = \beta^*$ (the "hourglass effect") .

In practice, the muon beam power and current are limiting parameters for energy frontier muon colliders, so the parameters are actually chosen to optimize the "specific luminosity":

$$l \equiv \frac{\mathcal{L}}{f_b \times N_0}. \qquad (2)$$

The luminosity is then determined from the choice of beam current that corresponds to the highest plausible beam powers.

Several further parameters in table 1 have been derived from the input parameters that determine the luminosity. These include, for example, the beam-beam tune disruption parameter, $\Delta\nu$. Other output parameters require additional modeling assumptions and/or further input parameters (2; 3). Examples include some of the output parameters for the final focus; these require both the input of a reference pole-tip magnetic field for the final focus quadrupoles ($B_{5\sigma}$) and a much simplified model for the final focus magnet lattice that is a linearized extrapolation from existing final focus lattice designs for lower energy muon colliders.

The physics parameters in table 1 include two examples of event sample sizes. As is discussed in references (4; 5) these give an indication of the physics potential corresponding to the specified luminosity and energy. Briefly, the number of $\mu\mu \to ee$ events gives a benchmark estimate of the discovery potential for elementary particles at the full CoM energy of the collider, while the production of hypothesized 100 GeV Higgs particles indicates roughly how the colliders might perform in studying physics at a lower energy scale.

B   Optimization of the 10 TeV and 100 TeV Parameter Sets

The Initial Choice of Energies

The two energies for the parameter sets, $E_{CoM} = 10$ TeV and 100 TeV, were chosen because they bracket that energy decade. The 10 TeV lower limit was chosen to be well above the highest energy that had been studied in detail, namely, $E_{CoM} = 4$ TeV for the Snowmass'96 workshop (6). Further, the neutrino radiation for very high luminosity $\mu^+\mu^-$ colliders at 10 TeV and above is high enough to rule out siting them at an existing laboratory, as is covered elsewhere in these proceedings (1). This necessitates a fresh outlook for the design optimization of the $\mu^+\mu^-$ colliders that is free from site-specific preconceptions involving existing laboratories, which was considered a good thing.

The choice of the upper energy limit was more technically constrained. For the 100 TeV parameter sets, the synchrotron radiation power had risen to become almost identical to the beam power, signaling a clear upper bound for the feasibility of circular $\mu^+\mu^-$ colliders.

To preview later discussion, it is noted that our understanding of the constraints on high energy muon colliders advanced during the workshop, as will be covered in section IV. An additional constraint on the maximum possible energies for circular muon colliders was discovered (7), due to beam heating arising from the quantum mechanical nature of the synchrotron radiation. On the other hand, the future prospects of many-TeV muon colliders were given a boost when the possible potential for linear colliders at even higher energies was uncovered (8).

Balancing Luminosity against Technical Difficulty

After deciding on the collision energies, it was then decided that the 10 TeV (set A) and the first of the 100 TeV parameter sets (set B) should assume only evolutionary changes in technology from the base-line parameters that have been previously posited for lower energy colliders (9). For example, the assumed 6-dimensional emittances are factors of 3.5 (10 TeV) or 50 (100 TeV) smaller than the value 170×10^{-12} m^3 that is normally used in Muon Collider Collaboration scenarios for first generation muon colliders. The smaller emittances assume that the performance of the muon cooling channel will be progressively improved through further design optimization, stronger magnets, higher gradient rf cavities and other technological advancements and innovations.

The second parameter set at 100 TeV (i.e., set C) encouraged study on some of the possibilities for using exotic technologies to improve the potential performance of future many-TeV $\mu^+\mu^-$ colliders. The additional assumed advances increased the luminosity by two orders of magnitude over the evolutionary parameter set at 100 TeV, to what would be a very impressive 1×10^{38} cm^{-2}.s^{-1}. (The luminosity should ideally rise as E_{CoM}^2, as is explained in reference (5).) The hypothesized technical advances included:

1. exotic cooling, to obtain a phase space density that is a further 3 orders of magnitude larger than the assumption for the evolutionary parameter set at 100 TeV

2. charge compensation at the interaction point (ip), to reduce the effective charge by a factor of 10. This assumption led rather directly to a corresponding increase in the luminosity by about a factor of 10.

3. more aggressive final focus parameters were included to allow for potential improvements in the final focus design, perhaps using exotic focusing technologies

4. the beam power was almost doubled from the evolutionary parameter set (B), to "top up" the luminosity to 1×10^{38} cm^{-2}.s^{-1}.

Final Focus Constraints

The final focus design may well present the most difficult design challenges that are relatively specific to high energy muon colliders. (This is to be contrasted with the muon cooling channel, which is a formidable challenge for all muon colliders.) References (2) and (3) have previously addressed the general design constraints and issues for final focus designs at many-TeV muon colliders.

To re-cap the discussion of references (2) and (3), higher energies demand progressively stronger focusing to generate the smaller spot sizes needed to increase the luminosity. Two simply defined parameters were used as benchmarks to obtain final focus specifications that might provide plausible starting assumptions for first attempts at magnet lattice designs. Firstly, an overall beam demagnification parameter is defined (2) in terms of one of the Courant-Snyder lattice parameters, β, as

$$M \equiv \sqrt{\frac{\beta_{\max}}{\beta^*}}. \tag{3}$$

This is a dimensionless parameter that gauges the strength of the focusing. The size of M should be closely correlated with fractional tolerances in magnet uniformity, residual chromaticity, etc., where the chromaticity is a measure of the change in response of the final focus to off-momentum particles. Secondly, a high residual chromaticity can be compensated for by decreasing the fractional momentum spread of the beams, δ. This suggests that another measure of the final focus difficulty might come from the product of the demagnification and momentum spread,

$$q \equiv M\delta, \tag{4}$$

where q has been referred to (2) as the "chromaticity quality factor".

In generating the parameter sets, the values of M and q were compared to those for existing e^+e^- and $\mu^+\mu^-$ final focus designs, as was discussed in references (2; 3). In practice, slightly more attention was paid to q than to M in obtaining the final parameters. It can be seen from table 1 that the two "evolutionary" parameter sets, A and B, were constrained to the value $q = 10$, which is very similar to the calculated value, $q = 11$, for the final focus lattice design of the 3 TeV $\mu^+\mu^-$ collider in reference (9). A more aggressive value, $q = 45$, was allowed for in the second parameter set at 100 TeV.

It is noted that the parameter sets at these high energies are always limited by $\Delta\nu$ and it is useful and straightforward to rewrite equations 1 and 2 in the form:

$$l \propto \frac{\Delta\nu}{\beta^*}, \tag{5}$$

which has no explicit dependence on emittance or bunch size for a given energy. The experience with optimizing the parameter sets was that this independence is true as an approximation only (3); residual dependences on limiting magnet apertures etc. meant that, in practice, it was almost always possible to slightly improve the specific luminosity by re-optimizing to parameter sets with smaller assumed emittances.

A value of $\Delta\nu = 0.10$ was assumed for all parameter sets. This was estimated by interpolating the results from a beam tracking study described in reference (6). Equation 5 and the discussion that follows indicate that the luminosity will scale approximately linearly with different assumed values for $\Delta\nu$.

Constraints on Energy Spread from Beamstrahlung

The "chromaticity quality factor" figure of merit, equation 4, favors decreasing the fractional momentum spread, δ, in order to ease the difficulty of the final focus, and this strategy was found to be effective in optimizing the luminosity for all three parameter sets. By the $E_{\text{CoM}} = 100$ TeV energy scale, however, the value of δ was found (3) to be limited from below by the rapidly rising beamstrahlung at collision. This occurred even though the fractional beamstrahlung energy loss, $(\Delta E)_{brem}$, remained at the level of parts-per-million per beam crossing, i.e., much less than the percent level expected at TeV-scale linear e^+e^- colliders. The difference is the need for multiple passes at $\mu^+\mu^-$ colliders, which compounds the sensitivity to beamstrahlung losses.

The average beamstrahlung energy losses can be replaced by rf acceleration, of course. However, the particle-by-particle variations will contribute to the spread in the beam momentum, and any such contributions from beamstrahlung must be limited to somewhat below the original momentum spread of the beam. The residual contributions to the beam energy spread should rise as the square root of the number of passes, since they will be statistically independent from turn to turn. Therefore, an appropriate criterion that was chosen to set lower limits on δ is:

$$\frac{(\Delta E)_{brem} \times \sqrt{n_{turn}^{eff}}}{\delta} \lesssim 1, \qquad (6)$$

where the effective (i.e. luminosity-weighted) number of turns, n_{turn}^{eff}, has values in the range $n_{turn}^{eff} \simeq 1000$. The evolutionary (B) and ultra-cool (C) parameter sets at $E_{\text{CoM}} = 100$ TeV had chosen values of 0.49 and 0.33 for the left hand side of equation 6, respectively.

As an aside, it is noted that reference (3) had suggested following the lead of proposed TeV-scale e^+e^- colliders by considering the option of using flat, rather than round, beam spots at the ip in order to reduce the beamstrahlung. This was tried, but all attempts led to disappointing luminosities and so round beam spots were retained for the parameter sets.

III STRAW-MAN ACCELERATION PARAMETERS

TABLE 2. Straw-man acceleration parameter sets for high energy muon colliders. The word "net" in the column "net E_{rf}" refers to the net energy gain per turn in the rf cavities after approximately subtracting synchrotron radiation losses from the 50 GeV and 250 GeV energy gains in the first and second recirculators, respectively. The parameter sets N_f^A, N_f^B and N_f^C are the numbers of muons per bunch at the exit of each FFAG corresponding to each of the three straw-man muon collider ring scenarios in table 1.

E_i [TeV]	E_f [TeV]	$\frac{E_f}{E_i}$	circum. [km]	B_{ave} [T]	net E_{rf} [GeV]	# turns	f_{decay} %	N_f^A [10^{12}]	N_f^B [10^{12}]	N_f^C [10^{12}]
	0.5							3.35	1.038	0.247
0.50	1.25	2.5	15	1.7	50	15	4.3%	3.21	0.993	0.236
1.25	2.50	2.0	15	3.5	50	25	3.3%	3.10	0.961	0.229
2.50	3.50	1.40	15	4.9	50	20	1.6%	3.05	0.945	0.225
3.50	4.55	1.30	15	6.4	50	21	1.3%	3.01	0.933	0.222
4.55	5.00	1.10	15	7.0	50	9	0.5%	3.00	0.929	0.221
5.0	12.5	2.5	100	2.6	250	30	5.7%		0.876	0.208
12.5	25.0	2.0	100	5.2	249	50	4.4%		0.838	0.199
25.0	35.0	1.40	100	7.3	246	41	2.2%		0.820	0.195
35.0	45.5	1.30	100	9.5	238	44	1.8%		0.805	0.192
45.5	50.0	1.10	100	10.5	229	20	0.7%		0.800	0.190

A Introduction

Table 2 gives straw-man acceleration scenarios that reproduce the final energy and bunch charge for each of the three straw-man muon collider ring scenarios given in table 1, labeled as A) 10 TeV with 10^{36} luminosity, B) 100 TeV with 10^{36} luminosity and C) 100 TeV with 10^{38} luminosity. The layout of each of the two recirculating complexes for table 2 is sketched schematically in figure 1.

The acceleration scenarios of table 2 and figure 1 will be described in subsection III D. For now, we note that the table contains only a minimal amount of information – much less than was provided for the collider ring – and, in practice, the acceleration parameters were much less critical than the collider ring parameters for determining the technical feasibility or otherwise of the collider scenarios. This viewpoint is supported by a much more detailed and knowledgeable acceleration scenario that is presented elsewhere in these proceedings (10).

Aside from the technical considerations, the acceleration is expected to dominate the cost of the colliders so its cost optimization will be very important and this was the main design criterion for the straw-man scenarios presented in table 2. To minimize the cost, the scenarios use configurations of recirculating linacs with "fixed field alternating gradient" (FFAG) magnet lattices.

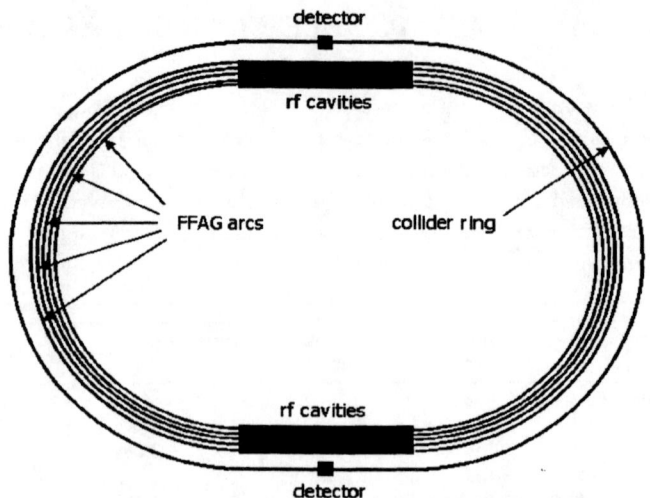

FIGURE 1. Accelerator layout for the acceleration scenario of table 2. The layout is schematic and is certainly not drawn to scale. A single tunnel contains 5 rings of FFAG arcs. All the arcs pass through the same rf cavities, shown here in 2 linacs on opposite sides of the tunnel. The collider ring is also shown in the same tunnel, indicating that this accelerator complex brings the beam up to collision energy, i.e., this could be the 0.5-5 TeV ring for the 10 TeV collider (parameter set A) or the 5-50 TeV ring for the 100 TeV colliders (sets B and C). Transfer lines between the rings are not shown. As an aside, 2 detectors are shown in the collider storage ring, although this was not assumed in the workshop. This would double the luminosity but would complicate the design of the storage ring.

The rest of this section is organized as follows. A very simplistic and non-technical introduction to FFAGs will be given in the next subsection. Some preliminaries on calculating decay losses during acceleration occupy the subsection after that before, in subsection III D, returning to describe the motivation for the parameter choices in table 2.

B FFAG Recirculating Arcs

The amount of expensive rf acceleration can be reduced many-fold relative to linear accelerators by bending the muons around for many passes through the same length of linac. The onus then shifts to minimizing the cost of the magnets in the recirculating arcs. In turn, it is then desirable that each of the arcs be able to accept a wide range of momenta so it can be reusable for many traverses. The most promising option for doing this appears to lie in a class of either quadrupole-loaded or combined function magnet lattices that are referred to as "fixed focus, alternating gradient" or "FFAG" lattices. Fast-ramping synchrotrons may also be considered (6; 9) but steady-state operation of FFAGs appears likely to be cheaper

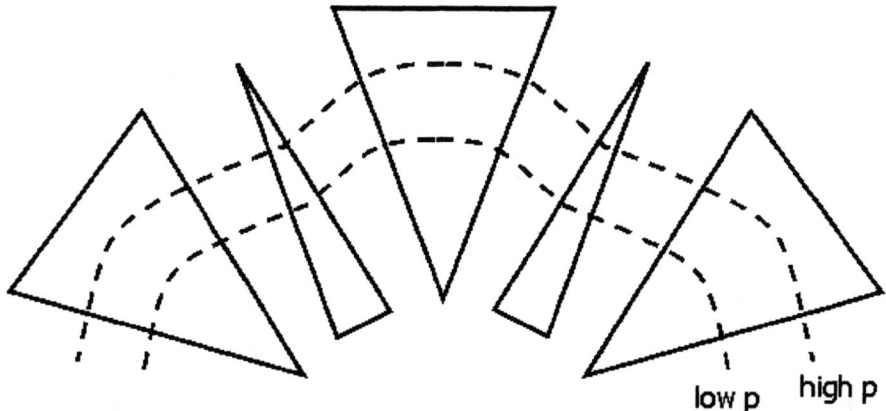

FIGURE 2. A very schematic illustration of the FFAG concept. Each triangle signifies a bending magnet with a non-uniform magnetic field. The acute point of each triangle signifies the direction of the bending magnetic field and the thickness at any radius signifies the magnetic field strength in this direction rather than the spatial extent of the magnet. (More generally, the increase in the field gradient will not be linear.) If the magnet spacings and magnetic field parameters are appropriate then the non-uniform bending fields automatically provide alternating gradient focusing in both transverse planes.

and might well be more reliable.

Figure 2 gives a very conceptual illustration of the basic idea of FFAGs. It can be seen that the alternating sign of the bending field results in net bending in the direction of the stronger dipoles a provides the scalloped beam trajectories that are characteristic of FFAGs. Further out trajectories see a progressively stronger average bending field and so are appropriate for transporting larger momenta in proportion to the average magnetic field strength.

FFAGs were first considered back in the 1950's (11) but, presumably, were not developed further at that time because the simpler alternative of slowly ramping synchrotrons was adequate for the acceleration of stable particles. Impressively, FFAG lattices have now been designed that transport as much as factors of 5 to 10 in muon momentum, although such extreme designs require very large apertures and the peak magnetic fields are several times the average bending field. Some initial design studies for more practical FFAG lattices are presented elsewhere in these proceedings (12; 13).

Perhaps the biggest technical problem with all FFAG scenarios is the difficulty in maintaining turn-by-turn an appropriate phase relationship with the rf acceleration, since the path lengths of the muon orbits within the FFAG lattice get progressively larger with increasing energy – as is conceptually illustrated in figure 2. It is a nice feature of many-TeV colliders that these problems become progressively less at higher energies because the increasing revolution period through the arcs gives more time for adjustments between passes through the linac.

C Rf Acceleration and Decay Losses

The amount of radio-frequency (rf) acceleration per turn will be determined by a trade-off between minimizing the expense and the tunnel length occupied by rf (favors less rf) and minimizing the number of turns and the decay losses (favors more rf).

A formula relating the decay losses to the rf and recirculator parameters can be derived directly from the decay equation for the change in the number of muons, N, with distance, x:

$$\frac{-1}{N}\frac{dN}{dx} = \frac{1}{\beta\gamma c\tau}, \qquad (7)$$

where c is the speed of light, the scaled muon velocity is essentially unity for the muon energies under consideration, $\beta = 1$, $\gamma \equiv \frac{E_\mu}{m_\mu c^2}$ is the conventional relativistic gamma factor and the muon mass and its lifetime, τ, are such that $\frac{mc^2}{c\tau} = 0.1604$ GeV.km^{-1}.

It follows easily that muon decay losses lead to ratios of initial to final bunch populations, $\frac{N_i}{N_f}$, that are related to the recirculator tunnel lengths in units of kilometers, $L^j[km]$, the number of GeV per turn of rf acceleration, $E_{rf}^j[GeV]$, and the ratio of final to initial energies in the recirculator, $\frac{E_i^j}{E_f^j}$, through

$$\ln\left(\frac{N_f}{N_i}\right) = 0.1604 \sum_{j=1,N} \frac{L^j[km]}{E_{rf}^j[GeV]} \ln\left(\frac{E_i^j}{E_f^j}\right), \qquad (8)$$

where $j = 1, N$ is the index for the jth of N recirculators. Equation 8 has made the approximation of averaging the acceleration to an assumed constant gradient over the length of the recirculator rather than the real situation where it will be concentrated in one or more rf linacs placed around the recirculator. This should introduce only small fractional errors in the calculated particle losses for the parameters given in table 2.

D Optimization of the Straw-man Acceleration Scenario

The straw-man acceleration scenario presented in table 2 starts at 500 GeV, working on the assumption that the acceleration to this energy range has already been developed and used for a previous TeV-scale $\mu^+\mu^-$ collider. The acceleration scenario for the $E_{CoM} = 10$ TeV collider (set A) then needs to provide exactly one decade of energy gain, accelerating the beams from 0.5 TeV up to their collision energy of 5 TeV. It economizes on expensive rf acceleration by utilizing only a relatively modest 50 GV of rf cavities.

The $E_{CoM} = 100$ TeV collider scenarios start with the $E_{CoM} = 10$ TeV acceleration scenario and add a further decade of acceleration to raise the beam energies to 50 TeV. A further 250 GV of rf cavities are utilized, which is, for example, much less rf than is required for the next generation of e^+e^- colliders and so should be easily compatible with the budget constraints on a 100 TeV collider.

In more detail, it can be seen from table 2 that the recirculating accelerator to 50 TeV is essentially a scaled copy of that to 5 TeV. Both recirculators use 5 rings of FFAG arcs and the fractional momentum increment in each of the 5 rings is the same between the first and second recirculator. As one difference, the average bending fields, B_{ave}, in the second recirculator are assumed to be a factor of 1.5 times higher than in the first.

Figure 1 is a schematic diagram for a possible layout of either of the two recirculators. As a specific suggestion of this layout, it assumes the 5 FFAG rings to be housed in the same tunnel. Further, this tunnel is assumed to be the collider tunnel, i.e., the 5+5 TeV collider for the first recirculator and the 50+50 TeV collider for the second.

The obvious motivation for the layout of figure 1 is to minimize the complexity and tunnel expense of the scenario. However, requiring all 5 FFAG rings in a recirculator to have the same radius as the collider ring has the obvious consequences of fixing the FFAG ring radii and of constraining the average bending magnetic fields in each ring according to the ranges of transported momenta in that ring. Table 2 gives a specific scenario for doing this.

The design of the later FFAG rings in each recirculator is clearly more constrained than those for the earlier rings because the average bending field must be closer to the (assumed high) average bending field of the collider ring. This is dealt with in table 2 by constraining the momentum swing to become progressively smaller for the later rings in the recirculators. Assumed energy ranges covered by the arcs range from a factor of 2.5 increase – for the lowest energy arcs in each recirculator – down to 10% energy gain for the highest energy arcs in each recirculator. These are really no more than guesses since, for example, the magnet apertures and ratios of peak-to-average magnetic fields required for this scenario are unknown.

Assumed average gradients for superconducting rf of 25 MV/m, as is assumed for the proposed TESLA e^+e^- collider, correspond to total rf lengths of 2 km (10 km) for the 10 TeV (100 TeV) colliders, which is 13.3% (10.0%) of the collider ring circumference. The example schematic layout of figure 1 shows the rf to be split equally between the two straight sections of tunnel on the opposing sides of the "race-track" collider ring, although this choice was somewhat arbitrary.

Decay losses were calculated according to equation 8. Non-decay losses were neglected. Synchrotron radiation energy losses – which range up to about 10% per turn at 50 TeV – have been included in a simple approximate manner. Table 2 shows the overall decay losses to be acceptably low, at 10.5% and 13.9% respectively, for each of the two decades of energy gain.

Having detailed the scenario, it should again be emphasized that the overall scenario, together with its specific choices and assumptions, was intended to do no more than provide the seed for more credible design studies from the accelerator physicists attending this workshop. Nothing but the qualitative assumptions of the scenario should be considered at all, and even these only at the reader's discretion. Of course, none of the specific numerical assumptions should be taken at all seriously, beyond perhaps obtaining a rough qualitative feel for such parameters as the amount of rf acceleration required and the magnitudes for the fractional decay losses.

Bearing the preceding paragraph in mind, we conclude this section by again referring the reader to the vastly more competent and detailed acceleration studies that emerged from the workshop: the overall acceleration scenarios of reference (10) and the FFAG design studies in references (12) and (13).

IV ASSESSMENT OF THE MUON COLLIDER PARAMETER SETS AT HEMC'99

TABLE 3. An assessment of the feasibility of high energy collider parameter sets, incorporating the advances in understanding from the HEMC'99 workshop. See text for details.

parameter set	A	B	C
center of mass energy, E_{CoM}	10 TeV	100 TeV	100 TeV
additional description	evol. extrap.	evol. extrap.	ultracold beam, etc.
Luminosity for Physics:	excellent	fair	excellent
Technology:			
acceleration	probably OK	OK	OK
detector backgrounds	probably OK	probably OK	probably OK
beam cooling	probably OK	probably OK	problematic
synch. radiation	probably OK	borderline	NOT FEASIBLE
final focus	challenging	problematic	problematic
overall technology:	challenging	problematic	NOT FEASIBLE
Cost:	challenging	problematic	problematic
neutrino rad./siting	dedicated new site	same site	same site
OVERALL	challenging	problematic	NOT FEASIBLE

This section reviews the studies and assessments at HEMC'99 of the collider ring and acceleration parameter sets of tables 1 and 2. It will concentrate on the muon collider design issues arising out of the parameter sets. The reader is also referred to the summary paper by Willis (14) for a more general overview of the findings of the workshop.

Table 3 summarizes the status of the acceleration and collider parameter sets after review at the workshop. As an important piece of contextual information, the assessment of parameter set A (10 TeV), assumes that a TeV-scale muon collider has already been built and successfully operated and the parameter set in each successive column assumes that the collider of the preceding column has already been built.

The following subsections have been grouped according to subject areas that follow fairly closely, but not exactly, the rows of table 3: on luminosity, acceleration, detectors, cooling, synchrotron radiation, final focus design and beam instabilities. A more general outlook and list of conclusions based on these observations will be deferred to the final section, section V.

A Assessment of Luminosities for Physics

The luminosity requirements for $\mu^+\mu^-$ colliders are discussed in some detail elsewhere in these proceedings (5). Ideally, collider luminosities should rise as $E_{CoM}{}^2$ and it is seen that, indeed, the luminosities for all three parameter sets in table 1 are higher than for any existing or (to the author's knowledge) other proposed collider.

Both parameter sets A and C have excellent luminosities, even considering their high energies, while the luminosity of parameter set B was still considered to be "fair" for a 100 TeV lepton collider. (See reference (5) for further discussion.)

B Assessment of the Acceleration Scenario

Studies at HEMC'99 focused more on the collider ring parameter sets of table 1 than on the acceleration scenario of table 2. The acceleration scenario was considered critical mostly to the extent that it would be expected to be the biggest single component of the overall cost of the collider. Unfortunately, the cost of the FFAG magnets was not able to be explicitly addressed in any detail due to the newness and developing nature of FFAG scenarios (12; 13) for muon colliders. A rather indirect source for some optimism on the acceleration costs could come from any assumed correlation with some relatively favorable cost estimates for the collider ring magnets, by Harrison (15), who roughly assessed the cost for the collider magnets for the 10 TeV scenario (set A) to be perhaps of order 400 million dollars.

Technically, muon acceleration tends to get easier at higher energies due to the increasing muon lifetime, smaller beam sizes and lower circulation frequencies in recirculating linacs. Hence, the technical feasibility of acceleration up to the energies in the table is automatically established to a large extent by the assumed previous success of the acceleration at a TeV-scale $\mu^+\mu^-$ collider. As a minor caveat to this, Harrison pointed out the increased load due to synchrotron radiation in the FFAGs. However, the collider ring magnets will need to handle the synchrotron radiation load for many times more turns than the FFAG arcs, so even this technical difficulty is concentrated more in the collider ring than the accelerating lattice.

Berg (10) pointed out that slightly increased technical difficulties might instead be expected for the *low energy* end of the acceleration for parameter sets A and, especially, B. This could result from the higher specified values for the longitudinal emittance in the many-TeV parameter sets: table 1 shows the longitudinal emittances for these parameter sets to be, respectively, similar to, and about twice as large as, the longitudinal emittance for the 3 TeV parameter set of reference (9).

C Detector Backgrounds

All muon collider detectors face challenging backgrounds resulting from the electron daughters of decaying muons near the interaction point. However, the amount of electromagnetic "junk" entering the detector is relatively independent of the collider energy since the power density of deposited electromagnetic energy depends primarily on the beam current rather than the beam energy. (For confirmation of this statement, see the values in the "power density into magnet liner" row of table 1.) Hence, such backgrounds are expected to be manageable for these many-TeV parameter sets under the stated assumption that the problem has already been solved at TeV-scale collider detectors. (A specific strategy for handling these backgrounds that was developed at the workshop is described in reference (16) of these proceedings.)

Muons entering the side of the detector, either from beam halo or Bethe-Heitler $\mu^+\mu^-$ pair production, are the one background that is expected to evolve markedly with energy. As muons become more relativistic they become less and less like minimum-ionizing particles and deposit larger amounts of energy "catastrophically" in, mainly, electromagnetic showers. This issue was not addressed at the workshop and it deserves further study.

D Beam Cooling

Parameter sets A and B assume only evolutionary improvements in the ionization cooling performance over that assumed (but far from demonstrated (9)!) for TeV-scale colliders so, by definition, the beam cooling should probably be OK if following on from the TeV-scale collider. Parameter set C is very different, assuming that some form of exotic cooling will be able to increase the phase space density of the muon beams by three orders of magnitude from that assumed for parameter set B.

Such ultra-cold muon beams are still looking plausible but have not yet progressed beyond that. The most promising of the exotic cooling methods is optical stochastic cooling (17). This method clearly has formidable technical challenges but no obvious show-stoppers. Other, very low energy, cooling methods were also presented at the workshop (18; 19). There is some concern that any cooling method using non-relativistic muons (i.e. with scaled velocity $\beta \ll 1$) may well not be feasible for preparing the high-charge muon bunches needed for colliders, due to space charge limitations.

It is noted that parameter set C provides a specific example of a general feature for ultra-cold muon beams. Since the collisions at many-TeV colliders would normally be tune-shift limited anyway, it is likely that improved cooling would also require ip compensation to substantially benefit the luminosity. We now discuss yet another barrier to the use of ultra-cold beams, at least at very high energies, from synchrotron radiation.

E Synchrotron Radiation

It has already been noted that the synchrotron radiation power in the 100 TeV colliders is already comparable to the beam power. During the workshop, Telnov (7) raised what might possibly be a stronger constraint from synchrotron radiation on the energy reach of circular muon colliders, namely, the quantum nature of synchrotron radiation may lead to heating, rather than damping, of the horizontal beam emittance if the beam energy is high enough and the emittance is already very small.

Telnov's observation clearly spells the end of parameter set C, with its ultra-cold beam at E_{CoM} = 100 TeV. The other parameter set at 100 TeV (set B) is also borderline, with an initial horizontal emittance that is larger by a factor of five (7) than the equilibrium emittance due to this effect, as calculated by Telnov using a simple approximate model.

The most likely possible loop-hole for parameter set B is that the heating effect is reduced for a very strongly focusing collider lattice. More specifically, Telnov's equation 2 shows the equilibrium emittance to be proportional to the average of the "H-function" around the collider ring, where

$$H \propto \frac{\beta^3}{\rho^2}, \tag{9}$$

for β the standard Courant-Snyder parameter and ρ the collider ring's radius. (Stronger focusing corresponds to smaller β values around the ring.)

To consider adjustments to parameter set B, equation 9 suggests that 100 TeV colliders with the emittances expected from ionization cooling still look to be feasible by increasing the ring radius, ρ, by, for example, a factor of two. This would lower both the equilibrium emittance by a factor of four and the radiated energy per turn by a factor of two, which is substantial compensation for halving the number of collisions per bunch.

A much more dramatic approach to beating the energy limits from synchrotron radiation has come from Zimmermann (8), in the form of single pass linear $\mu^+\mu^-$ colliders. Example parameter sets are included in Zimmermann's paper, and are commented on in more detail elsewhere in these proceedings (5).

F Final Focus Design

The final focus design extrapolations discussed in section II seemed to work well for the 10 TeV parameter set A. A magnet layout for the final focus from Johnstone (20) closely reproduced the predicted β_{max} in table 1. Further, the lattice design experts at the workshop seemed to appreciate the extremely challenging nature of the 10 TeV final focus parameters without everybody actually condemning them as being clearly unrealistic, i.e., an appropriate level of difficulty for a workshop of this nature! See reference (8) for more detailed studies and comments.

The 100 TeV parameter sets were less fortunate. Even the "evolutionary" parameter set B was immediately dismissed by the lattice experts as being incompatible with any final focus lattice designs using conventional magnets. It will be very useful to get further feedback on what exactly broke down in the simplistic energy extrapolation that was described in section II A. Hopefully, such feedback can then be used to obtain a better parameterization of the energy evolution in the final focus parameters. A more realistic and better established parameterization could then be used to predict the luminosity scaling with energy that might be expected using conventional final focus technologies.

Finally, two exotic final focus options were discussed that might go beyond conventional magnet designs: "dynamic focusing" (using auxiliary beams to focus the colliding beams) and plasma focusing. Discouragingly, both options looked much less plausible than when considered for single pass e^+e^- colliders, due to both the need for multiple passes and the larger bunch currents assumed for $\mu^+\mu^-$ collider parameters. Also disappointing are the obstacles to beam compensation at collision (as was assumed in parameter set C), which call into question the possibility of being able to do this – see reference (21) for discussion on this topic.

G Beam Instabilities in the Collider Ring

Papers by Keil (22) and Zimmermann (8) provide studies on beam instabilities. Keil provides a systematic assessment of the classes of instabilities, including parameter comparisons with the LHC collider ring. Zimmermann's tracking studies demonstrated that even circulating the beams for a single turn should not be taken for granted, let alone for of order 1000 turns over the lifetime of the muons.

As a connection to the physics capabilities of the collider ring that needs to be borne in mind, the common assumption (9) of collider rings that are isochronous is disfavored for retaining the beam polarization. (See also reference (23) for a discussion on the importance of polarization.) As a rough hand-waving explanation, the rate of polarization precession while circulating in the collider ring is proportional to the muon's energy. It is intuitively clear that the polarization will decay away more slowly if the energies of all the particles are allowed to slosh around the beam average energy – sometimes gaining in polarization precession (higher energy) over the bunch average precession and sometimes losing (lower energy than the bunch average). This is what happens in a collider ring with longitudinal focusing as opposed to isochronous rings. The same argument also favors small beam energy spreads.

V OUTLOOK AND CONCLUSIONS

The preceding section has reviewed the insights from HEMC'99 on the parameter sets of tables 1 and 2. More generally than this, HEMC'99 has provided the first speculative insights into (i) the ultimate physics potential for future colliders at the high energy frontier and (ii) the potential challenges to reaching very high energies with muon colliders. A personal interpretation of the workshop's findings through the energy decades is:

- **muon colliders to the TeV scale:** (added for completeness – these energies were not discussed in detail at the workshop) beam cooling is the dominant technical challenge. Other major challenges are the final focus region, backgrounds in the detector, cost-efficient acceleration and beam stability throughout the cooling, acceleration and storage in the collider ring. Neutrino radiation will impose significant design constraints and the beam currents may be well below those for the straw-man parameters in reference (9) (i.e. 6×10^{20} muons/sign/year in collision).

- **to advance to the 10 TeV scale:** neutrino radiation will probably dictate a new site. The final focus region of the collider and magnet cost reduction for acceleration may be the other major technical design issues.

- **to advance to the 100 TeV scale:** major breakthroughs are needed in magnet costs and in the final focus region.

- **to advance to the 1 PeV scale and beyond:** this is not absolutely ruled out in the far distant future using a linac and many technological breakthroughs, as illustrated by the parameter set in reference (8) and discussed further in reference (5).

It would certainly be very valuable to follow up on the understandings gained at this workshop. As a small first step, modified parameter sets for many-TeV muon colliders are being generated (24) that take into account the insights gained at HEMC'99. As a refinement to make interpolations easier, a parameter set at the intermediate center-of-mass energy of 30 TeV will be included.

More substantially, there is need for a new study and workshop. Preferably, this should include all three of the main accelerator technologies – pp, e^+e^- and $\mu^+\mu^-$ colliders. This is motivated (5) both for a more coherent understanding of the future of experimental high energy physics and in recognition that the three accelerator technologies are deeply intertwined. Planning is underway for such a study to take place in the Summer and Fall of 2001.

REFERENCES

[1] B.J. King, *Neutrino Radiation Challenges and Proposed Solutions for Many-TeV Muon Colliders*, these proceedings. Also available from
http://pubweb.bnl.gov/people/bking/heshop/hemc_papers.html.

[2] B.J. King, *Discussion on Muon Collider Parameters at Center of Mass Energies from 0.1 TeV to 100 TeV*, Proc. EPAC'98, BNL–65716. available from LANL preprint archive as *physics/9908016*.

[3] B.J. King, *Muon Colliders from 10 TeV to 100 TeV*, Proc. PAC'99, New York, 1999, pp. 3038-40, available from LANL preprint archive as *physics/9908018*.

[4] B.J. King, Muon Colliders: New Prospects for Precision Physics and the High Energy Frontier, Proc. Second Latin American Symposium on High Energy Physics, San Juan, Puerto Rico, 8-11 April, 1998, available from LANL preprint archive as *hep-ex/9908041*.

[5] B.J. King, *Prospects for Colliders and Collider Physics to the 1 PeV Energy Scale*, these proceedings. Also available from
http://pubweb.bnl.gov/people/bking/heshop/hemc_papers.html.

[6] The Muon Collider Collaboration, $\mu^+\mu^-$ *Collider: A Feasibility Study*, BNL-52503, Fermilab-Conf-96/092, LBNL-38946, July 1996.

[7] Valery Telnov, *Limit on Horizontal Emittance in High Energy Muon Colliders due to Synchrotron Radiation*, these proceedings. Also available from
http://pubweb.bnl.gov/people/bking/heshop/hemc_papers.html.

[8] F. Zimmermann, *Final Focus Challenges for Muon Colliders at Highest Energies*, ibid.

[9] The Muon Collider Collaboration, *Status of Muon Collider Research and Development and Future Plans*, Phys. Rev. ST Accel. Beams, 3 August, 1999.

[10] J. Scott Berg, *Acceleration for a High Energy Muon Collider*, these proceedings. Also available from http://pubweb.bnl.gov/people/bking/heshop/hemc_papers.html.

[11] K.R. Symon, *The FFAG Synchrotron – MARK 1*, MURA-KRS-6, November 12, 1954, pp. 1-19.

[12] Al Garren, *A Scaling Radial-Sector FFAG Lattice for a Muon Accelerator*, these proceedings. Also available from
http://pubweb.bnl.gov/people/bking/heshop/hemc_papers.html.

[13] Dejan Trbojevic, Ernest D. Courant and Al Garren, *FFAG Lattice Without Opposite Bends*, ibid.

[14] Bill Willis, *Muon Collider Workshop Summary*, ibid.

[15] Mike Harrison, *Magnet Challenges: Technology and Affordability*, oral presentation at this workshop. Transparency copies can be viewed at
http://pubweb.bnl.gov/people/bking/heshop/hemc_papers.html.

[16] P. Rehak et al., *Detector Challenges for $\mu^+\mu^-$ Colliders in the 10-100 TeV Range*, these proceedings. Also available from
http://pubweb.bnl.gov/people/bking/heshop/hemc_papers.html.

[17] A.A. Zholents, *The Potential for an Optical Stochastic Cooling After-Burner*, oral presentation at this workshop. Transparency copies can be viewed at
http://pubweb.bnl.gov/people/bking/heshop/hemc_papers.html.

[18] Paul Lebrun, *Comments on Frictional Cooling and the Zero Energy Options for Cooling Intense Muon Beams*, these proceedings. Also available from
http://pubweb.bnl.gov/people/bking/heshop/hemc_papers.html.

[19] Kanetada Nagamine, *Very Low-Energy Cooling Possibilities Towards Muon Colliders and Neutrino Factory*, ibid.

[20] Carol Johnstone, *Collider Ring Lattices*, oral presentation at this workshop. Transparency copies can be viewed at

http://pubweb.bnl.gov/people/bking/heshop/hemc_papers.html.
[21] Valery Telnov, *Some Problems in Plasma Suppression of Beam-Beam Interactions at Muon Colliders*, these proceedings. Also available from
http://pubweb.bnl.gov/people/bking/heshop/hemc_papers.html.
[22] Eberhard Keil, *Collective Single-Beam Effects*, ibid.
[23] Clemens A. Heusch, *Physics Opportunities with a High-Energy Collider of Same-Sign Muons*, ibid.
[24] B.J. King, parameters and paper in preparation for the EPAC 2000 conference.

Mighty MURINEs: Neutrino Physics at Very High Energy Muon Colliders [1]

Bruce J. King

Brookhaven National Laboratory
email: bking@bnl.gov
web page: http://pubweb.bnl.gov/people/bking

Abstract. An overview is given of the potential for neutrino physics studies through parasitic use of the intense high energy neutrino beams that would be produced at future many-TeV muon colliders. Neutrino experiments clearly cannot compete with the collider physics. Except at the very highest energy muon colliders, the main thrust of the neutrino physics program would be to improve on the measurements from preceding neutrino experiments at lower energy muon colliders, particularly in the fields of B physics, quark mixing and CP violation. Muon colliders at the 10 TeV energy scale might already produce of order 10^8 B hadrons per year in a favorable and unique enough experimental environment to have some analytical capabilities beyond any of the currently operating or proposed B factories. The most important of the quark mixing measurements at these energies might well be the improved measurements of the important CKM matrix elements $|V_{ub}|$ and $|V_{cb}|$ and, possibly, the first measurements of $|V_{td}|$ in the process of flavor changing neutral current interactions involving a top quark loop. Muon colliders at the highest center-of-mass energies that have been conjectured, 100–1000 TeV, would produce neutrino beams for neutrino-nucleon interaction experiments with maximum center-of-mass energies from 300–1000 GeV. Such energies are close to, or beyond, the discovery reach of all colliders before the turn-on of the LHC. In particular, they are comparable to the 314 GeV center-of-mass energy for electron-proton scattering at the currently operating HERA collider and so HERA provides a convenient benchmark for the physics potential. It is shown that these ultimate terrestrial neutrino experiments, should they eventually come to pass, would have several orders of magnitude more luminosity than HERA. This would potentially open up the possibility for high statistics studies of any exotic particles, such as leptoquarks, that might have been previously discovered at these energy scales.

---

[1] This work was performed under the auspices of the U.S. Department of Energy under contract no. DE-AC02-98CH10886.

I INTRODUCTION: THE ROLE OF MIGHTY MURINES

The dominant motivation for high energy muon colliders (HEMCs) is unquestionably to explore elementary particle physics at many-TeV energy scales. For the sake of completeness, however, this paper instead discusses what would be the most promising subsidiary fixed target physics program, namely, the parasitic use of the free and profuse neutrino beams at HEMCs to provide complementary precision studies of high energy physics (HEP) at lower energies. Perhaps, this might complement collider studies in fostering new and helpful insights into the properties of elementary particles.

Neutrino interactions have unique potential for precision HEP studies because they only participate in the weak interaction. Today's neutrino beams from pion decays lack the intensity to fully exploit this potential but future MUon RIng Neutrino Experiments (MURINEs), using neutrino beams from the decays of muons in a muon collider or other storage ring, hold the promise of neutrino beams that are several orders of magnitude more intense than today's beams (1). The first MURINEs may well be muon storage rings dedicated to neutrino production (2) ("neutrino factories") while the collider rings of any first generation muon colliders will also make excellent MURINEs. The topic of this paper is MURINEs at very high energies and these will be referred to as "Mighty MURINEs" [2].

At a minimum, Mighty MURINEs will improve on previous MURINEs in providing much useful bread-and-butter precision HEP to feed the hungry masses of HEP experimentalists with aversions to collider mega-experiments. More interestingly, there are a couple of plausible scenarios under which they might do much more; namely (i) if quark mixing and/or B physics offer more than predicted by our naive prejudices as parameterized in the standard model (SM) of elementary particles, and (ii) if leptoquarks or other exotica begin to emerge at or below the 100 GeV energy scale.

The following section surveys the experimental conditions and parameters that might be found at Mighty MURINEs. This is followed by an overview of the potential physics analyses and by three sections going into more detail on the most interesting topics: one each on exploiting Mighty MURINEs as B factories, on the possibilities for quark mixing studies and on the potential for heavy particle production up to the 100 GeV scale.

II EXPERIMENTAL OVERVIEW

A The Neutrino Beams

Neutrinos are emitted from the decay of muons in the collider ring:

---

[2] this is a play on words alluding to the venerable cartoon character Mighty Mouse through the dictionary definition of "murine" as "to do with mice".

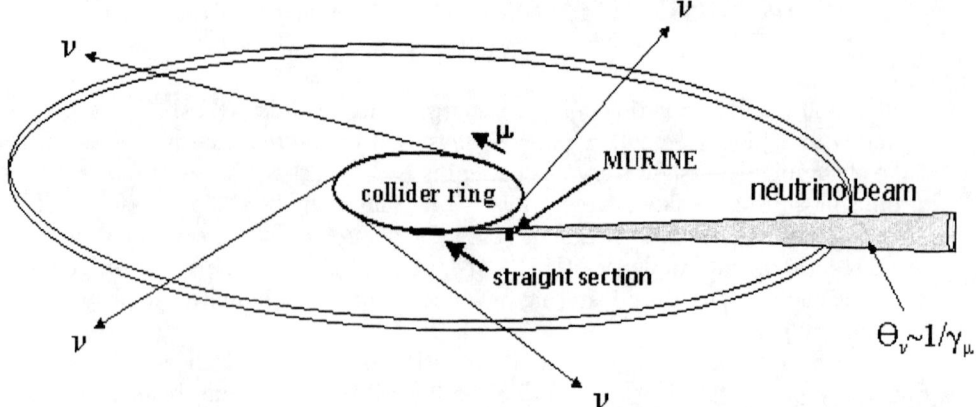

FIGURE 1. The decays of muons in a muon collider will produce a disk of neutrinos emanating out tangentially from the collider ring. The neutrinos from decays in straight sections will line up into beams suitable for experiments. The MURINEs will be sited in the center of the most intense beam and as close as is feasible to the production straight section.

$$\mu^- \to \nu_\mu + \overline{\nu_e} + e^-,$$
$$\mu^+ \to \overline{\nu_\mu} + \nu_e + e^+. \qquad (1)$$

As is illustrated in figure 1, the thin pencil beams of neutrinos for experiments will be produced from the most suitable long straight sections in the collider ring or, possibly, in the accelerating rings. These will be referred to as the production straight sections. The divergence of the neutrino beam is typically dominated by the decay opening angles of the neutrinos rather than the divergence of the parent muon beam. Relativistic kinematics boosts the forward hemisphere in the muon rest frame into a narrow cone in the laboratory frame with a characteristic opening half-angle, θ_ν, given in obvious notation by

$$\theta_\nu \simeq \sin\theta_\nu = 1/\gamma_\mu = \frac{m_\mu c^2}{E_\mu} \simeq \frac{10^{-4}}{E_\mu[\text{TeV}]}. \qquad (2)$$

For the example of 5 TeV muons, the neutrino beam will have an opening half-angle of approximately 0.02 mrad.

The large muon currents and tight collimation of the neutrinos results in such intense neutrino beams that potential radiation hazards (1; 3) are a serious design issue for the neutrino beam-line and even for the less intense neutrino fluxes emanating from the rest of the collider ring.

B Luminosities at Neutrino Experiments

For a cylindrical experimental target extending out from the beam center to an angle $\theta_\mu = 1/\gamma_\mu$, the luminosity, $\mathcal{L}$, is proportional to the product of the mass depth of the target, l, and the number of muon decays per second in the beam production straight section, according to:

$$\mathcal{L}[\text{cm}^{-2}.\text{s}^{-1}] = N_{\text{Avo}} \times f_{ss} \times n_\mu \, [\text{s}^{-1}] \times l[\text{g.cm}^{-2}], \tag{3}$$

where f_{ss} is the fraction of the collider ring circumference occupied by the production straight section, n_μ is the rate at which each sign of muons is injected into the collider ring (assuming they all circulate until decay rather than being eventually extracted and dumped) and the appropriate units are given in square brackets in this equation and all later equations in this paper. The proportionality constant is Avagadro's number, $N_{\text{Avo}} = 6.022 \times 10^{23}$, since exactly one neutrino per muon is emitted on average into the boosted forward hemisphere, i.e. each muon decay produces two neutrinos and half of them travel forwards in the muon rest frame.

Because Avagadro's number is so large, the luminosities at Mighty MURINEs will be enormous compared to those at collider experiments. The luminosities for a reasonable scenario using the workshop's straw-man parameter sets (4) are given in table 1 along with, for comparison, the final design goal luminosity for the HERA ep collider. It can be seen that, roughly speaking, Mighty MURINEs might achieve of order a million times the luminosity of HERA. An "accelerator year's" running – 10^7 seconds – at the 50 TeV MURINE's luminosity of 2×10^{37} cm^{-2}.s^{-1} would amount to an impressive integrated luminosity of 200 inverse *atto*barns per year while the even bigger straw-man luminosity at 5 TeV, 1×10^{39} cm^{-2}.s^{-1}, requires a luminosity prefix that is even less familiar to the HEP community: 10 inverse *zepto*barns per year.

C Center-of-Mass Energies for the Neutrino-Nucleon Interactions

It will be seen in section VI that HERA provides a useful comparison for some of the physics capabilities of Mighty MURINEs, particularly since the maximum center-of-mass energies, E_{CoM}, at the highest energy HEMCs might even be comparable to those at the HERA collider. The electron-proton E_{CoM} at the collider is given by relativistic kinematics as

$$E_{\text{CoM}}^{\text{HERA}} = 2\sqrt{E_p E_e}, \tag{4}$$

which is 314 GeV for the proton and electron energies of the year 2000 upgrade to HERA, $E_p = 820$ GeV and $E_e = 30$ GeV. For comparison, the MURINE's E_{CoM} is

$$E_{\text{CoM}}^{\text{MURINE}} = \sqrt{2E_\nu M_p c^2 + (M_p c^2)^2}, \tag{5}$$

where the proton mass corresponds to $M_p c^2 = 0.938$ GeV. The neutrino energy can range right up to the muon beam energy, $E_\nu^{\text{max}} = E_\mu$, and the energy spectrum seen by the detector is relatively hard (5), with an average neutrino energy within the $1/\gamma_\mu$ cone that is 49% of the muon beam energy. The comparative center-of-mass energies for Mighty MURINEs and HERA are summarized in table 1.

TABLE 1. Energy, luminosity and event rates for Mighty MURINEs at the 10 TeV and 100 TeV CoM muon colliders given in the HEMC'99 straw-man parameter sets. The center-of-mass energy, E_{CoM}, is given for the neutrino-nucleon system. The energy and luminosity at HERA are provided for comparison. The event rates assume fractional straight section lengths of $f_{ss} = 0.02, 0.01$ for the 5+5 TeV and 50+50 TeV parameter sets, respectively, and a detector mass-per-unit-area of $l = 1000$ g.cm^{-2} that intercepts the neutrino beam out to an angle $\theta_\mu = 1/\gamma_\mu$ subtended at the beam production straight section.

Facility	E_{CoM}	Luminosity, $\mathcal{L}$	events/year
5 TeV MURINE	0 to 97 GeV	1×10^{39} cm^{-2}.s^{-1}	1.7×10^{11}
50 TeV MURINE	0 to 306 GeV	2×10^{37} cm^{-2}.s^{-1}	4×10^{10}
HERA (2000 upgrade)	314 GeV	7×10^{31} cm^{-2}.s^{-1}	N.A.

D Cross Sections and Event Rates

The event rate in the neutrino detector is a product of the luminosity given in table 1 and the neutrino-nucleon scattering cross-section, which we now discuss.

The predominant interactions of neutrinos and anti-neutrinos at all energies above a few GeV are charged current (CC) and neutral current (NC) deep inelastic scattering (DIS) off nucleons (N, i.e. protons and neutrons) with the production of several hadrons (X):

$$\begin{aligned} \nu(\bar{\nu}) + N &\to \nu(\bar{\nu}) + X & (NC) \\ \nu + N &\to l^- + X & (\nu - CC) \\ \bar{\nu} + N &\to l^+ + X & (\bar{\nu} - CC), \end{aligned} \qquad (6)$$

where the charged lepton, l, is an electron if the neutrino is an electron neutrino and a muon for muon neutrinos. The cross-sections for these processes are approximately proportional to the neutrino energy, E_ν, with numerical values of (6):

$$\sigma_{\nu N} \text{ for } \begin{pmatrix} \nu - CC \\ \nu - NC \\ \bar{\nu} - CC \\ \bar{\nu} - NC \end{pmatrix} \simeq \begin{pmatrix} 0.72 \\ 0.23 \\ 0.38 \\ 0.13 \end{pmatrix} \times 10^{-35} \text{ cm}^2 \times E_\nu[\text{TeV}]. \qquad (7)$$

The number of events in the detector is easily seen to be given by:

$$N_{events} = \mathcal{L}[\text{cm}^{-2}.\text{s}^{-1}] \times 0.73 \times 10^{-35} \times 0.49 \times E_\mu[\text{TeV}] \times T[\text{s}], \qquad (8)$$

where T is the running time, $0.49 \times E_\mu$ is the average neutrino beam energy into the detector (5) and 0.73×10^{-35} is the total cross-section-divided-by-energy that is obtained from equation 7 after summing over the NC and CC interactions and averaging over neutrinos and anti-neutrinos.

The final column of table 1 shows the impressive event sample sizes predicted from equation 8: up to of order 10^{11} events per year in a reasonably sized neutrino target.

E High Performance Neutrino Detectors for Mighty MURINEs

The unprecedented event samples in small targets at MURINEs will undoubtedly also spark a revolution in neutrino detector design and performance, both to cope with event rates and to fully exploit the physics potential of the beams. An example of a novel general purpose neutrino detector that has been proposed previously (7) for MURINEs is shown in figure 2. The neutrino target is the cylinder at mid-height on the left hand side of the figure. It comprises a stack of equally-spaced CCD tracking planes, oriented perpendicular to the beam and with spacings of order 1 mm, that provides vertex tagging for events with hadrons containing charm or bottom quarks.

The general detector design of figure 2 should remain appropriate for Mighty MURINEs although it would likely be elongated to cope with the more boosted events, including perhaps lengthening the target to several meters to increase the target mass-per-unit-area and, correspondingly, the event rate.

At these higher energies, the target mass-per-unit-area could be increased still further by interspersing thin tungsten disks with the CCD planes. There are two reasons why such a dense, high-Z target becomes more practical than at lower energies: (i) multiple coulomb scattering becomes less important at higher energies, so the tracking resolution is degraded less, and (ii) the narrower pencil beam for Mighty MURINEs allows the disks to have smaller radii than at lower energies – smaller than the characteristic Moliere radius for electromagnetic showers – so it is speculated that electromagnetic showers will not develop to excessively pollute the events, despite the large number of radiation lengths along the axis of the target. (This assumption needs to be checked in more detailed follow-up studies.)

A specific scenario for the neutrino target that gives the 1000 g.cm^{-2} target mass assumed in table 1 is as follows: a 4 meter long target containing 4000 millimeter-long tracking subunits, where each tracking subunit contains a thin tungsten disk of thickness 118 microns (0.227 g.cm^{-2}) in front of a 100 micron thick CCD pixel detector (0.023 g.cm^{-2}). Each tungsten disk could have a radius of 2 cm to match the beam radius at approximately 1 km from production for a 5 TeV muon beam (as predicted from equation 2). The CCD detectors can be wider than the beam radius to also track particles moving outside the radial extent of the neutrino beam.

The vertex tagging performance of the target in figure 2 is expected (7) to be better than any other existing or planned high-rate detector for heavy quark physics, and should continue to improve with beam energy. Mighty MURINEs should attain close to 100 percent efficiency for both c and b (easier) vertex tagging in the target (excepting all-neutral decay modes, of course) since the average boosted lifetime for

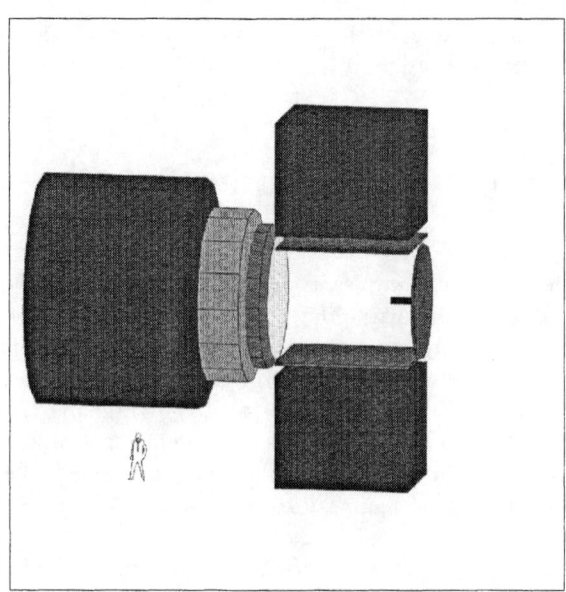

FIGURE 2. Schematic example of a general purpose neutrino detector, reproduced from reference (7). A human figure in the lower left corner illustrates its size. The neutrino target is the small horizontal cylinder at mid-height on the right hand side of the detector. Its radial extent corresponds roughly to the radial spread of the neutrino pencil beam, which is incident from the right hand side. Further details are given in the text.

TeV-scale charm and beauty hadrons – of order 10 cm – would span many planes of CCD's. The extremely favorable geometry for vertexing is illustrated in figure 3, where it is also compared with the vertexing geometry at a collider detector.

As with lower energy MURINEs, the detector backing the neutrino target should faithfully reconstruct both CC and NC event kinematics. The lower energy MURINEs provide essentially full particle identification through the muon toroids and dE/dx plus cherenkov radiation in the TPC. The particle ID for long-lived charged hadrons would become more difficult at higher energies although, speculatively, the cherenkov radiation in an elongated TPC might still give effective PID for particle energies up to a couple of hundred GeV – this requires further study. To somewhat compensate, the trajectories of most photons should be very well measured when they convert in the stack of tungsten disks that comprise most of the target mass. This will be particularly helpful for the reconstruction of neutral pions.

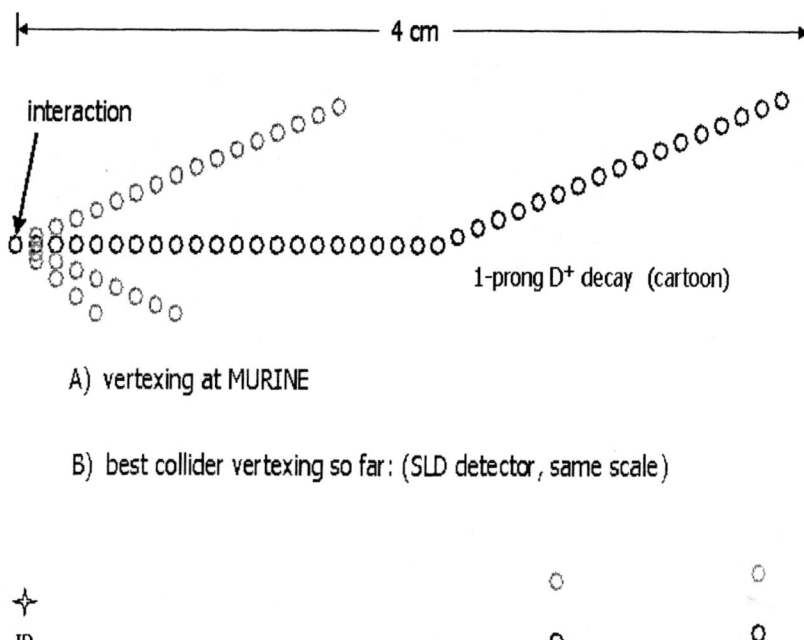

FIGURE 3. Conceptual illustration of the vertex tagging superiority at MURINEs over that with collider experiment geometries. MURINEs could have a vertex plane of CCD pixel detectors every millimeter. For comparison, the VXD3 vertexing detector at the SLD (8) experiment, which is universally regarded as the best existing vertex detector in a collider experiment, has its two innermost CCD tracking planes at 2.8 cm and 3.8 cm from the interaction point. A cartoon of a 1-prong D^+ decay has been drawn to illustrate the advantages of closely spaced vertex detectors. For clarity of illustration, the kink deflection angle has been drawn much larger than would be typical. The 2 cm distance to decay for the D^+ charmed meson corresponds to the average boosted lifetime for a 120 GeV D^+. Most charm and beauty hadrons at a Mighty MURINE, having much higher energies than this, will travel even further and hence traverse even more planes of tracking before decaying.

III MURINES: A NEW REALM FOR NEUTRINO PHYSICS

This subsection gives a brief non-technical overview of the high rate neutrino physics topics expected for MURINEs in general. Topics that might be expanded on with the higher energies of Mighty MURINEs are pointed out in preparation for more detailed discussion in the following sections.

A Neutrino Interactions with Quarks

Put simply, the DIS interactions of equation 6 involve a simple projectile (the neutrino), interesting interactions (the CC and NC weak interactions) and a complicated target (the nucleon). The bulk of the physics interest lies in the interactions with the quark constituents rather than in the properties of the neutrinos themselves. The complementary analyses that study the potential oscillations of neutrino flavors tend to be more the domain of lower energy MURINEs, at muon energies of order 100 GeV or below, and won't be discussed further here.

By the TeV-energy scale and above, the CC and NC interactions of equation 6 have become very well described as being the quasi-elastic (elastic) scattering of neutrinos off one of the many quarks (and anti-quarks), q, inside the nucleon:

$$\nu(\bar{\nu}) + q \rightarrow \nu(\bar{\nu}) + q \quad (NC) \quad (9)$$
$$\nu + q^{(-)} \rightarrow l^- + q^{(+)} \quad (\nu - CC) \quad (10)$$
$$\bar{\nu} + q^{(+)} \rightarrow l^+ + q^{(-)} \quad (\bar{\nu} - CC). \quad (11)$$

The CC and NC interactions are mediated through the exchange of a virtual W or Z boson, respectively. All quarks – up quarks (u), down quarks (d) and the smaller "seas" of the progressively heavier strange (s), charm (c) and even beauty (b) quarks – participate in NC scattering interactions of both neutrinos and anti-neutrinos. In contrast, charge conservation specifies the charge sign of the quarks participating in the CC processes as indicated by the labels: $q^{(-)} \in d, s, b, \bar{u}, \bar{c}$ and $q^{(+)} \in u, c, \bar{d}, \bar{s}, \bar{b}$. The hadrons seen in the detector are produced by the "hadronization" of the final state struck quark at the nuclear distance scale.

B The Intrinsic Richness of Neutrino Physics

Experimentally, the interaction type of almost all events can be distinguished with little ambiguity by the charge of the final state lepton (and its flavor: i.e. electron or muon): neutral (an unseen neutrino), negative or positive for the 3 respective processes in equation 6.

It is seen that equations 9 through 11 probe 3 different weightings of the quark flavors inside a nucleon, through weak interactions involving both the W and Z. For comparison, only a single and complementary weighting is probed by the best competitive process – the photon exchange interactions of *charged* lepton scattering experiments at HERA and fixed target facilities. Much of the uniqueness and richness of neutrino scattering physics derives from this variety of interaction processes.

C Physics Topics at MURINEs

Mighty MURINEs will extend and improve on the already considerable range of unique topics that will have been explored at lower energy MURINEs. The interested reader is referred to reference (5) for more details. Here we list those topics that will have already been well addressed in earlier MURINEs and comment on any added potential that might be available using the higher energies at Mighty MURINEs.

Probing Nucleon Structure

The redundant probes of the proton's and neutron's internal structure should provide some of the most precise measurements and tests of perturbative QCD - the theory of the strong interaction - and will also be invaluable input for many analyses at pp colliders. Neutrinos are also intrinsically 100% longitudinally polarized, so experiments with polarized targets could additionally map out the spin structure of the nucleon. Some of these analyses might obtain modest benefits from the higher energies and statistics at Mighty MURINEs.

Precision Electroweak Measurements

Besides using W and Z exchange as nuclear probes, the interactions themselves provide important precision tests of the standard model of elementary particles. Two measurements of total interaction cross sections will provide determinations of the fundamental weak mixing angle parameter of the electroweak theory, $\sin^2 \theta_W$, from (i) the ratio of NC to CC total cross sections and (ii) the absolute cross section for the rarer process of neutrino-electron scattering, which is 3 orders of magnitude less common than neutrino-nucleon scattering. In both cases, the fractional uncertainties in $\sin^2 \theta_W$ might approach the 10^{-4} level, which would be complementary and competitive to the best related measurements in collider experiments.

The first of the two measurements will already be systematically limited at MURINEs (5) so large gains should not be expected at Mighty MURINEs. The situation is not so clear for the electron scattering process, where the higher event statistics could still be beneficial. Speculatively, these higher statistics might also allow the use of liquid hydrogen targets with improved experimental capabilities.

Charm and Beauty Factories

MURINEs of all energies will be excellent charm factories, with of order 1% to 10% of the events containing a charmed hadron, depending on the beam energy. TeV-scale MURINEs and above will also produce enough B hadrons to be considered as beauty factories and Mighty MURINES might be very impressive B factories, as will be discussed in section IV.

Quark Mixing Studies

There is much additional interest in experimentally partitioning the CC event sample to obtain the partial cross sections for the various possible quark flavor transition combinations represented by the $q^{(-)}$ and $q^{(+)}$ symbols in equations 10 and 11. MURINEs, in general, should have the quark-tagging capability to separate the various quark flavor contributions, as was discussed in the preceding section. Mighty MURINES, with their extra capability for producing heavy final-state quarks, could make great strides beyond previous MURINEs for these "quark mixing" studies, as will be expanded on in section V.

Rare and Exotic Processes

The higher statistics and, particularly, energies available at Mighty MURINEs would clearly expand the scope for studies of rare processes and searches for exotic processes. This will be covered in section VI.

IV MIGHTY MURINES AS B FACTORIES

The charged current production of b quarks off the light quarks in the nucleon is heavily suppressed due to small off-diagonal CKM matrix elements. However, the fraction of neutrino-induced events containing B hadrons rises rapidly with energy (5) due to the decreasing threshold suppression for two higher-order processes involving gluons in the initial state:

1. $b\bar{b}$ pair production in neutral current interactions:

$$\nu N \to \nu b\bar{b} X. \tag{12}$$

2. charged current production of $c\bar{b}$ and $b\bar{c}$ from the charged current interactions of neutrinos or anti-neutrinos:

$$\nu N \to l^- \bar{b} c X \tag{13}$$

and

$$\bar{\nu} N \to l^+ b \bar{c} X, \tag{14}$$

respectively.

Preliminary estimates (9; 5) for the fraction of events from each of these processes are tabulated versus neutrino energy in table 2. The second of the two processes is seen to be less common than the first. To compensate, it provides an extremely pure and efficient tag to distinguish between b and anti-b quark production: b

TABLE 2. Fraction of events (9) producing B's in the final states $b\bar{b}$, $b\bar{c}$ or $c\bar{b}$, from neutrinos and anti-neutrinos of energies 1 TeV and 10 TeV. Estimates are preliminary.

ν or $\bar{\nu}$	E_ν	final state	fraction
ν	1 TeV	$b\bar{b}$	6×10^{-4}
$\bar{\nu}$	1 TeV	$b\bar{b}$	6×10^{-4}
ν	1 TeV	$c\bar{b}$	2×10^{-5}
$\bar{\nu}$	1 TeV	$b\bar{c}$	2×10^{-5}
ν	10 TeV	$b\bar{b}$	4×10^{-3}
$\bar{\nu}$	10 TeV	$b\bar{b}$	4×10^{-3}
ν	10 TeV	$c\bar{b}$	8×10^{-5}
$\bar{\nu}$	10 TeV	$b\bar{c}$	6×10^{-5}

production is always accompanied by a positive primary lepton (from anti-neutrino interactions) and anti-b production by a negative primary lepton (from neutrino interactions). This will be very helpful for studies of oscillations of B_0's and B_S's.

Combining the numbers in tables 1 and 2 predicts event rates of perhaps 10^8 to 10^9 B's per year at Mighty MURINEs. This is intermediate between the expectations of the e^+e^- B factory experiments ($\sim 10^7$ events/year) and the hadron B factories, HERA-B, BTeV and LHC-B (up to $\sim 10^{11}$ events/year, with up to a few times 10^9 events tagged for analysis). As already mentioned, however, the vertexing capabilities and other experimental conditions at Mighty MURINES should be superior in some aspects to those at the e^+e^- B factories and vastly superior to the very difficult experimental conditions at the hadronic B factories.

Three speculative examples of B analyses that would benefit from the unique experimental conditions at Mighty MURINEs are:

- the superior vertexing capabilities should be ideal for studying the expected fast oscillations of B_s's, perhaps following up on previous B factories with more precise measurements of the oscillation frequency and greater sensitivity to any asymmetry in the B_s and $\overline{B_s}$ decay rates

- some studies of the B baryons, Λ_b, Ξ_b^- and Ξ_b^0, which are not produced in e^+e^- B factories, may also plausibly be best performed at a Mighty MURINE

- it might have a chance (10) to measure the branching ratio for the all-neutral rare decay $B_d \to \pi^0\pi^0$, which is expected to be of order 10^{-6}. This would provide an estimate for the otherwise problematic "penguin-diagram pollution" in the analogous charged pion decay $B_d \to \pi^+\pi^-$, and this could go some way to resurrecting the charged decay mode as one of the central CKM processes at B factories. However, observing the neutral decay mode does not look feasible at any future B factories other than Mighty MURINEs.

The final process deserves some further explanation since the decay itself doesn't provide a vertex. However, the close to 100% vertex reconstruction efficiency could instead act as a veto to reduce the backgrounds from the pair-produced B's in neutral current interactions. The signature for the signal process would be a neutral current event with (i) a single vertex from the other B, (ii) 4 converted gammas reconstructing to 2 high energy π^0's that, in turn, reconstruct to the B_d mass and (iii) no suspicion of another B or charm vertex. Hence, the analysis – although admittedly still exceedingly difficult – would benefit from both the exceptional vertexing and neutral pion reconstruction at Mighty MURINEs.

Therefore, to summarize this section, the initial expectation is that Mighty MURINEs should be able to do follow-up precision studies in at least some of the most difficult areas of B physics, even after the other B factories have run.

V QUARK MIXING: MEASUREMENTS BEYOND THE B FACTORIES

This section enlarges on the theoretical interest in measurements of quark mixing at MURINEs and also provides detail on the central role that Mighty MURINEs could assume if they reached sufficient energies to begin producing top quarks.

A Theoretical Interest in the CKM Matrix

Quark mixing is one of the least understood and most intriguing parts of elementary particle physics, and the confinement of quarks inside hadrons also makes it one of the hardest areas to study. The CC weak interaction for quarks differs from this interaction for leptons by mixing quarks from different families, i.e. any positively charged quark, $q^{(+)} \in u, c, t$, has some probability of being converted into any of its negatively charged counterparts, $q^{(-)} \in d, s, b$, and vice versa, rather than being uniquely associated with its same-family counterpart (i.e. $d \leftrightarrow u$, $s \leftrightarrow c$ and $b \leftrightarrow t$). This feature is accommodated in the standard model of elementary particle physics through the unitary 3-by-3 Cabbibo-Kobayashi-Maskawa (CKM) matrix, V_{ij}, that connects the positively charged $q_i^{(+)}$'s with the three $q_j^{(-)}$'s. (The corresponding matrix for leptons is trivially the 3-by-3 identity matrix.)

Apart from verifying that the SM description for quark mixing is indeed correct, the CKM matrix has additional interest through its hypothesized association with CP violation: the puzzling phenomenon that some particle properties, such as decay rates, have been found to differ slightly from those of the corresponding anti-particles. CP violation could also have cosmological implications; it has been invoked as one possible explanation for the comparative scarcity of anti-matter in the universe. The presence of a complex phase in the CKM matrix is the largely *untested* standard model explanation/parameterization for CP violation.

It is a testament to the perceived importance of the CKM matrix and CP violation that much of today's experimental HEP effort is devoted such studies, including B and phi factory colliders, LHC-B, HERA-B, B-TeV, K-TeV and many others.

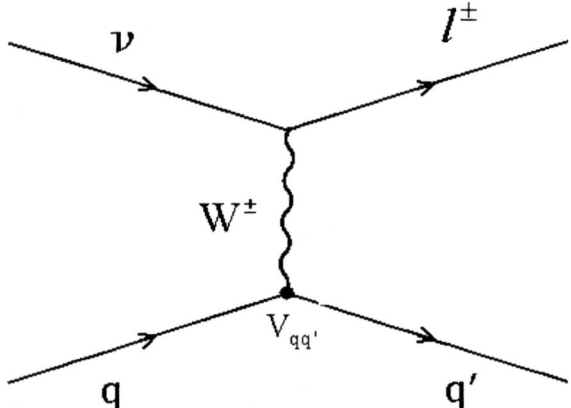

FIGURE 4. The Feynman diagram for charged current neutrino-quark scattering, showing that the CKM matrix element $V_{qq'}$ participates as an amplitude in the W-quark vertex.

Neutrino-nucleon scattering has impressive potential to augment these studies but, until the arrival of MURINEs, it will be held back by inadequate beam intensities.

B Quark Mixing Studies at MURINEs

The new neutrino studies at MURINEs will be much cleaner theoretically than most of the other experimental processes and will offer much complementary information.

Figure 4 is the Feynman diagram for the basic scattering process, showing that the CKM matrix element $V_{qq'}$ participates as an amplitude in the W-quark coupling. As a fundamental difference between νN DIS and all other types of CKM measurements, the scattering process involves the interaction of an external W boson probing the quarks inside a nucleon rather than an internal W interaction inside a hadron that, e.g., initiates a B decay. (In principle, the HERA ep collider could also do measurements involving an external W exchange, but these turn out not to be feasible in practice (11).) This is a substantial theoretical advantage for neutrino scattering because the "asymptotic freedom" property of QCD predicts quasi-free quarks with reduced influence from their hadronic environment at the higher 4-momentum-transfer (Q) scales available with an external W exchange.

The CKM measurements at MURINEs will be complementary to, say, the CKM measurements at B factories in that the measurements are of the magnitudes of individual CKM matrix elements rather than of interference terms involving pairs of elements. As can be inferred from figure 4, this arises because the differential cross-sections, $\frac{d\sigma}{dx}(q_i \to q_j)$, for the quark transitions are proportional to the absolute squares of the CKM elements:

$$\frac{d\sigma}{dx}(d \to c) \propto x\, d(x)|V_{cd}|^2 \times T(m_c, x) \qquad (15)$$

$$\frac{d\sigma}{dx}(s \to c) \propto x\,s(x)|V_{cs}|^2 \times T(m_c, x) \tag{16}$$

$$\frac{d\sigma}{dx}(u \to b) \propto x\,u(x)|V_{ub}|^2 \times T(m_b, x) \tag{17}$$

$$\frac{d\sigma}{dx}(c \to b) \propto x\,c(x)|V_{cb}|^2 \times T(m_b, x) \tag{18}$$

$$\frac{d\sigma}{dx}(d \to t) \propto x\,d(x)|V_{td}|^2 \times T(m_t, x) \tag{19}$$

$$\frac{d\sigma}{dx}(s \to t) \propto x\,s(x)|V_{ts}|^2 \times T(m_t, x) \tag{20}$$

$$\frac{d\sigma}{dx}(b \to t) \propto x\,b(x)|V_{tb}|^2 \times T(m_t, x), \tag{21}$$

where the Bjorken scaling variable, x, with $0 < x < 1$, is a relativistically invariant quantity that can be reconstructed for each event and, roughly speaking, measures the fraction of the nucleon's 4-momentum carried by the struck quark. The respective initial-state quark densities as functions of Bjorken x have been labeled $d(x)$, $s(x)$, $u(x)$, $c(x)$ and $b(x)$, and the $T(m_q, x)$'s are threshold suppression factors due to the masses, m_q, of the final-state quarks.

The $T(m_q, x)$ mass suppression factors are zero or much less than unity for all x below neutrino energies that can readily supply enough CoM energy to produce the massive final state quarks. From equation 5, the $T(m_q, x)$'s will asymptotically approach unity only for muon beam energies such that:

$$m_q^2 \ll 2M_p E_\mu/c^2 + M_p^2. \tag{22}$$

This places the following lower bounds on beam energies for the efficient production of charm, beauty and top quarks, respectively:

$$\begin{aligned} m_c &\sim 1.3 - 1.7\,\text{GeV}/c^2 &\Rightarrow& \quad E_\mu \gg 1\,\text{GeV} \\ m_b &\sim 5\,\text{GeV}/c^2 &\Rightarrow& \quad E_\mu \gg 13\,\text{GeV} \\ m_t &\sim 175\,\text{GeV}/c^2 &\Rightarrow& \quad E_\mu \gg 16\,\text{TeV}. \end{aligned} \tag{23}$$

The extraction of the CKM matrix elements from the MURINEs' experimental data will be analogous to, but vastly superior to, current neutrino measurements of $|V_{cd}|$, the only CKM matrix element that is currently best measured in neutrino-nucleon scattering (12). The experimentally determined event counts and kinematic distributions of the quark-tagged event samples provide measurements of the differential distributions for each of the final state quarks. The differential cross-sections, $\frac{d\sigma}{dx}(q_i \to q_j)$, and CKM matrix elements, $|V_{ij}|$, are derived from equations 15 through 21 using some auxiliary knowledge of the quark x-distributions within the nucleons and also a model for the mass threshold suppression terms, $T(m_q, x)$. In practice, this information should be obtainable largely from the data samples themselves: from CC and NC structure function measurements and the observed kinematic dependences in the heavy quark event sample.

C Expected Measurement Precisions at MURINEs

Today's measurements of $|V_{cd}|$ in νN scattering are already the most precise in any process, despite the coarse instrumentation of the neutrino detectors and the consequent requirement to use the semi-muonic subsample of charm decays for final state charm tagging. Even the lowest energy MURINEs under consideration (dedicated neutrino factories with $E_\mu \simeq 10$ GeV and up) will provide an opportunity (5) to extend to unique and precise measurements of the elements $|V_{cd}|$ and probably $|V_{cs}|$, now using vertex tagging of charm and with much improved knowledge of the quark distributions.

Further measurements of the more theoretically interesting elements $|V_{ub}|$ and $|V_{cb}|$ will become available at MURINES with muon energies of around 100 GeV or above, which can provide high enough neutrino energies for B production. The B-production analyses at these higher energy MURINEs should be experimentally rather similar to the charm analyses but would have vastly greater theoretical interest.

Both $|V_{ub}|$ and $|V_{cb}|$ determine the lengths of sides of the "unitarity triangle" that is predicted to exist if the CKM matrix is indeed unitary (13). The main goal of today's B factories is to measure the interior angles of this triangle to confirm that it is indeed a triangle, and the complementary input from a MURINE will be an enormous help in this verification process. In particular, the predicted (7; 5) 1-2 % accuracy in $|V_{ub}|^2$ is several times better than predicted accuracies in any future measurements of other processes, and will obviously provide a very strong constraint on the unitarity triangle.

Predicted experimental accuracies for a 500 GeV MURINE are summarized in table 3. These measurements would likely be improved still further with the higher event statistics and cleaner theoretical analysis available at a Mighty MURINE.

D Possible Measurements of V_{td} in Flavor Changing Neutral Current Interactions

The increased event statistics and neutrino energies at Mighty MURINEs might even allow the measurement of the further matrix element $|V_{td}|$ through the flavor changing neutral current (FCNC) interaction of figure 5.

This process is analogous to the predicted rare B decay $B \to X_d \nu \bar{\nu}$, shown in figure 6, where X_d represents inclusive production of hadrons containing a down quark. As related measurements, the very rare kaon decay processes $K^- \to \pi^- \nu \bar{\nu}$ and $K_L^0 \to \pi^0 \nu \bar{\nu}$ proceed through diagrams equivalent to 6 except with the incoming b quark replaced by an s quark and, correspondingly, V_{tb} replaced by V_{ts}. Therefore, the charged kaon decay has the potential to measure $|V_{ts}^* V_{td}|$ and its neutral counterpart actually measures the imaginary part of this quantity, $\mathrm{Im}(V_{ts}^* V_{td})$, due to $K^0 - \overline{K^0}$ interference. One event of the decay $K^- \to \pi^- \nu \bar{\nu}$ has been seen so

TABLE 3. Predicted CKM measurements for lower energy MURINES, reproduced from reference (14). The first row for each element, in bold face, is the absolute square of that matrix element, which is proportional to the experimental event rate – see equations 15 through 21. The second row for each element gives current percentage uncertainties in the absolute squares, without applying unitarity constraints, and speculative projections of the uncertainties after analyses from a 500 GeV MURINE. The measurements of $|V_{cd}|^2$ and $|V_{cs}|^2$ might be comparably good even for a 50 GeV MURINE but $|V_{ub}|^2$ and $|V_{cb}|^2$ would not be measured.

	d	s	b
u	**0.95** ±0.1%	**0.05** ±1.6%	**0.00001** ±50% → 1-2%
c	**0.05** ±15% → 0.2-0.5%	**0.95** ±35% → ~1%	**0.002** ±15% → 3-5%
t	**0.0001** ±25%	**0.001** ±40%	**1.0** ±30%

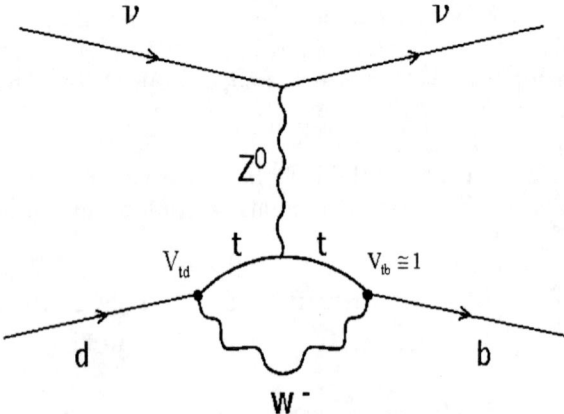

FIGURE 5. The Feynman diagram for B production through a flavor changing neutral current interaction involving a top quark loop.

far (15), consistent with its predicted tiny branching ratio of $(8.2 \pm 3.2) \times 10^{-11}$, and the even rarer neutral decay process has yet to be observed.

The search for the B decay signal looks to be extremely challenging even at future B factories, so it is unlikely to yield an accurate measurement of $|V_{td}|$. Therefore, a neutrino measurement of this quantity, at or below the 10% level, might still be valuable even after the B factories have run, augmenting the complementary measurements, perhaps eventually with comparable accuracy (16), of $|V_{ts}^* V_{td}|$ and

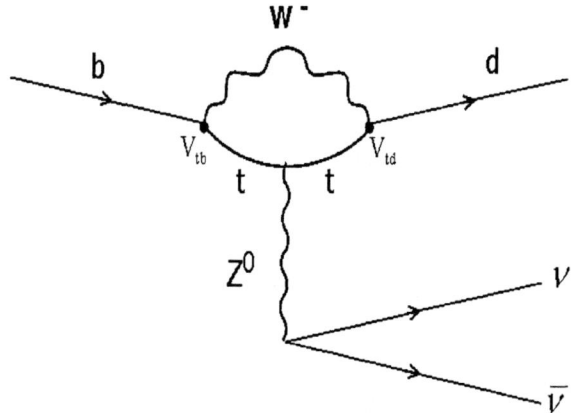

FIGURE 6. The Feynman diagram for the rare B decay $B \to X_d \nu \bar{\nu}$, which is analogous to the neutrino process of figure 5.

$\text{Im}(V_{ts}^* V_{td})$ expected in future generations of rare kaon decay experiments and the precise measurement of the ratio $|V_{ts}|/|V_{td}|$ that is to be eventually expected from B_d and B_s oscillations.

For neutrino energies well above the B production threshold, the process of figure 5 will occur at the level (5) of order 10^{-8} of the total neutrino-induced event sample unless some exotic physics process intervenes to increase the production rate. The signature is production of single B^- mesons from the valence d quarks at high x, whose rate should be directly proportional to the product of $|V_{td}|$ and the known valence quark density in nucleons.

The main background will come from b–anti-b production where the partner B meson containing the b anti-quark has escaped detection from either its primary decay vertex or through the decay vertex of its daughter charmed meson. The background process is easily separable from the signal on a statistical basis because it is symmetric in B^- versus B^+ mesons. However, the raw production rate (5) is roughly five orders of magnitude above the signal so the statistical viability of the analysis would require the raw background to be reduced by perhaps 4 to 5 orders of magnitude. This can be contemplated only because of (1) the very different event kinematics, with almost all of the background events at low Bjorken x and the signal mostly at high x, and (2) the unprecedented veto power for B decays that is expected in the vertexing detectors at MURINEs. Even so, a raw signal event sample of thousands of events might be needed for a 10% measurement of $|V_{td}|$ given the statistical dilution that background processes might entail. This would require several years running for the experimental parameters of table 1.

E CKM Measurements from the Production of Top Quarks at the Highest Energy Mighty MURINEs

Aside from top production in loops, a daunting leap of 3 orders of magnitude in beam energy would be required to move from the CKM elements involving B production to those involving top production, as is seen from comparing the second and third rows of equation 24. Uniquely precise direct measurements of $|V_{td}|^2$ and $|V_{ts}|^2$ and, possibly, $|V_{tb}|^2$ from the production of top quarks will become available if and when muon colliders eventually reach the 100 TeV CoM energy scale. (Note that muon collider energies even up to 1000 TeV, i.e. 1 PeV, have been speculated, using muon acceleration in linacs (17).)

Such impressive machines are prospects for the far distant future, and would be intended to zero in on a coherent understanding of the elementary building blocks of our universe. It should be stated that a major sea change from current theoretical prejudices would be required if the CKM matrix and its information on CP violation was to become central to the construction or verification of such a "theory of everything". Disregarding the current prejudices, top production at these highest energy Mighty MURINEs would move the experimental probing of the CKM matrix to a level of accuracy that appears to be inaccessible to any other type of experiment.

As will be explained further in the following section, a simple scaling from top production estimates calculated for HERA (18) predicts of order 10^5 top quark events for 1 inverse zeptobarn of integrated luminosity at muon energies slightly above 50 TeV. (A more accurate and detailed calculation is obviously required!)

Experimentally, top production should be relatively easy to tag with very high efficiency and purity. The two signatures are:

$$\nu_\mu N \to \mu^- (2 \text{ jets})(\text{b jet}) \qquad (24)$$

$$\nu_\mu N \to \mu^- l^+ \nu (\text{b jet}), \qquad (25)$$

with 68% and 32% BR's, respectively. Because of the large top mass, the final state jets can each have large acoplanarities, and the rarity of backgrounds with b quarks makes both signatures particularly distinctive. Additionally, in the first case the 2 other jets will reconstruct to the W mass while the presence of a second high-p_t, high energy lepton and large missing p_t from the neutrino will make the second signature even more striking.

No attempt will be made to even guess at the measurement accuracy. As a general comment, the beam energy will never be very far above the threshold for top quark production, so the feasibility and accuracy of the measurements would depend more strongly on the muon beam energy than the beam intensity. In almost all cases, the measurements of the CKM matrix elements involving top should be statistically limited because of the relatively small statistics (except at PeV-scale colliders!), their distinctive experimental signature and the accurately predictable threshold behavior. The sequentially decreasing populations at high x of the progressively

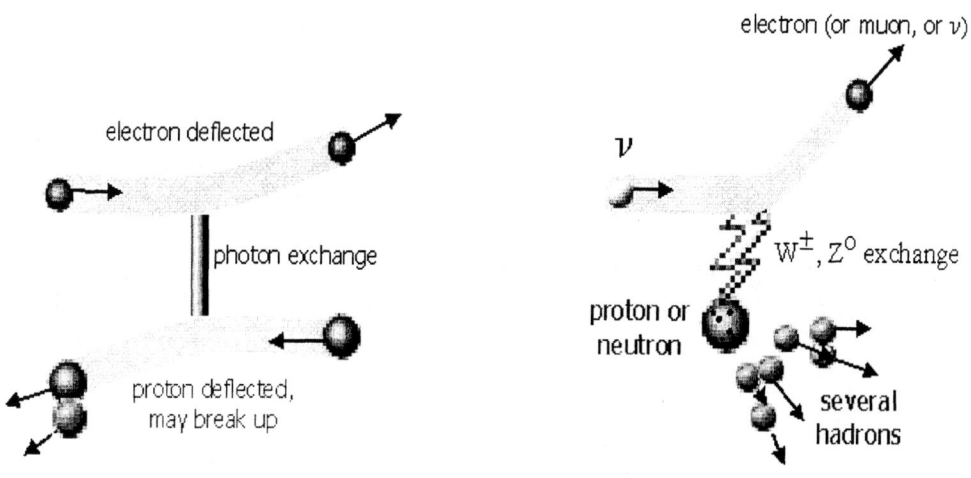

A) electromagnetic interaction at HERA   **B) weak interaction at MURINE**

FIGURE 7. Conceptual illustration of A) the relatively soft electromagnetic interactions, involving the exchange of a photon, that dominate the event sample at HERA, and B) the much harder weak interactions that will occur in Mighty MURINEs and that exist on the "hard scattering tail" at HERA.

heavier initial state quarks – d, s and b – should compensate or over-compensate the trend for the higher couplings to the top quark in the respective measurements of $|V_{td}|$, $|V_{ts}|$ and $|V_{tb}|$. Whether the first, the first two or all three matrix could be measured would presumably also depend strongly on the beam energy.

CKM measurements involving top would extend CKM studies beyond the paradigm of the unitarity triangle that is connected to B factory studies. For example, the conventional unitarity triangle is formed from the dot product of columns 1 and 3 of the CKM matrix (13). Measurements of the CKM elements involving the top quark would also provide enough experimental input to test the corresponding triangle involving *rows* 1 and 3 of the matrix, which is of comparable theoretical value in exploring CP violation. The analysis of experimental results would more likely be couched in more general theoretical terms, including unitarity tests and global fits to the 4 parameters – three magnitudes and a phase – that characterize the unitary 3-by-3 CKM matrix. Consistency of these fits would probe the SM hypothesis at a level that would not be possible without Mighty MURINEs.

VI HEAVY PARTICLE PRODUCTION – MIGHTY MURINES VERSUS HERA

The HERA ep collider is a convenient reference point for assessing the physics potential of Mighty MURINEs at the highest energy scales. As indicated in figure 7, MURINEs have the same *weak* interactions as HERA while avoiding the predominantly soft electromagnetic interactions that dominate the HERA event trigger rates but are less interesting because lower energy transfers probe physics at relatively lower energy scales. The event samples at MURINEs will correspond to the weak interactions in the "hard scattering tail" of the HERA event sample.

TABLE 4. Comparison of a 50–100 TeV mighty MURINE with HERA for the production of heavy particles. All of the numbers are order of magnitude only, and are based on simple and approximate scalings from reference (18) rather than on detailed calculations.

Particle/Process	HERA (1 inverse femtobarn)	Mighty MURINE (1 inverse zeptobarn)
b quark: $c \to b$	$O(10)$ events	$O(10^6)$ events
b quark: $u \to b$	$O(1)$ event	$O(10^5)$ events
top quark	$O(1)$ event	$O(10^5)$ events
W, Z bosons	tens of events	$O(10^5)$ events ?
120 GeV SM Higgs	$O(1)$ event	$O(10^5)$ events
exotica with small σ	luminosity limited	luminosity OK!

The HERA event samples involving weak interactions can be compared with MURINEs at energies where the high energy tail of the neutrino beam is comparable to the 314 GeV HERA center-of-mass energy. The maximum neutrino energy is the muon beam energy, which, according to equation 5, equals the HERA CoM energy for $E_\mu = 53$ TeV. At this energy or slightly above, very rough estimates of the event rates of similar or identical processes can be simply transcribed from HERA calculations after scaling by the 10^5 luminosity ratio shown in table 1. Such a comparison is shown in table 4. A range of MURINE energies has been given, in deference to the very approximate nature of the comparison. At the low end ($E_\mu = 50$ TeV), the MURINE event rates will probably be lower than the estimate, and the rates will normally be higher at the high end ($E_\mu = 100$ TeV).

The standard model physics processes involving weak interactions are the same in all cases except for the production of W and Z bosons, where HERA has the advantage due to processes involving photon exchange. The SM Higgs has not been found at the time of writing, so the 120 GeV mass is an example only.

The first three processes in table 4 have already been discussed in the preceding section. To be realistic, at the event rates shown it is very doubtful that W, Z and SM Higgs production could contribute anything useful beyond collider studies, despite the the astounding neutrino beam parameters and superior event reconstruction. Beyond this, possible exotic processes at the 100 GeV scale or below provide the only substantial potential for exciting discoveries. This motivation could become much stronger in the near future if, for example, one of the current leptoquark searches at HERA returned a discovery. It is noted that the leptoquarks produced at a Mighty MURINE might well be different – coupling to neutrinos rather than electrons – and so studies at MURINEs could potentially be complementary to those at a future ep collider with a higher E_{CoM} than HERA.

VII SUMMARY

The Mighty MURINE neutrino experiments that would come almost for free at any future many-TeV muon collider could improve on the pioneering advances from the previous MURINEs that would have existed at lower energy muon colliders. The most important improvements might well be on the unique and important measurements from previous lower energy MURINEs of $|V_{ub}|$ and $|V_{cb}|$, perhaps pushing the accuracy of both measurements below 1%. With total event statistics of a few times 10^{11} events, the rare production of B's through flavor changing neutral current interactions off valence d quarks might provide one of the best indirect determinations of $|V_{td}|$. More common channels for B production, particularly through neutral current interactions, might also provide some capabilities as a B factory with novel experimental strengths.

Upon crossing the threshold for top production, the even more interesting elements $|V_{td}|$, $|V_{ts}|$ and $|V_{tb}|$ could become successively available to uniquely precise measurements at the highest energy Mighty MURINEs. The addition of any or all of these three precise measurements would clearly advance our knowledge of the CKM matrix to a level where small perturbations from the Standard Model scenario could be searched for and, if found, could be studied. MURINEs would then truly play the central role in determining the CKM matrix parameters, with the best measurements of the magnitudes of perhaps seven of the nine elements (all but the two elements that are currently best measured: $|V_{ud}|$ and $|V_{us}|$) to add to the phase information from various other experimental processes.

If muon colliders ever reach the 100 TeV center of mass energy scale then their neutrino experiments will attain a center of mass energy reach comparable to the existing HERA ep collider, but at a luminosity that might be perhaps 5 orders of magnitude higher. HERA then becomes a convenient reference point for assessing their physics capabilities. Despite the promise of impressive luminosities, none of the standard model processes other than the CKM matrix appear to offer the chance of competitive physics potential to studies of the same processes at colliders. Therefore, only i) an enlarged theoretical importance for the CKM matrix or ii) the discovery, then or beforehand, of an exotic process that is accessible to Mighty MURINEs, would give Mighty MURINEs a chance for physics analyses of a comparable importance to those at the colliders. Leptoquarks that couple to neutrinos are the obvious candidate for such a new process.

REFERENCES

[1] B.J. King, *Assessment of the Prospects for Muon Colliders*, paper submitted in partial fulfillment of requirements for Ph.D., Columbia University, New York (1994), available from http://xxx.lanl.gov/ as **physics/9907027**.

[2] S. Geer, *Neutrino beams from Muon Storage Rings: Characteristics and Potential*, PRD 57, 6989 (1998).

[3] B.J. King, *Neutrino Radiation Challenges and Proposed Solutions for Many-TeV Muon Colliders*, these proceedings. Also available from
http://pubweb.bnl.gov/people/bking/heshop/hemc_papers.html.

[4] B.J. King, *Prospects for Colliders and Collider Physics to the 1 PeV Energy Scale*, ibid.

[5] I.I. Bigi et al., "The potential for High Rate Neutrino Physics at Muon Colliders and Other Muon Storage Rings", in preparation for publication in Physics Reports.

[6] See, for example, Chris Quigg, *Neutrino Interaction Cross Sections*, FERMILAB-Conf-97/158-T.

[7] B.J. King, *Neutrino Physics at a Muon Collider*, Proc. Workshop on Physics at the First Muon Collider and Front End of a Muon Collider, Fermilab, November 6-9, 1997, available from http://xxx.lanl.gov/ as **hep-ex/9907033**.

[8] K. Abe et al., Design and Performance of the SLD Vertex Detector: a 307 Mpixel Tracking System, NIM A 400 (1997) 287-343.

[9] Private communication with Tim Bolton.

[10] Ikaros Bigi pointed out the theoretical importance of this B decay process to the author and suggested considering the experimental possibilities to study it at MURINEs.

[11] R.M. Godbole et al., *The Kobayashi-Maskawa Matrix at HERA*, Proceedings of the HERA Workshop, Hamburg, October 12-14, 1987, ed. R.D. Peccei, publ. DESY; A. Ali et al., *Heavy Quark Physics at HERA*, ibid.

[12] H. Abramowicz et al., Z. Phys. **C21**, 27 (1982); S.A. Rabinowicz et al., Phys. Rev. Lett. **70**, 134 (1993); A.O. Bazarko et al., Z. Phys. **C65**, 189 (1995).

[13] C. Caso et al., *The Review of Particle Physics*, The European Physical Journal C3 (1998) 1 and 1999 off-year partial update for the 2000 edition available on the PDG WWW pages (URL: *http://pdg.lbl.gov/*).

[14] B.J. King, *High Rate Physics at Neutrino Factories*, Submitted to Proc. 23rd Johns Hopkins Workshop on Current Problems in Particle Theory, "Neutrinos in the Next Millenium", Johns Hopkins University, Baltimore MD, June 10-12, 1999, available from http://xxx.lanl.gov/ as **hep-ex/9911008**.

[15] S. Adler et al., *Evidence for the Decay* $K^- \to \pi^- \nu \bar{\nu}$ Phys. Rev. Lett. 79 (1997) 2204-2207, hep-ex/9708031; S. Adler et al., *Further Search for the Decay* $K^- \to \pi^- \nu \bar{\nu}$ hep-ex/0002015.

[16] Private communication with L.S. Littenberg.

[17] F. Zimmermann, *Final Focus Challenges for Muon Colliders at Highest Energies*, these proceedings. Also available from
http://pubweb.bnl.gov/people/bking/heshop/hemc_papers.html.

[18] Proceedings of the HERA Workshop, Hamburg, October 12-14, 1987, ed. R.D. Peccei, publ. DESY.

Neutrino Radiation Challenges and Proposed Solutions for Many-TeV Muon Colliders [1]

Bruce J. King

Brookhaven National Laboratory
email: bking@bnl.gov
web page: http://pubweb.bnl.gov/people/bking

Abstract. Neutrino radiation is expected to impose major design and siting constraints on many-TeV muon colliders. Previous predictions for radiation doses at TeV energy scales are briefly reviewed and then modified for extension to the many-TeV energy regime. The energy-cubed dependence of lower energy colliders is found to soften to an increase of slightly less than quadratic when averaged over the plane of the collider ring and slightly less than linear for the radiation hot spots downstream from straight sections in the collider ring. Despite this, the numerical values are judged to be sufficiently high that any many-TeV muon colliders will likely be constructed on large isolated sites specifically chosen to minimize or eliminate human exposure to the neutrino radiation. It is pointed out that such sites would be of an appropriate size scale to also house future proton-proton and electron-positron colliders at the high energy frontier, which naturally leads to conjecture on the possibilities for a new world laboratory for high energy physics. Radiation dose predictions are also presented for the speculative possibility of linear muon colliders. These have greatly reduced radiation constraints relative to circular muon colliders because radiation is only emitted in two pencil beams directed along the axes of the opposing linacs.

I INTRODUCTION

Neutrinos interact so rarely that, only 50 years ago, even their detection was not considered to be feasible. It is therefore quite surprising that the design of future muon colliders will usually be constrained by the need to limit hazards from neutrino radiation (1; 2). Neutrinos are produced copiously at muon colliders from the decays of the large currents of muons circulating in the collider ring:

$$\mu^- \to \nu_\mu + \overline{\nu_e} + e^-,$$
$$\mu^+ \to \overline{\nu_\mu} + \nu_e + e^+. \quad (1)$$

---

[1] This work was performed under the auspices of the U.S. Department of Energy under contract no. DE-AC02-98CH10886.

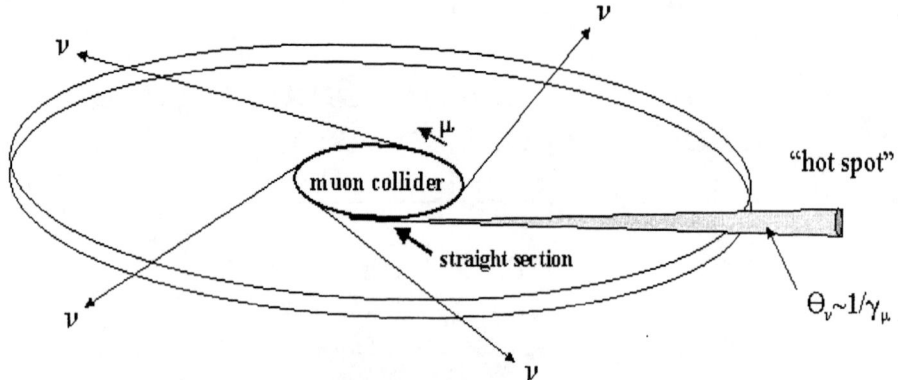

FIGURE 1. The decays of muons in a muon collider will produce a neutrino radiation disk emanating out tangentially from the collider ring. Radiation hot spots in the disk will occur directly downstream from straight sections in the collider ring.

The neutrino direction is tightly collimated to within a characteristic angle, θ_ν, of the decaying muon's direction, where:

$$\theta_\nu = 1/\gamma_\mu = \frac{m_\mu c^2}{E_\mu} \simeq \frac{10^{-4}}{E_\mu[\text{TeV}]}, \qquad (2)$$

for γ_μ the relativistic boost factor of the muon, m_μ the muon rest mass, c the speed of light and E_μ the muon energy. (Units are given in square brackets in the equations throughout this paper.) The combined effect of all the muon decays will be a disk of neutrinos emanating out in the plane of the collider ring (2), as shown in figure 1. Straight sections in the ring will cause radiation hot spots in the disk (1) where all of the decays in the straight section line up into a pencil beam that is superimposed on the disk, again with a characteristic opening half-angle for the cone of $1/\gamma_\mu$. As a notable contrast to all other radiation hazards, the neutrino attenuation length is too long for the beam to be appreciably attenuated by any practical amount of shielding material, including even the expanse of ground between the collider ring and where the radiation disk breaks ground.

Example parameter sets for muon colliders that illustrate the radiation hazard are given in table 1. The entries in the table will be referred to and explained throughout this paper. For now, we note that the radiation doses may be compared with the U.S. Federal off-site limit of 1 mSv/year or, in alternative units, 100 mrem/year. (The limit is comparable to typical background radiation levels of 0.4 to 4 mSv/year (3).) The radiation hazard is seen to rise sharply with muon collider energy, increasing from a tiny fraction of the legal limit for the lower energy colliders to well above the limit for the collider scenarios at 10 TeV and 100 TeV. This behavior will be quantified in the two sections that follow; the next section will characterize the radiation dose for colliders up to the TeV energy scale and the section after that will discuss some mitigating factors that come into play for the many-TeV muon colliders that are the subject of this workshop. Possible solutions for the hazard at many-TeV energies will be discussed in the next-to-last section of the paper before rounding out with a summary section.

TABLE 1. Straw-man muon collider specifications and the corresponding neutrino radiation parameters for muon colliders at 0.1, 4, 100 and 100 TeV. The first two muon collider scenarios were taken from references (4) and (5), respectively, while the final two scenarios were straw-man parameter sets for this workshop (6). The calculation of the neutrino radiation parameters is discussed in the text.

center of mass energy, E_{CoM}	0.1	4 TeV	10 TeV	100 TeV
additional description	H^0 factory	"lite"	2nd gen.	3rd gen.
collider luminosity, $\mathcal{L}$ [cm^{-2}.s^{-1}]	1×10^{31}	6×10^{33}	1×10^{36}	1×10^{36}
collider int. lum., $\int \mathcal{L}$ [fb^{-1}/yr]	0.1	60	10 000	10 000
muon beam energy, E_μ [TeV]	0.05	2	5	50
muon decays/yr, N_μ^+ [10^{20}]	6	0.08	8	0.4
collider reference depth, d [m]	10	300	100	100
ν beam distance to surface, L [km]	11	62	36	36
ν beam radius at surface [m]	24	3.3	0.8	0.08
ave. rad. dose in plane [mSv/yr]	2×10^{-5}	5×10^{-4}	2.3	10
str. sec. len. for 10x ave. rad. [m]	1.9	1.1	1.0	4.2

II THE NEUTRINO RADIATION HAZARD FOR MUON COLLIDERS UP TO TEV ENERGY SCALES

This section reviews reference (2) in characterizing the potential neutrino radiation hazard and giving numerical estimates for the radiation dose in the so-called equilibrium situation that pertains to the beams from muon colliders at the TeV energy scale and below.

Neutrinos at energies beyond a few GeV interact predominantly through deep inelastic scattering off nucleons. The radiation hazard arises from the showers of penetrating charged particles produced through neutrino interactions with any material bathed by the neutrino radiation disk, as is indicated in figure 2. Starting from the initial interaction products, an avalanche effect of secondary, tertiary, etc. interactions, produces the vast majority of the charged particles. For TeV-scale neutrinos, neutrino interactions in people themselves may therefore only account for as little as 0.1% of their radiation dose (7) because the primary hadrons from the interaction will typically exit the person before interacting to commence the shower of charged particles.

The development of particle showers makes the dose very dependent on the local surroundings. Radiation hazard calculations must conservatively consider the worst case configuration where a person is (i) completely bathed in the radiation disk or the pencil beam downstream from a straight section and (ii) is surrounded

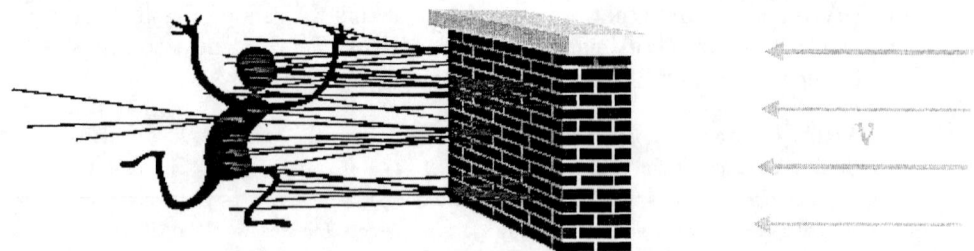

FIGURE 2. The radiation hazard arises primarily from particle showers initiated by neutrino interactions in material near the person.

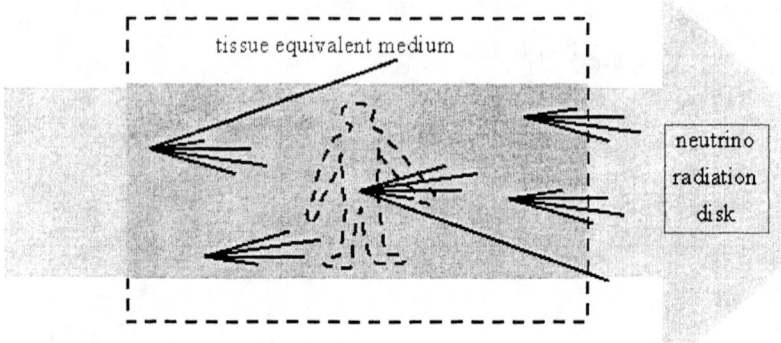

FIGURE 3. Conceptual illustration of the "equilibrium approximation" geometry used for quantitative worst-case radiation calculations. The entire person is bathed by the neutrino radiation disk in the plane of the collider ring and perhaps additionally by the radiation cone downstream from a straight section in the collider ring. The person is also enclosed in a "tissue equivalent medium" that has sufficient density to localize the charged particle showers from neutrino interactions so that little of the showers' energy spills out beyond the radiation disk.

by material that will initiate showers from neutrino interactions. A further requirement for the validity of these calculations is that the characteristic density of the surrounding material must be sufficient to contain the showers within the pencil beam or radiation disk. Such geometries can be contrasted with the more common situation where the showers will spread out transversely beyond the beam and, hence, dilute the dose received by someone within the disk. Instead, this containment criterion will be satisfied only by materials where the nuclear interaction length that characterizes shower development is shorter than the beam radius, as will generally be the case for solids or liquids but not for air or other gaseous media. As examples (3), water and quartz have interaction lengths of 85 cm and 43 cm, respectively, while the interaction length of air, 700 meters, is much larger than the typical few-meter beam radii expected at TeV-scale muon colliders (see table 1).

The worst-case situation can be conveniently modeled for numerical calculations (2) by considering a person completely enclosed within a "tissue equivalent medium" i.e. a medium with the approximate density of water. (The "scuba diver" configuration.) This geometry is illustrated in figure 3. For this simplified model and rather artificial geometry, it is clear that the energy–per–unit–mass absorbed by a person is constrained simply by conservation of energy to be approximately equal to the summed energy of the neutrino interactions in the person. This applies even though most of the deposited energy is from shower products of interactions upstream from the person. This approximate equality is referred to as the equilibrium approximation. For more realistic and general geometries with inhomogeneous distributions of mass, it can be argued (2) that the equilibrium approximation is either valid or, alternatively, conservatively overestimates the radiation dose.

The calculation of radiation doses is straightforward (2) within this equilibrium assumption because neutrino interaction cross sections are well known and the approximate neutrino flux within the pencil beams can be simply predicted from the known decay kinematics and relativistic kinematics. Here we merely reproduce, in a slightly reorganized form, the formula of reference (2) for the average whole-body radiation dose, D^{ave}, in the plane of the collider ring:

$$D^{ave}[mSv] \simeq 3.7 \times N_\mu^+[10^{20}] \times \frac{(E_\mu[TeV])^3}{(L[km])^2}. \tag{3}$$

N_μ^+ is the number of muons decaying in the collider ring, per year and per charge sign and given in appropriate units of 10^{20} decays. L is the tangential distance from the ring to where the dose is measured – typically where the radiation disk exits the ground.

The additional radiation hot-spot from a straight section of length l^{ss} is given by (2):

$$D^{ss}[mSv] = 1.1 \times 10^5 \times N_\mu[10^{20}] \times f^{ss} \times \frac{(E_\mu[TeV])^4}{(L[km])^2}, \tag{4}$$

where the fraction, f^{ss}, of the ring circumference, C, corresponds to l^{ss} through:

$$f^{ss} = \frac{l^{ss}}{C}. \tag{5}$$

Equation 4 can be rewritten in terms of l^{ss} and the average bending magnetic field in the collider ring, B^{ave}, as

$$D^{ss}[mSv] \simeq 5.3 \times N_\mu^+[10^{20}] \times l^{ss}[m] \times B^{ave}[T] \times \frac{(E_\mu[TeV])^3}{(L[km])^2}, \tag{6}$$

by making use of the relation

$$C[km] = \frac{2\pi \cdot E_\mu[TeV]}{0.3 \cdot B^{ave}[T]}. \tag{7}$$

Equations 3, 4 and 6 are not claimed to be accurate at much better than order-of-magnitude level and detailed follow-up Monte Carlo simulations have confirmed their predictions to this level of accuracy (8).

The ratio of equations 3 and 6 immediately gives the length of straight section, l^{equiv}, that approximately doubles the in-plane average radiation dose:

$$l^{equiv}[\text{meters}] \simeq \frac{0.7}{B_{ave}[T]}. \tag{8}$$

This is only of order 10 cm for the typical average bending fields expected in muon collider storage rings.

The energy-cubed dependences of equations 3 and 6 deserve further comment. The dependence in equation 3 is the product of three linear factors: 1) the inverse width of the disk, which goes as $1/\theta_\nu$, 2) the neutrino cross-section, $\sigma_{\nu N}$, and 3) the average neutrino energy, $<E_\nu>$, that is deposited per interaction:

$$\text{disk average dose, } D^{ave} \sim \frac{1}{\theta_\nu} \cdot \sigma_{\nu N} \cdot <E_\nu> \propto E_\mu \cdot E_\mu \cdot E_\mu = E_\mu^3. \tag{9}$$

For the straight sections, the inverse disk width is replaced by the inverse cross-sectional area of the pencil beam, which goes as $1/\theta_\nu^2$. Also, the proportionality of equation 4 on f^{ss} brings in a factor of $1/E_\mu$ for a given value of l^{ss} in equation 6, using equations 5 and 7. The energy scaling of equation 6 can therefore be broken down into:

$$\text{str. section dose, } D^{ss} \sim (\frac{1}{\theta_\nu})^2 \cdot \sigma_{\nu N} \cdot <E_\nu> \cdot f^{ss} \propto E_\mu^2 \cdot E_\mu \cdot E_\mu/E_\mu = E_\mu^3. \tag{10}$$

Equations 3 and 6 predict the numerical radiation doses in table 1 for the 0.1 TeV and 4 TeV collider examples. The values used for L assume a collider located at a specified depth, d, under a site with a smooth surface having the average curvature of the Earth, so that

$$L_{exit} = (2 \times d \times R_E)^{1/2}, \tag{11}$$

where the Earth's radius has the value $R_E = 6.4 \times 10^6$ m and the equation very reasonably assumes that $d \ll R_E$.

The preceding discussion in this section has provided the background information for the numerical estimates of table 1 for the 0.1 TeV and 4 TeV parameter sets. The tabulated radiation predictions for the 10 TeV and 100 TeV collider parameters anticipate some mitigating factors at multi-TeV energies that will now be discussed.

III MITIGATING FACTORS AT MANY-TEV ENERGIES

Equations 3 and 6 are too pessimistic at the many-TeV energies addressed at this workshop, for two reasons. The smaller of the two effects is a partial leveling

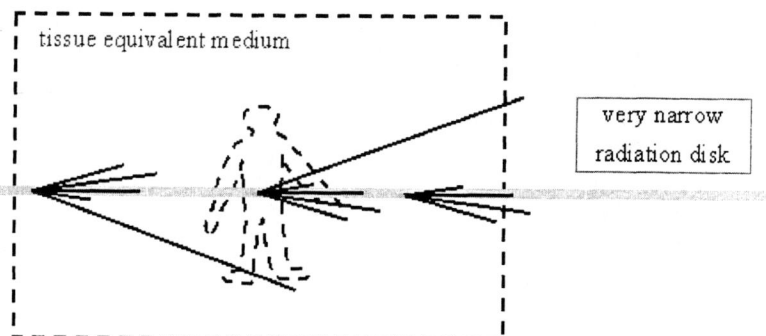

FIGURE 4. The worst-case geometry for radiation exposure at many-TeV muon colliders, to be contrasted with the "equilibrium approximation" geometry of the preceding figure.

TABLE 2. Parameterization of the cross section factor $X(E_\mu)$ that describes the fall-off from a linear cross section rise with energy. By definition, $\alpha \equiv \log_{10}(E_\mu[TeV])$. The parameterizations are simple logarithmic energy interpolations between the cross sections given in reference (9). The numerical factors in the expressions, 1.453, 1.323, 1.029, 0.512 and 0.175 are the total summed neutrino-nucleon and antineutrino-nucleon cross-sections-divided-by-energy (9) at the neutrino energies of 100 GeV, 1 TeV, 10 TeV, 100 TeV and 1 PeV, respectively, given in units of 10^{-38} cm^2/GeV. As an adequate approximation to avoid convolutions with neutrino energy spectra, the muon energies in the table have been set equal to the corresponding neutrino energies. By definition, $X(E_\mu) \equiv 1$ for $E_\mu = 100$ GeV, which is the reference energy used for the radiation dose parameterizations of equations 3 and 6.

muon energy range	expression for $X(E_\mu)$
$E_\mu < 1$ TeV	$(-1.453 \times \alpha + 1.323 \times (\alpha+1))/1.453$
1 TeV$< E_\mu <$ 10 TeV	$(1.323 \times (1-\alpha) + 1.029 \times \alpha)/1.453$
10 TeV$< E_\mu <$ 100 TeV	$(1.029 \times (2-\alpha) + 0.512 \times (\alpha-1))/1.453$
100 TeV$< E_\mu <$ 1000 TeV	$(0.512 \times (3-\alpha) + 0.175 \times (\alpha-2))/1.453$
1000 TeV$< E_\mu$	$(0.175/1.453) \times 3^{3-\alpha}$

off in the neutrino cross section. Rather than a continuation of the linear rise up to TeV energy scales, it is predicted that (9), for example, the neutrino cross section at 100 TeV is only 33 times that at 1 TeV, instead of a 100-fold increase.

More significantly, the beam radius, $\theta_\nu L$, ceases to be large compared to the size of a person or to the width of the shower it produces. This prevents the possibility of a person ever experiencing the geometry of figure 3 that is needed for the equilibrium approximation to apply. Instead, a modified worst-case geometry that often applies for many-TeV muon colliders is illustrated in figure 4.

The formulae 3 and 6 can be modified to apply roughly to the case of a narrow, many-TeV neutrino radiation disk or pencil beam by considering how the equilibrium approximation breaks down for a beam radius that, due to either increasing E_μ or decreasing L, becomes comparable to or smaller than a hadron shower radius. Instead of depositing the dose over a progressively smaller vertical band (in the case of a radiation disk), or a spot (in the case of the pencil beam), the shower radius imposes a limiting transverse size scale. For definiteness, the calculations for table 1 spread the disk-average dose out evenly over a half-height of 0.5 meters – which is comparable to the 0.43 m interaction length of granite that was mentioned previously – and spread the hot-spot dose over a circle of the same radius.

The explicit many-TeV equations corresponding to the lower energy equations 3 and 6 become, respectively:

$$D^{ave}_{many-TeV}[mSv] \simeq 3.7 \times N^+_\mu[10^{20}] \times \frac{(E_\mu[TeV])^3}{(L[km])^2} \times X(E_\mu) \times F(E_\mu, L) \quad (12)$$

and

$$D^{ss}_{many-TeV}[mSv] \simeq 5.3 \times N^+_\mu[10^{20}] \times l^{ss}[m] \times B^{ave}[T] \times \frac{(E_\mu[TeV])^3}{(L[km])^2}$$
$$\times X(E_\mu) \times (F(E_\mu, L))^2, \quad (13)$$

where $X(E_\mu)$ and $F(E_\mu, L)$ are the high energy suppression factors for the cross section leveling and small spot size, respectively.

The cross-section suppression factor, $X(E_\mu)$, has, by definition, the value unity for $E_\mu = 100$ GeV, which was the energy chosen to calculate the numerical coefficients for equations 3 and 6. Its slow decrease with increasing energy has been approximated by simple logarithmic energy interpolations between the cross sections given in reference (9) and the form and coefficients for the interpolations are given in table 2.

The appropriate form of $F(E_\mu, L)$, incorporating the assumed minimum transverse shower dimension of 0.5 m that was discussed above, is:

$$F(E_\mu, L) = \min\left(1, \frac{\theta_\nu[rad]L[km]}{5 \times 10^{-4}}\right) = \min\left(1, \frac{L[km]}{5 \times E_\mu[TeV]}\right), \quad (14)$$

where equation 2 has also been used to obtain the second form of the expression.

Substituting in the explicit form of equation 14 for the case where $L[km] < 5 \times E_\mu[TeV]$ returns the narrow-beam form of equations 12 and 13:

$$D^{ave}_{many-TeV}[mSv] \to 0.74 \times N^+_\mu[10^{20}] \times \frac{(E_\mu[TeV])^2}{L[km]} \times X(E_\mu) \quad (15)$$

and

$$D^{ss}_{many-TeV}[mSv] \to 0.21 \times N^+_\mu[10^{20}] \times l^{ss}[m] \times B^{ave}[T] \times E_\mu[TeV] \times X(E_\mu). \quad (16)$$

The second of these equations is independent of the distance L in this limit, as it intuitively must be: the beam is now narrow enough that a person will intercept essentially the entire beam, so the dose should become independent of distance in this limit.

We can now review the energy scaling of the radiation dose at many-TeV energies, to compare the power law dependences with those for the lower energy colliders that are given in equations 9 and 10. The 0.5 m fixed half-height removes the dependence of equation 9 on the disk height, $\theta_\nu L$. Combined with the partial leveling of the cross-section, this softens the radiation rise with energy relative to slightly less than quadratic in energy:

$$\text{many-TeV disk average dose},\ D^{ave}_{many-TeV} \sim \sigma_{\nu N} \cdot <E_\nu> \propto E_\mu^{<1} \cdot E_\mu = E_\mu^{<2}. \tag{17}$$

Similarly, the many-TeV dose from straight sections loses both powers of E_μ that came from the $1/(\gamma_\mu L)^2$ factor in equation 10, giving:

$$\text{many-TeV str. sec. dose},\ D^{ss}_{many-TeV} \sim \sigma_{\nu N} \cdot <E_\nu> \cdot f^{ss} \propto E_\mu^{<1} \cdot E_\mu/E_\mu = E_\mu^{<1}, \tag{18}$$

i.e. a less-than-linear rise with energy for radiation hot spots from a straight section of fixed length l^{ss}.

The assumed shower radius of 0.5 meters is clearly a somewhat arbitrary choice, so it should be borne in mind that equations 12 and 13 will be even more approximate than the lower energy predictions of equations 3 and 6. These predictions at many-TeV energies should also be interpreted even more as worst case scenarios than those at lower energies, since the material surrounding the person is now required to have a density comparable to granite in order to confine the hadron showers to the assumed transverse dimensions of 0.5 meters.

No detailed follow-up Monte Carlo simulations for various material geometries have yet been performed for the Many-TeV scenarios discussed in this section. Such simulations would be very valuable in confirming and refining the rough numerical estimates obtained from applying the above high-energy modifications to equations 3 and 6.

IV PROPOSED SOLUTIONS

Several means to reduce the radiation hazard have been proposed previously in reference (2):

1. minimize straight sections in the collider ring, e.g. by superimposing some bending field on all focusing magnets

2. improve the luminosity per unit current, from better beam cooling etc.

3. use fenced-off radiation enclosures downstream from the largest straight sections

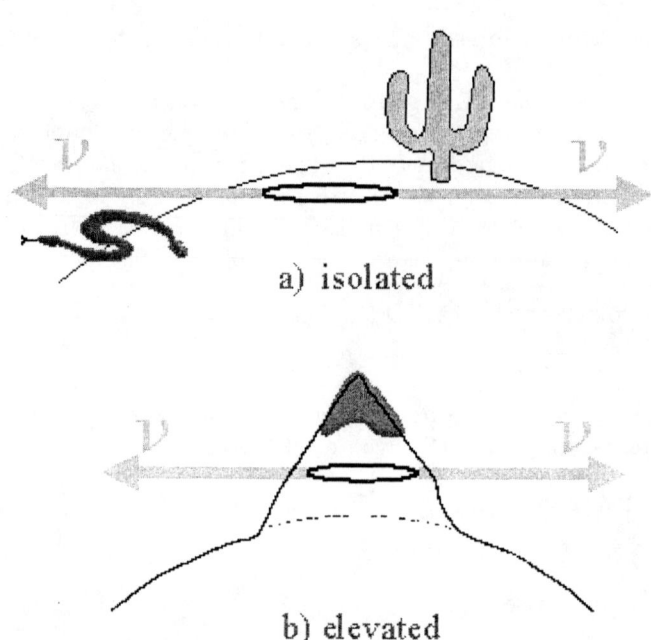

FIGURE 5. Two site options that avoid exposing people to the neutrino radiation disk in the plane of the muon collider ring.

4. bury the muon collider ring deeper underground to increase the distance before the neutrino disk exits the ground. Optionally, the ring can also be tilted and oriented to take advantage of natural geological features

5. choose to build the muon collider on a site where human exposure to the radiation disk will be minimized or, ideally, nobody at all will be exposed to the neutrino radiation.

The orders-of-magnitude reductions desired for many-TeV colliders appear difficult to achieve through any combination of items 1 through 4 alone so the final option – a specially chosen site – may well be unavoidable for muon colliders at the highest energies.

Two classes of siting options are shown in figure 5. The first is an isolated site, where nobody is exposed to the radiation disk before it exits into the atmosphere due to the local curvature of the Earth. The height above ground, h, at distance L from a collider ring close to the surface of a spherical Earth is given by rearranging equation 11 to:

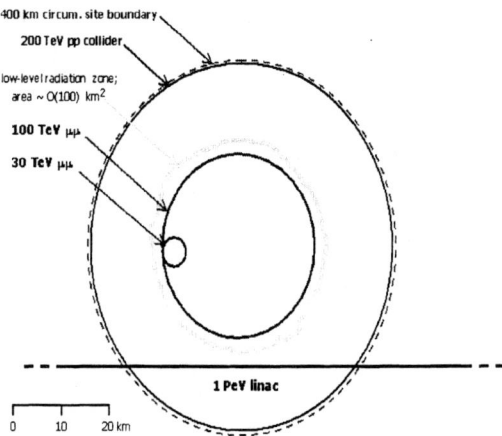

FIGURE 6. The ultimate high energy physics laboratory ? See further discussion in reference (10). Figure reproduced from reference (10).

$$h = \frac{L^2}{2R_E}. \tag{19}$$

As an example that is discussed further below, a distance to the site boundary of $L = 64$ km corresponds to $h = 320$ m, which might be considered a very comfortable clearance height at the site boundary for an isolated region. Some zoning restrictions could also be placed on tall structures near the laboratory site if this was additionally required.

The second option in figure 5, siting the laboratory on elevated land, would take advantage of the local topology to provide either a smaller site or higher clearances for the radiation disk at the site boundary. In practice, both siting options would likely be combined by choosing the most elevated site in an isolated region.

A third siting option, placing the collider in a valley or other depressed region to extend the distance before the disk exits the ground, is discussed elsewhere in these proceedings (11). This can be considered to be a variation on item 4 of the above list. As speculation, this siting option might well be practical up to perhaps the 10 TeV energy scale but would likely give an inadequate dose reduction for muon colliders at the highest energies and luminosities.

The transient radiation doses to people in planes and birds flying through the disk would clearly be negligible for either of the siting options of figure 5 since the doses in table 1 must be accumulated over a full accelerator year of 10^7 seconds and also, as explained previously, the conservatively calculated doses are anyway about three orders of magnitude above the full-year doses for a person or bird when they are not surrounded by dense material.

FIGURE 7. An example to illustrate the size of a HEP laboratory with a 400 km site boundary circumference. A circle of this diameter has been drawn in the Great Victoria Desert (just above the "A" in the label "South Australia"), showing that the outline of a laboratory of this size would even be visible on a map of Australia. A 1000 km long strip of land for the 1 PeV linear collider would perhaps not be included inside the original laboratory boundary. The choice of country and positioning of the site are for illustration only. Figure reproduced from reference (10).

Because of the dilute beam halo from showers induced in the atmosphere, the radiation dose at the site boundary would not be strictly zero even if the neutrino radiation disk was well above ground level. Speculatively, the halo might extend down below the neutrino disk by perhaps of order the interaction length of air, $\lambda_{air} = 700$ m. On naively comparing with equations 14 and 12, the disk average dose might be expected to be down by of order $(0.5 \text{ m})/\lambda_{air}$, which is about three orders of magnitude, from the in-disk prediction of equation 12. Similarly, comparison with equation 13 suggests that the additional doses from straight sections might speculatively be predicted to be diluted by the square of this ratio, i.e. by about six orders of magnitude. (These predictions are essentially just repeating the

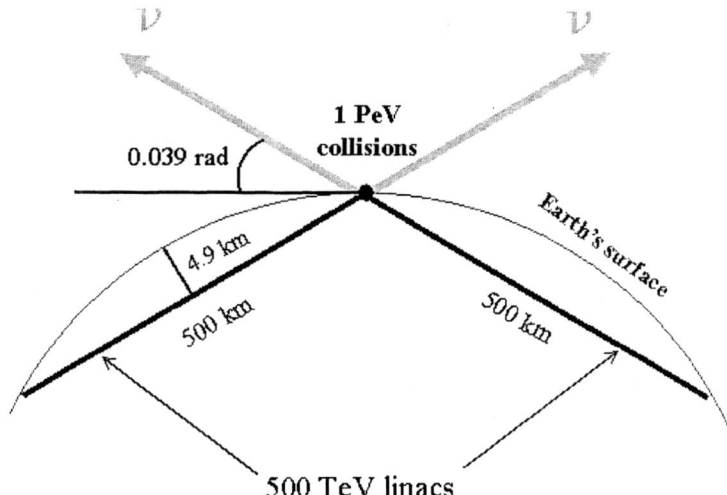

FIGURE 8. A PeV-scale linear muon collider (12) would shoot two neutrino beams upwards and towards the sky at angles to the horizontal of perhaps a few tens of milliradians, depending on the length of the linacs.

arguments leading up to equations 15 and 16, but now with a limiting radial extent of 700 meters rather than 0.5 meters.) If these tentative predictions are valid then both the average dose and the straight section dose would be safely below radiation limits. However, the predictions are very speculative and detailed Monte Carlo showering calculations are called for to predict the true level of the beam halo for this geometry.

On the positive side, a site diameter of order 100 km is also an appropriate size scale for jointly housing the largest potentially plausible proton or electron colliders along with the many-TeV muon colliders. This is the case in the "proof-of-plausibility" scenario for future energy frontier colliders that is presented in reference (10) and that might lead to a site layout that is illustrated conceptually in figure 6. As a detail on the site layout, the various muon acceleration and collider rings could plausibly be placed in the same plane to minimize the radiation zone to perhaps of order 100 square kilometers.

Just to illustrate the size scale, figure 7 shows a circle with a 400 km circumference (i.e. a 64 km radius) drawn in an unpopulated desert region of Australia. The content of both figures 6 and 7 is further discussed in reference (10). Note that the figures also include a linear collider, whose radiation hazards we now discuss.

V NEUTRINO RADIATION AT MANY-TEV LINEAR MUON COLLIDERS

It was speculated during this workshop that the ultimate potential energy reach for muon colliders could be extended by using linear $\mu^+\mu^-$ colliders, and straw-man parameter sets were presented for such single pass colliders (12).

From the standpoint of neutrino radiation, a linear muon collider has two advantages over circular muon colliders. Firstly, the radiation is confined to two pencil beams which would naturally be oriented at an upward tilt of perhaps tens of milliradians, as illustrated in figure 8. Secondly, the spent muons after each single-pass collision can be immediately ranged out in a beam dump rather than surviving until they decay into high energy neutrinos. This gives a large reduction in the radiation dose within the pencil beams.

We now derive the dose for the unpleasant and very artificial situation of a person living full-time in dense material immediately downstream from a linac, in the center of one of the neutrino beams. This will then be used to assess doses for more realistic situations.

The calculation proceeds by considering the total dose as an integral over the dose contributions, $\delta D(E)$, received in each energy interval, $[E, E + dE]$, as the muon bunch accelerates to the beam energy, E_μ:

$$D^{linac} = \int_0^{E_\mu} dE\, \delta D(E). \tag{20}$$

We already have the means to estimate $\delta D(E)$, since it can be compared to the dose from a straight section in a circular collider. After scaling by the relative fraction of muons decaying in the two cases, one obtains:

$$\delta D(E) = D^{ss}_{many-TeV}(E) \times \frac{1}{f^{ss}} \times \frac{df}{dE}(E)\, dE, \tag{21}$$

where $\frac{df}{dE}(E)\, dE$ is the fraction of muons that decay in the energy interval $[E, E + dE]$. This is given by:

$$\frac{df}{dE}(E) = \frac{1}{\gamma_\mu \beta c\tau \times g}, \tag{22}$$

where the product $\gamma_\mu \beta c\tau = 660$ is the characteristic decay length of ultra-relativistic muons, with $\beta c\tau = 660$ meters, and $g = dE/dz$ is the acceleration gradient.

Substituting in the expressions and constants from equations 16, 2, 5 and 7 gives, after some algebra and on taking care with units:

$$D^{linac}[mSv] = \frac{0.67 \times N_\mu[10^{20}]}{g[GeV/m]} \times \int_0^{E_\mu[TeV]} dE\, E\, X(E). \tag{23}$$

To solve this equation analytically, it is an adequate approximation to replace the energy-weighted integral of $X(E)$ by its value at, say, $E = E_\mu/2$, giving finally:

$$D^{linac}[mSv] \sim \frac{0.33\, N_\mu[10^{20}]}{g[GeV/m]} \times X(E_\mu/2) \times (E_\mu[TeV])^2. \tag{24}$$

As a numerical example, the straw-man parameters for the 1 PeV muon linear collider of reference (12) assume $E_\mu = 500$ TeV and $N_\mu = 0.064 \times 10^{20}$ in a 10^7 second year. Substituting in these values, with the additional assumption that $g = 1$ GV/m, gives:

$$D^{linac}\,[2 \times 500 \text{ TeV example}] = 1400 \text{ mSv/year}. \tag{25}$$

To put this figure for a whole-year dose in perspective, it is approximately 30 times the recommended maximum dose for a U.S. radiation worker, i.e. 50 mSv/year (3), so such a radiation worker would be able to work directly in the beam for of order 100 hours of accelerator running time. This shows that, while the dose would be well above legal limits for long-term occupancy, any doses from transient exposure would still be relatively small. In particular, the calculation gives reassurance on the negligible dose that would be received by a bird or plane flying through the beam.

VI SUMMARY

Neutrino radiation is a very serious problem and design constraint for muon colliders, particularly at very high center-of-mass energies. A characterization of the neutrino hazard has been presented that quotes the numerical formulae from reference (2) applying for TeV energy scales and below, and that extends the numerical predictions up to many-TeV energies.

It has been shown that the radiation dose in the plane of the muon collider ring rises quickly as the cube of the beam energy up to around the TeV collider energy scale before leveling off for many-TeV colliders to a slightly less than quadratic rise with energy on average and slightly less than linear for the radiation hot spots downstream from a fixed length of straight section.

To solve the neutrino radiation problem, many-TeV muon colliders and their associated neutrino radiation disks may be located within a new world laboratory with site diameter of order 100 km and located at a carefully chosen site in either the U.S., Canada, Australia, Northern Europe or elsewhere where large isolated tracts of land can be found with resources for a large-scale high technology laboratory.

REFERENCES

[1] B.J. King, *Assessment of the prospects for muon colliders*, paper submitted in partial fulfillment of requirements for Ph.D., Columbia University, New York (1994), available from LANL preprint archive as *physics/9907026*.

[2] B.J. King, *Potential Hazards from Neutrino Radiation at Muon Colliders*, available from LANL preprint archive as *physics/9908017*. An abbreviated version of this paper is available as Proc. PAC'99, New York, 1999, pp. 318-320.

[3] C. Caso et al., *The Review of Particle Physics*, The European Physical Journal **C3** (1998) 1.

[4] The Muon Collider Collaboration, *Status of Muon Collider Research and Development and Future Plans*, Phys. Rev. ST Accel. Beams, 3 August, 1999.

[5] B.J. King, *Discussion on Muon Collider Parameters at Center of Mass Energies from 0.1 TeV to 100 TeV*, Proc. EPAC'98, BNL-65716. available from LANL preprint archive as *physics/9908016*.

[6] B.J. King, *Parameter Sets for 10 TeV and 100 TeV Muon Colliders, and their Study at the HEMC'99 Workshop*, these proceedings. Also available from
http://pubweb.bnl.gov/people/bking/heshop/hemc_papers.html.

[7] J.D. Cossairt, N.L. Grossman, E.T. Marshall, *Fermilab-Conf-96/324* (1996) and *Fermilab-Pub-97/101* (1997); Private communications with N.V. Mokhov.

[8] C.J. Johnstone and N.V. Mokhov, Proc. PAC'97 (1997) 414; Nikolai Mokhov and Andreas Van Ginneken, *Neutrino Radiation at Muon Colliders and Storage Rings*, to appear in Proc. ICRS-9 Int. Conf. on Radiation Shielding, Tsukuba, Ibaraki, Japan, 1999.

[9] See, for example, Chris Quigg, *Neutrino Interaction Cross Sections*, FERMILAB-Conf-97/158-T.

[10] B.J. King, *Prospects for Colliders and Collider Physics to the 1 PeV Energy Scale*, these proceedings. Also available from
http://pubweb.bnl.gov/people/bking/heshop/hemc_papers.html.

[11] Colin Johnson, Siting Options for a High Energy Muon Collider, oral presentation at this workshop. Transparency copies can be viewed at
http://pubweb.bnl.gov/people/bking/heshop/hemc_papers.html.

[12] F. Zimmermann, *Final Focus Challenges for Muon Colliders at Highest Energies*, these proceedings. Also available from
http://pubweb.bnl.gov/people/bking/heshop/hemc_papers.html.

Technicolor Signatures at the High Energy Muon Collider

Kenneth Lane[1,2]

Department of Physics
Boston University
590 Commonwealth Avenue
Boston, Massachussets 02215

Abstract. I discuss high mass signatures of technicolor that would be observable at a very high energy muon collider. Most intriguing is the spectrum of spin–one technihadrons, ρ_T, ω_T and A_{1T}, which may extend to 100 TeV and beyond in a walking technicolor theory.

1. INTRODUCTION

It is a real pleasure to talk at a workshop in which the theorists are down–to–earth participants and the machine physicists are wild–eyed dreamers. Here is an e-mail exchange between my session organizer and me:

- Joe –

  I just realized that the workshop title refers to muon colliders at 10-100 TeV (!). I don't have a hell of a lot in the way of TC signals at those energies. How seriously should I take that energy range as a charge??

  Ken

- You can completely ignore the 10 TeV stuff - that is for the accelerator people (i.e. what is the highest energy muon collider one could ever have any hope of building).

  –Joe

Accordingly, I prepared a talk that discusses TC signatures at 1–4 TeV: The technivector mesons ρ_T and ω_T of the minimal, one–doublet TC model [1]; the Z' and

---

[1] lane@buphyc.bu.edu
[2] Talk delivered at the workshop "Studies on Colliders and Collider Physics at the Highest Energies: Muon Colliders at 10 TeV to 100 TeV", Montauk, Long Island, NY, 27 September–1 October 1999.

higher–dimensional electroweak singlet technifermions of topcolor–assisted technicolor [2]; and the electroweak–$SU(2)$ singlet fermions of the top seesaw model [3].

In the course of this, however, I recalled an old idea that would give the HEMC physics to do all the way from 1 TeV to 100 TeV. This has to do with the fact that walking technicolor [4], an essential ingredient of any viable TC model, implies that the spectrum of technivector mesons cannot be QCD–like [5–7]. It must extend in some sense to 100 TeV and beyond. This idea is so intriguing that I will emphasize it here. I hope someone will be able to decide whether it makes sense.

The rest of this paper is organized as follows: Section 2 presents a summary of the dynamical approach to electroweak and flavor symmetry breaking: technicolor [1], extended technicolor (ETC) [8,9], and all that. This scenario's signatures at the HEMC are discussed in Section 3, with emphasis on the technivector spectrum in walking technicolor models.

2. OVERVIEW OF TECHNICOLOR

Technicolor—a strong interaction of fermions and gauge bosons at the scale $\Lambda_{TC} \sim 1\,\mathrm{TeV}$—induces the breakdown of electroweak symmetry to electromagnetism *without* elementary scalar bosons [1]. Technicolor has a strong precedent in QCD. There, the chiral symmetry of massless quarks is spontaneously broken by strong QCD interactions, resulting in the appearance of massless Goldstone bosons, π, K, η. [3] In fact, if there were no Higgs bosons, this chiral symmetry breaking would itself cause the breakdown of $SU(2) \otimes U(1)$ to electromagnetism. Furthermore, the W and Z masses would be given by $M_W^2 = M_Z^2 \cos^2\theta_W = \frac{1}{8} g^2 N_F f_\pi^2$, where g is the weak $SU(2)$ coupling, and N_F the number of massless quark flavors. Alas, the pion decay constant f_π is only 93 MeV and the W and Z three orders of magnitude too light.

In its simplest form, technicolor is a scaled up version of QCD, with massless technifermions whose chiral symmetry is spontaneously broken at Λ_{TC}. If left and right-handed technifermions are assigned to weak $SU(2)$ doublets and singlets, respectively, then $M_W = M_Z \cos\theta_W = \frac{1}{2} g F_\pi$, where $F_\pi = 246\,\mathrm{GeV}$ is the weak *technipion* decay constant. [4]

In the standard model and its extensions, the masses of quarks and leptons are produced by their Yukawa couplings to the Higgs bosons—couplings of arbitrary magnitude and phase that are put in by hand. This option is not available in technicolor because there are no elementary scalars. Instead, this explicit breaking

---

[3] The hard masses of quarks explicitly break chiral symmetry and give mass to π, K, η, which are then referred to as pseudo-Goldstone bosons.

[4] In the minimal model with one doublet (U,D) of technifermions, there are just three technipions. They are the linear combinations of massless Goldstone bosons that become, via the Higgs mechanism, the longitudinal components $W_L^\pm$ and Z_L^0 of the weak gauge bosons. In non-minimal technicolor, the technipions include the longitudinal weak bosons as well as additional Goldstone bosons associated with spontaneous technifermion chiral symmetry breaking. The latter must and do acquire mass—from the extended technicolor interactions discussed below.

of quark and lepton chiral symmetries must arise from *gauge interactions alone*. The most economical approach employs extended technicolor [8,9]. In its proper formulation [9], the ETC gauge group contains technicolor, color, and flavor as subgroups and there are very stringent restrictions on the representations to which technifermions, quarks, and leptons belong: Specifically, they must be combined into the same few large representations of ETC. Otherwise, unbroken chiral symmetries lead to axion–like particles. Quark and lepton hard masses are generated by their coupling (with strength g_{ETC}) to technifermions via ETC gauge bosons of generic mass M_{ETC}:

$$m_q(M_{ETC}) \simeq m_\ell(M_{ETC}) \simeq \frac{g_{ETC}^2}{M_{ETC}^2}\langle \bar{T}T\rangle_{ETC}, \qquad (1)$$

where $\langle \bar{T}T\rangle_{ETC}$ and $m_{q,\ell}(M_{ETC})$ are, respectively, the technifermion condensate and quark and lepton masses renormalized at the scale M_{ETC}.

Technicolor is an asymptotically free gauge interaction. If it is like QCD, with its running coupling α_{TC} rapidly becoming small above its characteristic scale $\Lambda_{TC} \sim 1\,\mathrm{TeV}$, then $\langle \bar{T}T\rangle_{ETC} \simeq \langle \bar{T}T\rangle_{TC} \simeq \Lambda_{TC}^3$. To obtain quark masses of a few GeV thus requires $M_{ETC}/g_{ETC} \lesssim 30\,\mathrm{TeV}$. This is excluded: Extended technicolor boson exchanges also generate four-quark interactions which, generically, include $|\Delta S| = 2$ and $|\Delta B| = 2$ operators. For these not to conflict with K^0-$\bar{K}^0$ and B_d^0-$\bar{B}_d^0$ mixing measurements, M_{ETC}/g_{ETC} must exceed *several hundred* TeV [9]. This implies quark and lepton masses no larger than a few MeV, and technipion masses no more than a few GeV.

Because of this conflict between constraints on flavor-changing neutral currents and the magnitude of ETC-generated quark, lepton and technipion masses, classical QCD–like technicolor was superseded long ago by "walking" technicolor [4]. Here, the strong technicolor coupling α_{TC} runs very slowly, or walks, for a large range of momenta, possibly all the way up to the ETC scale of *several hundred* TeV. The slowly-running coupling enhances $\langle \bar{T}T\rangle_{ETC}/\langle \bar{T}T\rangle_{TC}$ by almost a factor of M_{ETC}/Λ_{TC}. This, in turn, allows quark and lepton masses as large as a few GeV and $M_{\pi_T} \gtrsim 100\,\mathrm{GeV}$ to be generated from ETC interactions at $M_{ETC} = \mathcal{O}(100\,\mathrm{TeV})$.

In almost all respects, walking technicolor models are very different from QCD with a few fundamental $SU(3)$ representations. One example is that integrals of weak-current spectral functions and their moments converge much more slowly than they do in QCD. The consequence of this for the HEMC will be discussed in Section 3. Meanwhile, this and other calculational tools based on naive extrapolation from QCD and on large-N_{TC} arguments are suspect. It is not yet possible to predict with confidence the influence of technicolor degrees of freedom on precisely-measured electroweak quantities—the S, T, U parameters to name a frequently discussed example [10].

Another major development in technicolor was motivated by the discovery of the top quark at Fermilab [11]. Theorists have concluded that ETC models cannot explain the top quark's large mass without running afoul of either experimental constraints from the ρ parameter and the $Z \to \bar{b}b$ decay rate [12]—the ETC mass must be about 1 TeV to produce $m_t = 175\,\text{GeV}$; see Eq. (1)—or of cherished notions of naturalness—M_{ETC} may be higher, but the coupling g_{ETC} then must be fine-tuned near to a critical value. This state of affairs led to the proposal of "topcolor-assisted technicolor" (TC2) [2].

In TC2, as in many top-condensate models of electroweak symmetry breaking [13], almost all of the top quark mass arises from a new strong "topcolor" interaction [14]. To maintain electroweak symmetry between (left-handed) top and bottom quarks and yet not generate $m_b \simeq m_t$, the topcolor gauge group under which (t,b) transform is usually taken to be $SU(3) \otimes U(1)$. The $U(1)$ provides the difference that causes only top quarks to condense. Then, in order that topcolor interactions be natural—i.e., that their energy scale not be far above m_t—without introducing large weak isospin violation, it is necessary that electroweak symmetry breaking is still due mostly to technicolor interactions [2].

Extended technicolor interactions are still needed in TC2 models to generate the masses of light quarks and the bottom quark, to contribute a few GeV to m_t,[5] and to give mass to technipions. The scale of ETC interactions still must be hundreds of TeV to suppress flavor-changing neutral currents and, so, the technicolor coupling still must walk. In TC2 there is no need for large technifermion isospin splitting associated with the top-bottom mass difference. Thus, for example, ω_T and ρ_T partners are nearly degenerate $\bar{U}U \pm \bar{D}D$ states.

Another, more recent, variant of topcolor models is the "top seesaw" mechanism [3]. Its motivation is to realize the original top–condensate idea of the Higgs boson as a fermion–antifermion bound state. This failed for the top quark because it turned out to be too light! In top seesaw models, an electroweak singlet fermion F acquires a dynamical mass of *several* TeV. Through mixing of F with the top quark, it gives the latter a much smaller mass (the seesaw) and the scalar $\bar{F}F$ bound state acquires a component with an electroweak symmetry breaking vacuum expectation value.

This completes our brief summary of technicolor. We turn now to the technicolor signatures for which a high energy muon collider is well-suited.

3. TECHNICOLOR SIGNATURES AT THE HEMC

The principal signals of technicolor are discussed in a number of places [15]. Most of them are accessible at low energies—at the Tevatron in Run II, certainly at the LHC, and, possibly, even at LEP. In the minimal technicolor model, with just one technifermion doublet, the only prominent signals in a TeV–scale collider are modest enhancements in longitudinally-polarized weak boson production. These are the s–channel color-singlet technirho resonances near 1.5–2 TeV: $\rho_T^0 \to W_L^+ W_L^-$

---

[5] Massless Goldstone "top-pions" arise from top-quark condensation. This ETC contribution to m_t is needed to give them a mass in the range of 150–250 GeV.

and $\rho_T^\pm \to W_L^\pm Z_L^0$. The $\mathcal{O}(\alpha^2)$ cross sections of these processes are quite small at such masses. This and the difficulty of reconstructing weak-boson pairs with reasonable efficiency make observing these enhancements a challenge. These states would be more easily seen in a lepton collider—if one can be built with $\sqrt{s} = $ 1.5–2 TeV at an affordable cost. Nonminimal technicolor models are much more accessible in a hadron collider because they have a rich spectrum of lower mass technirho vector mesons and technipion states into which they may decay.

If technicolor is the basis for electroweak symmetry breaking, it will have been discovered once the LHC has acquired and analyzed 10 fb^{-1} of data. The question we address here is what the HEMC can do to add to our understanding of this new dynamics.

3.1 The Technivector Spectrum of Walking Technicolor

The slow decrease with energy of the coupling α_{TC} in walking technicolor means that the $\mu^+\mu^-$ cross section approaches asymptotia only near the extended technicolor scale, probably even above the reach of the HEMC. This is most directly seen by considering the integrals in Weinberg's spectral function sum rules for the weak-isospin vector and axial vector currents [16]. These sum rules are

$$\int_0^\infty ds\,[\rho_V(s) - \rho_A(s)] = F_\pi^2$$
$$\int_0^\infty ds\,s\,[\rho_V(s) - \rho_A(s)] = 0\,, \qquad (2)$$

where $F_\pi = 246\,\mathrm{GeV}$. Here, the spectral functions ρ_V and ρ_A are analogs for the weak-isospin currents of the ratio of cross sections, $R(s) = \sigma(e^+e^- \to \mathrm{hadrons})/\sigma(e^+e^- \to \mu^+\mu^+)$. In QCD, the sum rules corresponding to Eq. (2) are saturated by the lowest lying spin-one resonances, ρ and A_1, and the sum rules converge rapidly above the A_1 mass. Similarly, in technicolor without a walking coupling, the sum rules would be saturated by the lowest ρ_T and A_{1T} and the difference $\rho_V - \rho_A \sim 1/s^3$ for $s \gtrsim M_{A_{1T}}^2 \sim 1\,\mathrm{TeV}^2$. In walking technicolor, the slow running of $\alpha_{TC}(s)$ implies that $\rho_V - \rho_A \sim 1/s^2$ below $s \sim M_{ETC}^2$ and $1/s^3$ above. Thus, *the spectral functions cannot be saturated by a single pair of low-lying resonances.* Either there must be a tower of resonances above ρ_T and A_{1T}, all of which contribute significantly to the spectral integrals (see Ref. [5,6]; also Ref. [7] for an explicit attempt to realize this), or the spectral functions are smooth but anomalously slowly decreasing up to M_{ETC}. The same alternative applies to the $\mu^+\mu^-$ cross section. Moreover, the isoscalar state ω_T and its excitations appear there. Thus, exploration of the 1–100 TeV region of $\mu^+\mu^-$ annihilation is bound to reveal crucial information on the dynamics of a walking gauge theory, dynamics on which we theorists can only speculate.

In the minimal one–doublet model of technicolor, it has always been assumed that the lowest lying ρ_T, ω_T, and A_{1T} decay mainly into two and three longitudinally–polarized weak bosons, $W_L^\pm$ and Z_L^0. In the minimal model, however, $M_{\rho_T} \sim$

$M_{A_1T} = 1\text{--}2\,\text{TeV}$, and this is so far above $2M_W$ that it is possible that decay modes with more than two or three weak bosons are important if not dominant. [6] Thus, in the minimal walking technicolor model, there may be a tower of vector and axial vector mesons in the s–channel of $\mu^+\mu^-$ annihilation which decay to many W and Z bosons. It is an open question how narrow and discernible these resonances will be.

In nonminimal models, the spectrum of technihadrons is quite rich and the scale of their masses is lower (roughly as the square root of the number of technifermion doublets). There are technipions π_T as well as weak bosons for the ρ_T, ω_T, and A_{1T} to decay into. These π_T may be color singlets and, if colored technifermions exist, octets and triplets ("leptoquarks"). Technipions are expected to have masses in the range 100—500 GeV and to decay into the heaviest fermion pairs allowed. The large value of $\langle\bar{T}T\rangle_{ETC}/\langle\bar{T}T\rangle_{TC}$ in walking technicolor significantly enhances technipion masses. Thus, for example, $\rho_T \to \pi_T\pi_T$ decay channels may be closed for the lowest–lying state. Instead, $\rho_T \to W_L W_L$, $W_L \pi_T$, and $\gamma\pi_T$ [15]. The excited states should be able to decay into pairs of technipions. The ρ_T, ω_T, and A_{1T} that lie above multi–π_T threshold are likely to be wider than their counterparts in the minimal model. Still, the structure of $\mu^+\mu^-$ annihilation up to 100 TeV will provide valuable insight to walking gauge dynamics.

3.2 Topcolor–Technicolor Signals

As I said above, topcolor–assisted technicolor generally employs an extra "hypercharge" $U(1)$ to help induce a large condensate for the top, but not the bottom quark. This additional $U(1)$ is broken, leading to a Z' boson which is strongly coupled to at least the third generation. In the models of Ref. [17], it is strongly coupled to all fermions. Some of the lower energy phenomenology of this Z' was studied in Refs. [18,19]. Its nominal mass, in the range 1–4 TeV, and potentially strong coupling to muons make it a target of opportunity for the HEMC. [7] Unfortunately, its strong couplings and many decay channels to ordinary fermions and technifermions may also make the Z' so broad that it is difficult discover and study in any collider.

An intriguing feature of this Z' is that it must acquire its mass from condensation of a technifermion ψ [17]. The Z' mass of several TeV implies that the ψ–fermion's mass is 1–2 TeV. Thus, ψ must transform according to a higher–than–fundamental representation of the technicolor gauge group. In order that its condensation not break electroweak $SU(2)\otimes U(1)$, ψ must either be a singlet or transform vectorially under this symmetry. The obvious way to access it is via $Z' \to \bar{\psi}\psi$ in the s–channel

---

[6] The QCD 2^3S_1 state $\rho'(1700)$ decays predominantly to four, not two pions, presumably because the two–pion mode is suppressed by an exponential form factor and/or a node in the decay amplitude.

[7] Top seesaw models also have an extra $U(1)$ gauge symmetry, broken spontaneously. There, the Z' boson mass is expected to be roughly 5 TeV.

of the HEMC. The phenomenology of these higher representation technifermions has not been studied in detail. One crucial question is whether ψ is stable. If not, how does it decay? If it is, what are the cosmological consequences?

Finally, there is the $SU(2)$ singlet, charge–2/3 quark F of top seesaw models. This fermion also has a mass of several TeV and may be pair produced via γ, Z, Z' at the HEMC. It decays by virtue of its mixing with the top quark as $F \to t \to Wb$, a striking signature indeed.

4. CONCLUSIONS AND ACKNOWLEDGEMENTS

The HEMC technicolor signatures that I have presented here are, quite obviously, at a primitive stage of development. I think all of them deserve further thought because they bear directly on unfamiliar dynamics such as walking technicolor and strongly–coupled topcolor. Corresponding uncertainties face the design of the HEMC. Again, the particle theorists and the accelerator theorists are in the same boat. The need to go on to higher energies remains and it always will. This was said very well by an Amherst poet long ago:

"Faith" is a fine invention
When Gentlemen can see —

But *Microscopes* are prudent
In an Emergency.

— *Emily Dickinson, 1860*

I thank the organizers, especially Bruce King and Joe Lykken for inviting me to this stimulating workshop and for the wonderful opportunities to explore Montauk and Block Island. Kathleen Tuohy ran a perfect workshop and I send her my gratitude. I am grateful to my fellow participants in the joint Physics and Detector Working Group. They provided the mental stimulation that led to my contribution. I am also indebted to Sekhar Chivukula for discussions about top seesaw models and for reading this manuscript. This research was supported in part by the Department of Energy under Grant No. DE–FG02–91ER40676.

REFERENCES

1. S. Weinberg, *Phys. Rev.* D **19**, 1277 (1979);
   L. Susskind, *Phys. Rev.* D **20**, 2619 (1979).
2. C. T. Hill, *Phys. Lett.* B **345**, 483 (1995).
3. B. A. Dobrescu and C. T. Hill, *Phys. Rev. Lett.* **81**, 2634 (1998), hep-ph/9712319;
   R. S. Chivukula et al., *Phys. Rev.* D **59**, 075003 (1999), hep-ph/9809470;
   G. Burdman and N. Evans, *Phys. Rev.* D **59**, 115005 (1999), hep-ph/9811357.

4. B. Holdom, *Phys. Rev.* D **24**, 1441 (1981); *Phys. Lett.* B **150**, 301 (1985);
   T. Appelquist, D. Karabali and L. C. R. Wijewardhana, *Phys. Rev. Lett.* **57**, 957 (1986);
   T. Appelquist and L. C. R. Wijewardhana, *Phys. Rev.* D **36**, 568 (1987);
   K. Yamawaki, M. Bando and K. Matumoto, *Phys. Rev. Lett.* **56**, 1335 (1986);
   T. Akiba and T. Yanagida, *Phys. Lett.* B **169**, 432 (1986).
5. K. Lane, *An Introduction to Technicolor*, Lectures given at the 1993 Theoretical Advanced Studies Institute, University of Colorado, Boulder, published in "The Building Blocks of Creation", edited by S. Raby and T. Walker, p. 381, World Scientific (1994).
6. K. Lane, *Technicolor and Precision Tests of the Electroweak Interactions*, Proceedings of the 27th International Conference on High Energy Physics, edited by P. J. Bussey and I. G. Knowles, Vol. II, p. 543, Glasgow, June 20–27, 1994.
7. M. Knecht and E. de Rafael, *Phys. Lett.* B **424**, 335 (1998).
8. S. Dimopoulos and L. Susskind, *Nucl. Phys.* B **155**, 237 (1979).
9. E. Eichten and K. Lane, *Phys. Lett.* B **90**, 125 (1980).
10. B. W. Lynn, M. E. Peskin and R. G. Stuart, in *Trieste Electroweak 1985*, 213 (1985);
    M. E. Peskin and T. Takeuchi, *Phys. Rev. Lett.* **65**, 964 (1990); A. Longhitano, *Phys. Rev.* D **22**, 1166 (1980); *Nucl. Phys.* B **188**, 118 (1981);
    R. Renken and M. Peskin, *Nucl. Phys.* B **211**, 93 (1983);
    M. Golden and L. Randall, *Nucl. Phys.* B **361**, 3 (1990);
    B. Holdom and J. Terning, *Phys. Lett.* B **247**, 88 (1990);
    A. Dobado, D. Espriu and M J. Herrero, *Phys. Lett.* B **255**, 405 (1990);
    H. Georgi, *Nucl. Phys.* B **363**, 301 (1991).
11. F. Abe, et al., The CDF Collaboration, *Phys. Rev. Lett.* **73**, 225 (1994); *Phys. Rev.* D **50**, 2966 (1994); *Phys. Rev. Lett.* **74**, 2626 (1995);
    S. Abachi, et al., The DØ Collaboration, *Phys. Rev. Lett.* **74**, 2632 (1995).
12. R. S. Chivukula, S. B. Selipsky, and E. H. Simmons, *Phys. Rev. Lett.* **69**, 575 (1992);
    R. S. Chivukula, E. H. Simmons, and J. Terning, *Phys. Lett.* B **331**, 383 (1994), and references therein.
13. Y. Nambu, in *New Theories in Physics*, Proceedings of the XI International Symposium on Elementary Particle Physics, Kazimierz, Poland, 1988, edited by Z. Adjuk, S. Pokorski and A. Trautmann (World Scientific, Singapore, 1989); Enrico Fermi Institute Report EFI 89-08 (unpublished);
    V. A. Miransky, M. Tanabashi and K. Yamawaki, *Phys. Lett.* B **221**, 171 (1989); *Mod. Phys. Lett.* **A4**, 1043 (1989);
    W. A. Bardeen, C. T. Hill and M. Lindner, *Phys. Rev.* D **D41**, 1647 (1990).
14. C. T. Hill, *Phys. Lett.* B **266**, 419 (1991);
    S. P. Martin, *Phys. Rev.* D **45**, 4283 (1992); *ibid* **D46**, 2197 (1992); *Nucl. Phys.* B **398**, 359 (1993);
    M. Lindner and D. Ross, *Nucl. Phys.* B **B370**, 30 (1992);
    R. Bönisch, *Phys. Lett.* B **268**, 394 (1991);
    C. T. Hill, D. Kennedy, T. Onogi, H. L. Yu, *Phys. Rev.* D **47**, 2940 (1993).
15. K. Lane, *The Scalar Sector of the Electroweak Interactions*, Proceedings of the 1982 DPF Summer Study on Elementary Particle Physics and Future Facilities, edited by

R. Donaldson, R. Gustafson and F. Paige (Fermilab 1983), p. 222;

E. Eichten, I. Hinchliffe, K. Lane and C. Quigg, *Rev. Mod. Phys.* **56**, 579 (1984);

K. Lane and M. V. Ramana, *Phys. Rev.* D **44**, 2678 (1991);

M. Golden, et al., *Strongly Interacting Electroweak Sector: Model–Independent Approaches*, in *Electroweak Symmetry Breaking and New Physics at the TeV Scale*, T. L. Barklow, editor, et al., pp292–351, hep-ph/9511206;

T. L. Barklow, et al., *Strong Coupling Electroweak Symmetry Breaking*, 1996 DPF/DPB Summer Study on New Directions for High Energy Physics (Snowmass 96), Snowmass, CO, 25 June–12 July 1996,hep-ph/9704217;

E. Eichten and K. Lane, *Electroweak and Flavor Dynamics at Hadron Colliders, I and II*, 1996 DPF/DPB Summer Study on New Directions for High Energy Physics (Snowmass 96), Snowmass, CO, 25 June–12 July 1996, hep-ph/9609297 and hep-ph/9609298;

K. Lane, *Non-Supersymmetric Extensions of the Standard Model*, plenary talk at the 28th International Conference on High Energy Physics, edited by Z. Ajduk and A. K. Wroblewski, Vol. I, p. 367, Warsaw, July 25–31, 1996, hep-ph/9610463;

K. Lane, *Phys. Rev.* D **60**, 075007 (1999), hep-ph/9903372.

16. S. Weinberg, *Phys. Rev. Lett.* **18**, 507 (1967);

    K. G. Wilson, *Phys. Rev.* **179**, 1499 (1969);

    C. Bernard, A. Duncan, J. Lo Secco and S. Weinberg, *Phys. Rev.* D **12**, 792 (1975).

17. K. Lane and E. Eichten, *Phys. Lett.* B **352**, 382 (1995);

    K. Lane, *Phys. Rev.* D **54**, 2204 (1996).

18. Y. Su, G.-F. Bonini, and K. Lane, *Phys. Rev. Lett.* **79**, 4075 (1997), hep-ph/9706267.

19. T. Rador, *Phys. Rev.* D **59**, 095012 (1999), hep-ph/9810252.

Comments on Frictional Cooling and the Zero Energy Options for Cooling Intense Muon Beams

Paul Lebrun

Fermi National Accelerator Laboratory
P.O. Box 500, Batavia, Illinois 60510[1]

Abstract. It is shown that the proposed frictional cooling method is not directly applicable to intense ($\approx 10^{12}$) muon bunches, mostly due to space charge constraints. Other difficulties stem from the fact that the initial emittance must be quite small, compared to the nominal muon collider emittance. Excessive heat due to energy deposition in the foils, from the primary muon beam or from secondary electrons could also destroy the thin foils used as moderator. Other "zero energy" schemes are considered, separately for μ^- and μ^+. All of them lead us to the study of exotic electrons-ions-muons plasma.

INTRODUCTION

Ionization cooling has been so far adopted by our collaboration as the method of choice to reach transverse emittance adequate for a muon collider [1]. While other methods based on low energy beams have been used in low and medium energy applications, ionization cooling has yet to be demonstrated in a real experiment. However, our emphasis on ionization cooling is justified for the following reasons: (i) None of the schemes based on low energy muon can accept the large emittances typical of pion produced muon beams. This is particularly true for the longitudinal emittances, since the typical energy of such muons is many hundreds of MeV, with $\Delta P/P \approx 1.$, and the required kinetic energy for frictional cooling is a ≈ 10 KeV.

(ii) Detailed simulations of ionization cooling channels have been successful. Transverse cooling in solenoidal field based channels is largely understood. In some cases, cooling in 6D phase space has been obtained in our computer models. Preliminary engineering studies have done, leading to various constraints on these channels which have been implemented in our simulation codes. However, we now know that realistic channels reaching transverse (normalized) emittances below $500\, mm.mrad$ will be very hard to achieve, as they require either more than

---

[1] Operated by University Research Association Inc. under Contract No. DE-AC02-76CH03000 with the United States Departement of Energy

15 *T.* solenoids (Alternate Solenoid configuration) or short (10 to 20 cm, or up to 2 time the coil radii) running at $\approx 8\,T.$, opposite to each other (SFOFO), and placed close to each others, so that low beta can be reached. This means that the field in the coil pack is at or above the critical limit where super-conductivity will be lost. Finally, we have to come up with a scheme to implement longitudinal to transverse emittance exchange.

Thus, low energy cooling must be viewed as a complementary approach to ionization cooling. In particular, we are interested in a scheme that would take a long (≈ 100 ns) bunch, possibly with a substructure (ns. sub-bunches), with a transverse emittance of $\approx 750\,mm.\,mrad$ [2], and cool it to the emittance suitable for a muon collider. First, such beams will have smaller transverse emittances. This will allow us to reduce the beam intensity for a given luminosity at the I.P., use smaller aperture machine for accelerators and of collider rings and possibility to implement transverse stacking to recombine longitudinal bunches together [2]. Second, the 100 ns. bunch could be compressed longitudinally, if the muons can be stopped and extracted from the exotic plasma fast enough.

In this paper, I'll briefly review the seminal experiment which demonstrated frictional cooling. This apparatus can not be used as such for intense beam: the beam will simply blow up due to space charge and the foil will likely be destroyed due to excessive heating, from the primary beam and secondary emissions due to the intense electric field produced by the space charge. Other low intensity cooling method will suffer from similar problems. In all cases, we might end up with a low temperature, highly unstable exotic plasma. The means by which such a state is realized will depend on the charge of the muons, leading to different μ^+ and μ^- beams.

FRICTIONAL COOLING

By low energy muons, we refer to non-relativistic muons, typically a few MeV/c momentum or ≈ 10 KeV of kinetic energy ($\beta \approx .02$). Reaching such low energies without dilution of the 6D phase space is certainly non-trivial. The friction cooling experiment [3] described below has been achieved starting with a surface muon beam, obtained from the PE5 beam line at PSI. In addition, Wien filters or electrostatic separators have been used to further select slow muons. The normalized incident beam emittance at 10 KeV, reaching the setup described in references [3,5] was about

$$5\,mm \times 1000.\,mrad \times 0.015 = 75\,\pi\,mm\,mrad.$$

It is unlikely that such techniques will be directly applicable to the muon collider: the ratio μ/π is simply too low. ($\approx 10^{-8}$ to 10^{-9}) As stated above, we plan to start with our baseline pion/muon capture and decay channel, followed by ionization

---

[2] Lots of approximate signs in this paper! By this, I mean the number is o.k. within 20 to 50 %!

cooling. At the end of such a channel currently considered for the neutrino factory, the normalized transverse emittance will be about $\approx$ 2000 to 1500 $\pi\, mm.\, mrad$, a factor 20 higher than the initial emittance quoted above. Thus, this assumes that we can use larger foils in the friction cooler (e.g. $\approx$ 10 cm. radius instead of 2.). The 6D invariant emittance will be about 100 $\pi\, mm^3$ (sub-bunch emittance) [4], at a momentum of 186 MeV/C, with $\Delta P/P \approx 10\%$. Thus, one has to decrease the momentum by two order of magnitude without substantial phase space dilution.

Passive absorbers alone won't do the trick, because straggling and the non-linear behavior of energy loss are such that the significant fractions of the beam are likely to pass through such absorbers without loosing enough energy, or stop in them. Not to mention additional multiple scattering. The following schemes can be mentioned:

- r.f. deceleration followed by an inverse radio frequency quadrupole. (e.g., imagine a conventional ion injector, working backward).

- Use of a cyclotron trap. [6]

- 5D Emittance exchange: based on a succession of Wien filters and passive absorbers, followed by recombinations by transverse stacking.

- 5D Emittance exchange: the passive absorbers are wedges placed at high dispersion points in a bend solenoid based channel. Phase space dilution occurs transversely due to multiple scattering. The bunch length increases as well, which is not a problem, while P and ΔP decreases, which is our goal.

- Same as above, in helicoidal channel.

None of these ideas have been worked out. The first one can probably be rejected quickly given the large emittance we start with, and the constraint on the channel length (at $\beta \approx .02$, one muon life time is 12.5 m.). The beam optic and extraction for the second one looks difficult The others looks more promising, although the transverse re-heating might be prohibitive. However, let us give us the benefit of the doubt, and assume we can overcome these difficulties.

Description of the method and experiment

At 10 KeV kinetic energy, in thin carbon foils, the electron cloud maintain the muons motion, because the electron and muon velocity are about equal. This means that, if a muon slows down too much, he will be re-accelerated. If a longitudinal boost is continuously applied, cooling can be obtained. Experimental details are given in reference [3]. With a stack of 10 to 12 foils, each $\approx 5\, \mu gr/cm^2$, and a static electric field of about 1.4 KV/cm to restore the longitudinal momentum, cooling merit factor of about 2.5 transversely and 3.7 longitudinally have been achieved [5]. The beam loss due to muon absorbed in the foils was approximately 30 %.

Difficulties for intense muon bunches

The counting rate was about 10 to 20 muons per second. (Assuming the PSI cyclotron gave 1 mA on target). Thus, for all practical purpose, the muons are "alone" in the apparatus, as they cool. Not quite, though: as the muons traverse the foil, secondary emission occurs, followed by acceleration in the d.c. electric field. As for the muons, they trapped in the confining magnetic field (5 Tesla). However, there is no collective motion because there are so few of them.

Let us consider now dumping about 10^{12} muons in that magnetic bottle. The r.m.s σ_x, σ_y of the beam spot is of the order of 2 cm. Even with a relatively long magnetic trap of 30 cm, capturing a longitudinal bunch length of 10 cm., the quasi-static electric field (E_r, E_z) will reach (0.17, 2.7) MV/m. The longitudinal component is about 20 times greater than the applied electric field which keeps the beam moving in the right direction. The next problem could be the heat dissipated in the thin foils due to the primary muons and the secondary emission: each muon looses approximately half of KeV in each foil. A 10^{12} bunch will dissipate $\approx 7.2\,10^{-5}$ Joule in the $5\,\mu$ gr. of Carbon, raising it's temperature by $\approx$ 20 degree C. This does not include secondary emissions. May be this is not a problem, although these foils are quite fragile, and may age rather quickly in such an environment.

The space charge limit is a real issue. For the scheme to work properly, the charge of the bunch must be reduced by roughly two order of magnitude. The luminosity of the muon collider will then decrease by 4 orders of magnitude. We might have gain one order in magnitude due to the cooling, which is not enough... This is not an issue bringing the beam in the friction cooling channel, because the macro bunch length is much longer. However, no matter what the cooling scheme is once the bunch is compressed longitudinally (e.g., quasi stopped), the space charge issues is very much relevant to the extraction and re-acceleration. I will come back to that point.

THE ZERO ENERGY OPTIONS AND EXOTIC PLASMA

The friction cooling method comes short on three counts: (i) The transverse emittances coming from the ionization channel into the friction cooler do not match, by a factor 10 to 20 [3] (ii) Since the beam does not stop in the foil (at least the 60 % fraction one intends to keep), but merely drifts at a reduced velocity (percent of the speed of light), the longitudinal compression (geometrical) is not yet optimal. (iii) Severe space charge will limit this longitudinal compression entering the magnetic bottle and moving into it, where the foils/re-accelerators resides.

---

[3] This depends on which ionization cooling final stage (neutrino factory vs 15 T. Alternate Solenoid)

One somehow must learn how to neutralized the bunch, while keeping the muon "charged", in a small physical volume: $\approx 100.\,cm^3$, at most. If the entire macro-bunch is stopped in a thin foil (tens of micro-gram up to 1 milligrams), in presence of such extreme charge densities, it is very likely that the foils will suffer damage and can not be used over any reasonable number of booster cycle. So be it, we will find a way to replace the foils every 1/15 of a second.

If the charge is to be compensated, say with low energy beam of protons or electrons, we are then compelled to consider our final cooling system not as made of solid or gaseous moderators with an externally applied re-acceleration voltage, but as an exotic plasma: This medium is composed of ions being ejected from the moderators, KeV electrons or protons to provide charge compensation and the muons. We now briefly review such methods, where in each cases, during a fraction of micro-second, the muons are literally "thermal" and slowly moving in solids. in less than $\approx$ 100 ns., the physical state must become a plasma. Evidently, it is best to study these exotic μ^+ or μ^- plasma separately.

Thermal muonium from metallic foils

Some of these concerns might also be applicable to the cooling method based on muonium produced in hot tungsten foils [7]. There also, charge densities will be high. Yet, the scheme proposed by Prof. Nagamine is certainly compelling. May I take this opportunity to ask the following naive questions or remarks on this preferred method. The crude sketch given on figure 1 illustrates a cooler based on a magnetic bottle, a tungsten absorber where muonium are produced, an a scheme to extract them and ionization. In more details:

1. Once again, the transverse emittance coming from the ionization channel and moderator will not quite match the spot size of the laser (mm^2. vs few cm^2). Hence achieving resonant ionization of the muonium will be costly, if achievable.

2. Same concern in the longitudinal direction: it is very unlikely that the dispersive channel will reduce the momentum down to a $\approx 4.\,MeV/c$ with a $\Delta P/P \approx 6\%$. A crude guess-estimate is that we might reach a momentum of 40 MeV/c with a $\Delta P/P$ of 30 %[4]. The last moderator will gives a beam of $\approx 20 MeV/c$ with a $\Delta P/P \approx 50\%$. Hence the need to use multiple foils. However, $20 \times 100\,\mu m. = 2\,mgr/cm^2$. might not be enough to stop such muons. Hence, either we further increase the number of foils (and lasers!) or we place these foils in a magnetic trap, such that the high muons bounce multiple times until the stop in the foils. The limitation by the leak rate of the bottle (muon going straight on the axis can escape) and by the muon life time. Both limits

---

[4] Without detailed simulation, of a scheme discussed above, one might easily be off by a factor 2 in these quantities

probably impose a maximum number of bounces or foil traversal in the bottle, of about ≈ 10.

3. If the muonium formation inside the tungsten is more (or as) efficient at low temperature [8], why not consider a scheme where the Tungsten will be maintained at a more modest temperature (few hundreds degrees as oppose to ≈ 2,000 degree C.)

4. But then, the muonium won't diffuse out of the foils. As longitudinal compression has to occur, this is good: they better stay put in the foil for ≈ 100 ns → 300 ns until the complete macro-bunch is captured.

5. We now have to heat the foil, quickly (< 100ns.), and ionize the muonium once it is out of the foil. Conventional heating method will be too slow. It also better be rather efficient: the total amount of Tungsten to heat to thousands of degree is approaching a fraction of a gram.

6. Can we organize the tungsten as a number of cold needles, falling through a grid of electrodes, and send a high power pulse (tens to 100 kJoules, MW peak power), so that the needles are "Z-pinched"? Such Z-pinches are the preferred method to heat tungsten at KeV-like temperature.

7. As the electron temperature in these imploded needles get higher than ≈ 100 eV, muonium trapped inside will ionize. Assuming the muons are in thermal equilibrium with these electrons, there velocity will be of the order of ≈ 40. $cm/\mu sec.$, fast enough to move away from the needles before the plasma recombines.

8. We are left with a few cm^3 volume where the electron-tungsten ions do form a highly unstable, non-uniform plasma. Note that the muons are too diluted to contribute to the collective motion of the plasma (they are do not form a plasma per say, they embedded in a conventional plasma).

9. However, if might still make sense to compute the plasma frequency for muons. After about 100 ns, they will occupy the entire volume of this irregular plasma (the tungsten atoms or ions are 2,000 times slower). Assuming a volume a few cm^3 and a $1.0\,10^{12}$, the plasma frequency for the muons will of the order of a GHz. If not compensated (or partially compensated), the electric field due to their collective static charge will be of the order of 10 MV/m. Can such frequency and field parameters could be matched to the r.f. field used on the extracting electrode, or the buncher?

The μ^- cooler

A collider works best if the emittances of the colliding beam are about the same. Thus, similar emittances must be reached for the μ^-. This has been quoted as "a

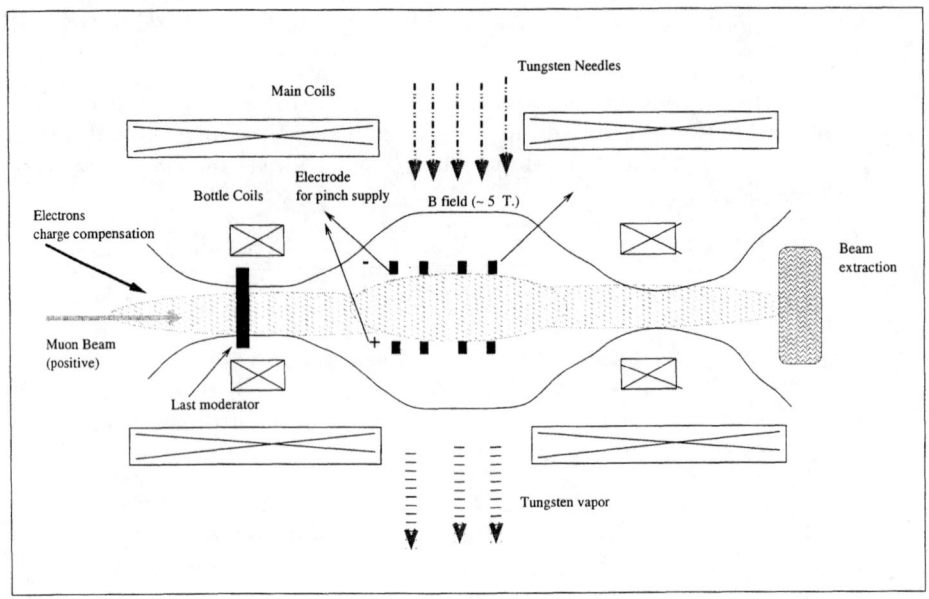
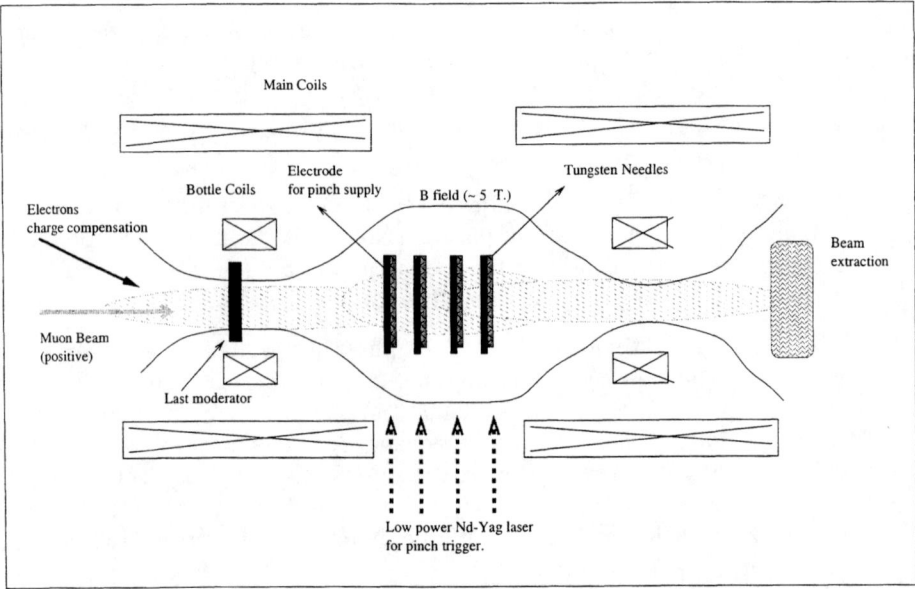

FIGURE 1. Naive sketch of the Tungsten - μ^+ cooler. Top : elevation, bottom plain view. The diameter of the inner coils would be approximately 10 cm. and the length of the bottle is $\approx 30\,cm$.

problem" during this workshop [7] A few ideas have been proposed, on which I wish to comment[5].

1 He3-mu exotic ions

In the magnetic bottle described above, one could remove the Tungsten foils or needle and replace with an H3 gas or liquid jet. μ^- are stopped and quickly captured. The $He3 - \mu^-$ exotic atom is left with a net positive charge, as the electrons are ejected in the capture process. If a plasma forms quickly thereafter, recombination between electrons and the exotic atom does not occur, due to the low density of the medium. One then extract these exotic ions (hopefully with reasonable efficiency) and accelerate them. The emittance of the muon remains small in the process. A foil intercepting this beam strips the negative muons. In the rest frame of the exotic atom, this stripping process give $\approx 20 KeV$ kinetic energy to the muons. Such energy transfer can occur at sufficiently high velocity via transition radiation from the stripping foil. In this process, the momentum transfer to the μ^- is of the order of 2 MeV/c .

In order to limit the emittance growth due to this momentum transfer, the exotic atom must be boosted at sufficiently high energy. At $\approx 10.$ GeV/c the $He - \mu^-$ and therefore the μ^- will be boosted at $\beta \approx .76$ and the liberated μ^- will keep there mere 125 MeV/c momentum, and a $\Delta P/P$ of 2 to 3 % is expected. (neglecting the $\Delta P/P$ of the $He3 - \mu^-$ exotic ions. Accelerating these ions up to these energy if less than a few micro-second (without much gain from an appreciable Lorentz boost!) is a tall order, unfortunately. Hopefully, one can do better.

2 Z-pinched D2-Muons

Instead of using Helium or Hydrogen[6], let us consider deuterium. Muon catalyzed fusion for cooling negative muons has not only been proposed, in fact, it has been experimentally demonstrated [9]. However, in order to avoid having the μ^- merely forming exotic $D2 - \mu^-$ molecules in the solid deuterium without escaping in his lifetime, catalyzing the fusion and doing it again until it decays, the thickness of the D2 film must be limited to 30 μgr. Such thin films will not stop of $\approx 20 MeV/c$ μ^- (They have a range in liquid Hydrogen is 2 mm.)

Remember now that we could cross this D2 target a few times, thank to the magnetic trap and the relatively long decay length of these μ^-. However, it still won't bring us down to the required thickness. Thus, the negative muons are trapped in the solid D2 for a long time, unless the D2 temperature get raised very quickly after the complete stop of the muon beam.

---

[5] One should be limited to only one such "new" idea per paper, sorry for being greedy

[6] One could consider $p - \mu^-$ exotic neutral atoms, and let these capture an electron, so that they can be extracted and accelerated. However, the probability to capture this electron in such low density plasma is too small

Once again, let us consider Z-Pinches. This technique has been used to raised the D2 temperature high enough so that nuclear fusion occurs (without muon catalysis) [10]. However, in our application, there is no need to reach such high temperature, one simply has to avoid that the negative muons bind to an other D2 molecule by reducing the density. Cryogenic Deuterium fibre have been manufactured in-situ, either from condensed droplets (or "snow") or continuous extrusion [11]. These experiments were conducted on single fibres, for ease of implementation. Implosion velocity of the fibers of $1.\,cm/\mu sec.$ are considered low. In a gas embedded compressional Z-pinch, explosion velocity of $0.4\,cm/\mu sec$ have been recorded. These phenomena occur in the right time scale for us: 10 to 100 ns. [12].

As soon as the density is reduced, negative muons liberated during the fusion process are no longer able to find other D2 molecules. The emerging plasma can them be heated further and the charge can be compensated, by dumping low energy proton beam into it. (or simply, by ionizing enough D2, and collecting the rapidly moving electrons. On is then left with a negative muons embedded in the typical ion source.

Caveats

The present author is interested at learning where such naive schemes would fail. Such ideas are indeed very speculative. For instance, how do we drop or shoot the Tungsten needles evenly so that all positive muons are stopped, and the Z-pinch occurs in each needles. (The hope is the impedance of a formed plasma is not too low compared to Ohmic resistance of the adjacent cold needle, or the trigger provide a uniform pre-ionization on all the needle tips.). Same concern for our D2 fibres or "snow" flakes: relatively uniform heating must occur so that the local density drops sufficiently rapidly. Dumping tens of kJoules of power at 15 Hz in leads us awfully close to the MW range. (To avoid evaporating the D2 fiber to due heat released while the μ^- beam stops, one would need enough D2 mass..). An other concern: in such exotic ion sources: e.g., radiated heat from the electrodes onto the cold D2 fibers could also melt them prematurely...)

Once such relatively high volume, non-uniform, plasma are formed, how do we extract efficiently the $\mu^{\pm}$ from it? And how fast? A goal for a muon collider would be to achieve a macro-bunch length of $\approx 30\,cm. \rightarrow 1.\,m.$, with a transverse normalized emittance of $\approx 3 \rightarrow 10\,\pi\,mm.\,mrad$ (beam spot size of a few cm, with angles of about a radiant and $\beta \approx 10^{-3}$). with 10^{11} particles in such bunches. Space Charge is still an issue at the extraction and early acceleration stage. Finally, such beams must be accelerated at relativistic speed fast enough. Tens of MV/m are required in the RFQ. This means higher frequencies ($\approx$ GHZ) are needed, or superconducting RFQs [13][7]. Such High Frequencies would match the muon plasma

---

[7] The idea of SRFQ is mentioned for completeness. Actually, the main advantage would be to allow for d.c. operation, irrelevant in our case. Acceleration of 2.16 MV/m have been obtained with such SRFQ's, not that much higher than normal conducting structures at the same

frequency better (unclear if this is an advantage), but would require smaller beam spot size to fit in the structure. And, unfortunately, multi-beam, high frequency, RFQ have not been invented yet (to my knowledge). So, we have work ahead of us. However, the "zero energy" option for muon cooling should not be abandoned too quickly.

ACKNOWLEDGEMENTS

I would like to thank Alexander Zholents for prompting me to study again these issues. I also would like to thank Prof. D. Taqqu and F. Kotzman from PSI for the detailed and pertinent e-mail on various aspects of the friction cooling.

REFERENCES

1. C. Ankenbrandt et al (Muon Collider Collaboration) Phys. Rev. ST Accel. Beams 2, 081001 (1999) Fermilab-Pub-98-179
2. *An Emittance Exchange Idea Using Transverse Bunch Stacking* Charles H. Kim, Muon Collider Note 70
3. M. MuhlBauer et al, Nucl. Phys. B (Proc. Suppl.) **51A** (1996) 135-142
4. See for instance (and reference quoted therein) J. Monroe et al, MuCool Note # 68 (1999).
5. David Taqqu and Franz Kotzman, private communication. See also Markus Muhlbauer, *Kuehlung eines Strahls niederenergetischer Myonen durch Moderation und Beschleunigung: Reibungskuehlung* Publisher: Herbert Utz Verlag, Wissenschaft, Munich, 1997 Tel xx49 89 3077 8821, FAX xx49 89 3077 9694 ISBN 3-89675-181-6
6. See for instance http://www1.psi.ch/www_f1_hn/lke/cycltrap.html or L.M. Simons, Phys. Scripta T22 (1988) 90.
7. K. Nagamine, talk presented at this workshop. See also Y. Miyake, *Generation of Ultra Slow Muons and VUV Laser Source*, talk presented at the workshop on Laser application to Muon Science, March 1998, KEK proceedings 98-8, JHF-98-3, November 1998
8. K. Nagamine, private communication, UCLA Spring 99 workshop
9. P. Strasser et al, Phys. Lett. **B 368** (1996) 32-38
10. See for instance M.G. Haines, *An overview of the DZP Project at Imperial College*, talk at the 4th conference on Dense Z-Pinches, 1997, Vancouver Canada. AIP conference Proceedings 409, p 27.
11. R. Aliaga-Rossel and J. Bayley, *A novel Cryogenic Fibre Maker for Continuous extrusion*, talk at the 4th conference on Dense Z-Pinches, 1997, Vancouver Canada. AIP conference Proceedings 409, p. 55.

---

frequencies

12. L. Soto, *Comparative Studies on a Gas Embedded Compresional Z-Pinch in H2 and D2*, talk at the 4th conference on Dense Z-Pinches, 1997, Vancouver Canada. AIP conference Proceedings 409, p. 47.
13. See for instance A. Pisent *et al*, *Status of the Superconducting RFQ project for the Legnaro new positive ion injector*, EPAC 96, conf. proceeding, page 768.

Constraints on plasma compensation of beam-beam effects in muon colliders

Konstantin V. Lotov

Budker Institute of Nuclear Physics, Novosibirsk, 630090, Russia

Abstract. We derive necessary conditions for the plasma compensation to work in muon colliders. For this, we analyze the suppression of beam fields by the plasma, collisional diffusion of the return plasma current, possible beam filamentation, and dynamics of plasma ions. We show that a good compensation requires very short beams and allows no freedom in choice of the plasma density.

I INTRODUCTION

A plasma can sustain extremely large electric fields. Because of this ability, various applications of plasmas to high-energy accelerators are intensively studied [1–3]. Among them are the wakefield acceleration, passive plasma lens, plasma guiding of beams, photon acceleration, and plasma compensation of beam fields at the interaction point of colliders. Here we consider the plasma compensation as applied to multi-TeV ultimate muon colliders discussed in these Proceedings.

Earlier, the plasma compensation was first studied as a possible mean of beamstrahlung suppression in linear electron-positron colliders [4,5], and it was realized soon that the plasma density required for a good compensation is too high in future colliders, higher than the density of conduction electrons in solids. Later, the suppression of beam-beam effects in circular colliders was considered [6], but degradation of the beam lifetime due to plasma turns out to be unacceptable for proton and electron machines. Here we derive necessary conditions for the plasma compensation to work in muon colliders and show that these conditions are difficult to meet. Nevertheless, they are no more fantastical than the ultimate muon collider itself.

II SUPPRESSION OF THE BEAM FIELDS

The plasma can efficiently neutralize the electric and magnetic fields of the beam if the following conditions are fulfilled [4–6]:

$$n_e \gg n_b, \qquad (1)$$

$$k_p \sigma_r \gg 1. \quad (2)$$

Here n_e is the electron density in the plasma, n_b is the beam density (the sum of beam densities in the case of several beams), σ_r is the rms beam radius, and k_p is the reciprocal plasma skin depth related to the plasma electron frequency ω_p as follows:

$$k_p = \frac{\omega_p}{c} = \sqrt{\frac{4\pi n_e e^2}{mc^2}}, \quad (3)$$

where m is the electron mass, c is the light velocity, and e is the elementary charge.

It is convenient to characterize the plasma compensation by the ratio of tune shifts with (ξ) and without (ξ_0) the plasma. The smaller ξ/ξ_0 the better compensation is. For round Gaussian beams and the optimum plasma thickness, this ratio is [7]

$$\frac{\xi}{\xi_0} \approx \frac{1}{k_p^2 \sigma_r^2} \left(1 + \frac{8\ln(k_p^2 \sigma_r^2 - 1) + 1}{4\sqrt{\pi \ln(k_p^2 \sigma_r^2 - 1)}}\right). \quad (4)$$

This formula is correct up to beam densities of the order of n_e.

The plasma neutralizes the electric field of the beam much better than its magnetic field. It is uncompensated magnetic field that makes the dominant contribution to ξ. The electric field is v/c times smaller [7], where v is the longitudinal velocity of plasma electrons. However, the electric field always pushes plasma ions radially away from the beam axis [7].

Let us rewrite inequalities (1), (2) in terms of number of particles in each colliding beam (N_b), beam size (σ_r, σ_z), plasma ion density (n_i), ion charge state ($Z = n_e/n_i$), and "ion" skin depth

$$\lambda_{pi} = \frac{c\sqrt{Z}}{\omega_p} = \sqrt{\frac{mc^2}{4\pi n_i e^2}}.$$

For Gaussian beams, the maximum beam density

$$n_b = \frac{N_b}{(2\pi)^{3/2} \sigma_z \sigma_r^2}, \quad (5)$$

and

$$\frac{n_b}{n_e} = \sqrt{\frac{2}{\pi}} \frac{N_b r_e \lambda_{pi}^2}{Z \sigma_r^2 \sigma_z}, \quad (6)$$

where r_e is the classical electron radius. Then (1) becomes

$$\sigma_z \sigma_r^2 \gg \sqrt{\frac{2}{\pi}} \frac{N_b}{Z} r_e \lambda_{pi}^2, \quad (7)$$

and (2) turns to

$$\sigma_r \gg \lambda_{pi}/\sqrt{Z}. \quad (8)$$

III COLLISIONAL DIFFUSION OF THE RETURN CURRENT

When the beam (or two opposite beams in our case) enters the plasma, it inductively generates the plasma return current. This current provides an approximate local compensation and the exact integral compensation of the beam current. Due to electron-ion collisions in the plasma, the area of the return current broadens, and the local compensation becomes worse. This is the well known phenomena of the diffusion of a magnetic field into a stationary conductor (see, e.g., [8]). Electron-electron collisions do not change the return current and affect the field diffusion only via heating of plasma electrons. Quantitatively, the magnetic diffusion is described by the equation

$$\frac{\partial B}{\partial t} = \frac{c^2}{4\pi\sigma}\triangle B, \qquad (9)$$

where B is the azimuthal magnetic field, σ is the plasma conductivity, and $\triangle$ is the Laplacian. The plasma conductivity can be expressed in terms of electron-ion collision frequency ν_{ei} or electron velocity v:

$$\sigma = \frac{n_e e^2}{m\nu_{ei}} = \frac{mv^3}{4\pi\Lambda Z e^2}, \qquad (10)$$

where Λ is the Coulomb logarithm.

Assume we want the plasma to reduce the magnetic field of the beam K times. Then, as follows from (9), the beam should be short:

$$\frac{1}{K} > \frac{1}{B}\frac{\partial B}{\partial t} \cdot \frac{\sigma_z}{c} = \frac{a_d \Lambda Z e^2 c \sigma_z}{mv^3 \sigma_r^2}. \qquad (11)$$

Here the constant $a_d \sim 1$ comprises possible errors introduced when we estimate spacial derivatives of B. The velocity of plasma electrons can be written as

$$v = b_d\, c\, n_b/n_e. \qquad (12)$$

The factor $b_d \in (0,1)$ appears since at the beam periphery plasma electrons move slower than at the beam center. Substituting (12) into (11), we can rewrite the condition of admissible magnetic diffusion in the form of limitation on beam dimensions:

$$\sigma_r \sigma_z < \left(\frac{2\sqrt{2}}{\pi\sqrt{\pi}a_d K\Lambda}\right)^{1/4} \frac{(b_d N_b)^{3/4}}{Z} r_e^{1/2} \lambda_{pi}^{3/2}. \qquad (13)$$

We retain all numerical factors in formulae to avoid accumulation of errors.

IV BEAM FILAMENTATION

Cold beams in the plasma are subject to filamentation. This phenomenon has long being known as Weibel instability and was studied in detail as applied to plasma wakefield acceleration. It was found [9] that the beam is stable if the transverse component of its velocity satisfies the condition

$$v_{b\perp} > c\sqrt{\frac{mn_b}{\gamma_b m_\mu n_e}}, \tag{14}$$

where γ and m_μ are the relativistic factor and rest mass of the beam particles, correspondingly. Assume that the beta-function at the interaction point is equal to σ_z. Then

$$\frac{v_{b\perp}}{c} \sim \frac{\sigma_r}{\sigma_z}, \tag{15}$$

and the stability condition reads as

$$\frac{\sigma_r^4}{\sigma_z} > \frac{a_f m N_b}{\gamma_b m_\mu Z} r_e \lambda_{pi}^2, \tag{16}$$

where $a_f \sim 1$ is a numerical factor.

V MOTION OF PLASMA IONS

In the presence of an ultrarelativistic beam, a small radial electric field appears in the plasma. This field balances the magnetic force exerted on plasma electrons moving axially in the incompletely neutralized magnetic field of the beam [7]. The electric field always pushes plasma ions out of the beam region. When the ion density at the beam axis reduces to zero, any compensation of the beam fields (both electric and magnetic) disappears. Here we write out the limitation to beam parameters imposed by the dynamics of plasma ions.

The typical value of the radial electric field is

$$E \sim \frac{v}{c} B \sim \frac{v}{c} \cdot \frac{en_b \sigma_r}{(k_p \sigma_r)^2} \sim \frac{a_i e n_b^2}{k_p^2 \sigma_r n_e}, \tag{17}$$

where the numerical factor $a_i \sim 1$ reflects an uncertainty in determination of E. This field shifts plasma ions radially by the distance $\sim \sigma_r$ in the time

$$\tau_i \sim \sqrt{\frac{M_i \sigma_r}{ZeE}}, \tag{18}$$

where M_i is the ion mass. For the compensation to take place, we need

$$\sigma_z < b_i c \tau_i, \tag{19}$$

where $b_i \sim 1$. Substituting (17), (18) and (6) into (19), we obtain

$$\sigma_r > \left(\frac{a_i m}{2\pi b_i^2 Z M_i}\right)^{1/6} N_b^{1/3} r_e^{1/3} \lambda_{pi}^{2/3}. \tag{20}$$

Thus, for a good plasma compensation, the beam should be wide enough.

VI UNIVERSAL CONSTRAINTS

Let us put inequalities (7), (8), (13), (16), and (20) together and choose the most important ones. We take as a reference point the following parameters:

$$N_b = 5 \cdot 10^{12}, \qquad m_\mu \gamma_b/m \approx 10^7, \qquad n_e = n_i \approx 5 \cdot 10^{22} \text{ cm}^{-3}, \tag{21}$$

$$\lambda_{pi} \approx 2.4 \cdot 10^{-6} \text{ cm}, \qquad M_i/m \approx 5.5 \cdot 10^3, \qquad Z = 1, \qquad K = 10, \tag{22}$$

which corresponds to a 5 TeV muon beam and conduction electrons of liquid lithium as the plasma. The above inequalities then become, correspondingly,

$$\sigma_z \sigma_r^2 / \lambda_{pi}^3 \gg 4.7 \cdot 10^5, \tag{23}$$

$$\sigma_r / \lambda_{pi} \gg 1, \tag{24}$$

$$\sigma_z \sigma_r / \lambda_{pi}^2 < 2.5 \cdot 10^5, \tag{25}$$

$$\sigma_r^4 / (\sigma_z \lambda_{pi}^3) > 0.06, \tag{26}$$

$$\sigma_r / \lambda_{pi} > 23. \tag{27}$$

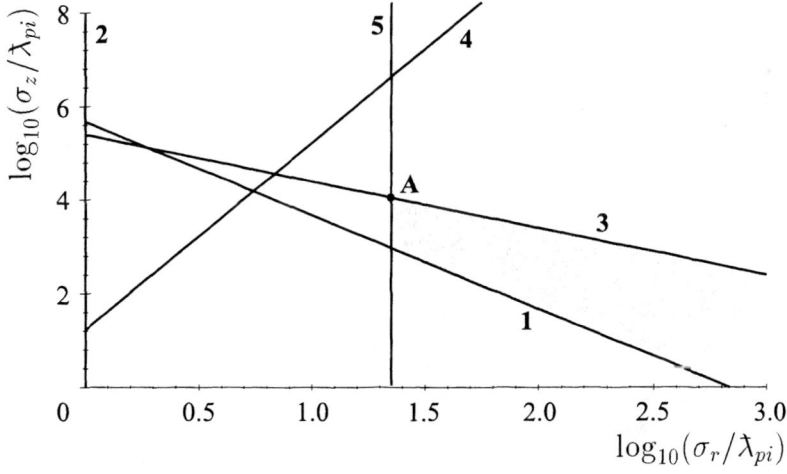

FIGURE 1. Graphical representation of inequalities: 1–(23), 2–(24), 3–(25), 4–(26), 5–(27). In the shaded area all the inequalities are fulfilled.

In derivation of (23)–(27) we put $\Lambda = 3$, $b_d = 0.5$, and $a_d = a_f = a_i = b_i = 1$. The areas determined by inequalities (23)–(27) are shown in Fig.1 in logarithmic scale. It is seen that, for the plasma compensation to work, the beam should be very short. The maximum beam length (point **A** in Fig.1) can be found by substitution of the minimum σ_r [determined by (20)] into (13). It equals

$$\sigma_{z,max} \approx \frac{4\, b_i^{1/3} b_d^{3/4}}{a_i^{1/6} a_d^{1/4} \Lambda^{1/4}} \cdot \frac{M^{1/6} N_b^{5/12} r_e^{1/6} \lambda_{pi}^{5/6}}{Z^{5/6} K^{1/4}}, \tag{28}$$

where M is the ion mass number.

For the above parameters, $\sigma_{z,max} \approx 0.04$ cm. This value is much smaller than any conceivable bunch length. Using heavier metals as the plasma can not save the situation because of the very weak dependence of $\sigma_{z,max}$ on M. Multiple ionization of the ions, which is possible at typical energies of plasma electrons, makes $\sigma_{z,max}$ appreciably shorter. More accurate calculations of the magnetic diffusion (a_d), electric field in the plasma (a_i), ion dynamics (b_i), or Coulomb logarithm (Λ) will not noticeably change the expression (28), because the dependence of $\sigma_{z,max}$ on the corresponding coefficients is weak. Possibly, more accurate analysis of the plasma conductivity at different radii (b_d) can change the numerical factor in (28), but unlikely more than an order of magnitude.

As we see, the only way to make the plasma compensation work in muon colliders is to abandon the liquid metal plasma in favor of lower density plasmas. Inverting (28) and neglecting numerical factors of the order of unity,

$$\lambda_{pi} \sim \frac{Z K^{3/10} \sigma_{z,max}^{6/5}}{M^{1/5} N_b^{1/2} r_e^{1/5}}, \tag{29}$$

we find that, for $\sigma_{z,max} = 0.3$ cm and all other parameters from (21)–(22), the required ion density

$$n_i \sim 10^{20}\,\text{cm}^{-3} \qquad (\lambda_{pi} \sim 5 \cdot 10^{-5}\,\text{cm}). \tag{30}$$

For a fixed bunch length and a variable plasma ion density determined by (29), we find from (20) and (4)

$$\sigma_r \sim \frac{0.2\, Z^{1/2} K^{1/5} r_e^{1/5} \sigma_{z,max}^{4/5}}{M^{3/10}}, \tag{31}$$

and

$$\frac{\xi}{\xi_0} \sim \frac{25\, K^{1/5} M^{1/5} \sigma_{z,max}^{4/5}}{N_b\, r_e^{4/5}}. \tag{32}$$

Thus, the longer is the bunch, the wider should it be, and the worse is its compensation. Substituting $\sigma_{z,max} = 0.3$ cm into (31)–(32), we obtain

$$\sigma_r \sim 3\,\mu\text{m}, \qquad \xi/\xi_0 \sim 0.04. \tag{33}$$

We see that the bunches longer than several millimeters are unacceptable since they are to be too wide and their compensation is poor.

VII ACKNOWLEDGEMENTS

The author is grateful to A. N. Skrinsky, V. I. Telnov, and V. V. Parkhomchuk for helpful discussions.

REFERENCES

1. **E. Esarey, P. Sprangle, J. Krall, and A. Ting**, *Overview of plasma-based accelerator concepts.* — IEEE Trans. Plasma Sci., v. 24 (1996), p. 252–288.
2. **A. Ogata and K. Nakajima**, *Recent progress and perspectives of laser-plasma accelerators.* — Laser and Particle Beams, v. 16 (1998), p. 381–396.
3. **J. S. Wurtele**, *The role of plasma in advanced accelerators.*— Phys. Fluids B, v. 5 (1993), p. 2363–2370.
4. **D. H. Whittum, A. M. Sessler, J. J. Stewart, and S. S. Yu**, *Plasma suppression of beamstrahlung.* — Part. Accel., v. 34 (1990), p. 89–104.
5. **A. M. Sessler and D. H. Whittum**, *Suppression of beamstrahlung by means of a plasma.* — In: Advanced Accelerator Concepts, AIP Conference Proceedings, edited by J. S. Wurtele, v. 279, p. 939–944, (AIP Press, New York, 1993).
6. **G. V. Stupakov and P. Chen**, *Plasma suppression of beam-beam interaction in circular colliders.* — Phys. Rev. Lett., v. 76 (1996), p. 3715–3718.
7. **K. V. Lotov, A. N. Skrinsky, and A. V. Yashin**, *Plasma suppression of beam-beam interaction in a muon collider.* — Nucl. Instr. Methods A, to be published.
8. **R. B. Miller**, *An Introduction to the Physics of Intense Charged Particle Beams.* — New York – London, Plenum Press, 1982.
9. **J. J. Su, T. Katsouleas, J. M. Dawson, P. Chen, M. Jones, and R. Keinigs**, *Stability of the driving bunch in the plasma wakefield accelerator.* — IEEE Trans. Plasma Sci., v. PS-15 (1987), p. 192–198.

New physics and the post-LHC era

Joseph D. Lykken

Theoretical Physics Dept.
Fermilab, P.O. Box 500, Batavia, IL 60510

WHERE ARE WE GOING AND HOW DO WE GET THERE?

While we are correct to spend sleepness nights agonizing over the optimal path to our long term future, we should also rejoice that our immediate future looks extremely bright. The B factories coming on line now will provide fundamental new inputs to our global view of particle physics, including, most likely, some surprises. The same can be said for the many neutrino experiments in progress or under construction, and for a number of other low energy projects. As a bonus, a flood of new astrophysical data will impact on a number of important ideas and problems circulating in particle physics.

Beginning in 2001, the next run of the Tevatron will extend our reach for the Higgs [1], supersymmetry [2], B physics, top physics, electroweak physics, and new strong dynamics, to name but a few. Major discoveries are very possible. If we are fortunate, we might even get the first experimental hints of extra spatial dimensions [3-7], quantum gravity, or strings [8-11].

Obviously new discoveries from any of these arenas will help bootstrap funding and resources for future experiments and new facilities. In addition, discoveries –or even hints– of physics beyond the Standard Model will crystalize our thinking about what future facilities are needed.

The same can be said of the LHC era, scheduled to begin around 2006. With the LHC we will probe the TeV scale, $\sqrt{s_x} \sim 1 - 5$ TeV. A possible linear collider could also probe $\sqrt{s} \sim 1$ TeV. We expect to see a lot of new physics at these machines. One of our main jobs in studying this physics is to look for hints about new physics thresholds at even higher energies.

If we want to be ready to exploit these exciting discoveries, we better get serious **now** about post-LHC machines. At the same time, we must keep in mind that physics input along the way will help us to make better choices; thus the more flexibility we can incorporate into our planning, the more likely we will be able to respond appropriately to new discoveries.

To be concrete I have summarized below the various planned or proposed future accelerator projects, according to my opinion of what we might have and when.

2006 – 2012: The LHC Era

- **LHC** : $\sqrt{s}$=14 TeV, $\mathcal{L}$=10^{33} to 10^{34}.
- **Linear Collider (LC)** [12-14] : $\sqrt{s}$=350 GeV to 1 TeV, $\mathcal{L}$=10^{34}ish.
- **Muon storage ring ν factory** [15-18] : 1 millimole of muons per year.
- **upgraded Tevatron?** [19,20] : $\sqrt{s}$= 4 – 6 TeV, $\mathcal{L}$=5×10^{32}.

2013 – 2025: Within the Energy Frontier

- **stretch LC** : $\sqrt{s}$=1.5 TeV.
- $\gamma\gamma$, e^-e^- : piggyback on LC.
- **First Muon Collider** [21] : Higgs factory? Heavy Higgs factory?

2013 – 2025: Extending the Energy Frontier

- **upgraded LHC?** : $\sqrt{s}$=?.
- **CLIC** [22] : $\sqrt{s}$=3 – 5 TeV, $\mathcal{L}$=10^{35}.
- **site-filling HEMC** : $\sqrt{s}$=3 – 4 TeV, $\mathcal{L}$=10^{35}.

Or, make a bigger energy jump:

- **VLHC** [23] : $\sqrt{s}$=100 – 200 TeV, $\mathcal{L}$=10^{35}.
- **ultimate HEMC** : $\sqrt{s} \geq$10 TeV, $\mathcal{L}$ =?.

The dates of course are only estimates, but the groupings are important. It is important in thinking about future machines to make a clear distinction between those which are intended to operate in the LHC era, and those which are clearly post-LHC successors. It is also important to discriminate between those proposals which extend the energy frontier, and those which would operate within the energy frontier defined by the LHC. I have also made note of possible upgrades of the Tevatron, LHC, and linear e^+e^- collider (LC), since recycling is likely to become increasingly popular in a difficult funding climate.

PHYSICS QUESTIONS FOR LHC ERA EXPERIMENTS

The physics agenda for the LHC era is both clear and robust. Although this agenda involves the exploration of new physics at the TeV scale, it can be outlined in a way that avoids most theoretical bias about the nature of that new physics:

- The Standard Model is an effective theory for physics below some high energy cutoff Λ. What is the value of Λ?

- What are the relevant degrees of freedom for the new effective theory at energies above Λ?

- What are the symmetries of this new effective theory?

- What symmetries and organizing principles of the Standard Model turn out to be artifacts of the "low energy" approximation?

- Do the symmetries and organizing principles of the new effective theory explain parameters/hierarchies of the SM, e.g. the flavor problem?

- Does the new effective theory give any hints of physics at even higher scales?

WHAT COULD BE OUT THERE?

Most theoretical speculation about the new effective theory at high energies involves *adding* things to the Standard Model:

- Add particles (e.g. superpartners, techniparticles, messenger sector, Kaluza-Klein or string resonances).

- Add new symmetries or organizing principles (e.g. supersymmetry)

- Add new gauge interactions, dynamics, either strongly coupled (e.g. technicolor) or weakly coupled (e.g. Z' boson).

However it is just as likely that at higher energy scales we have more radical changes:

- Qualitatively new degrees of freedom (e.g. strings, membranes, extra dimensions).

- Symmetries are broken (e.g. B and L violation).

- Sacred principles are violated!

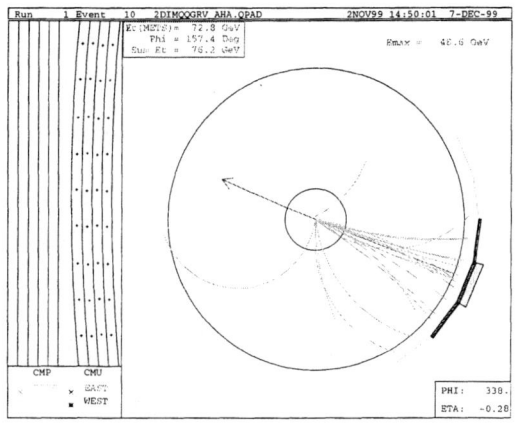

FIGURE 1. Extra dimensions at the Tevatron: a monojet recoiling off a Kaluza-Klein graviton. Simulated event from M. Spiropulu, CDF.

This would not be the first time that sacred cows got ground into hamburger. A major challenge for future experiments is to determine the energy scale and conditions for which the following theoretical assumptions break down:

- The fundamental dynamical entities are point-like particles.

- Relativistic quantum field theory (locality, microcausality, CPT).

- General relativity.

- Quantum mechanics.

It is interesting to note in this regard that in 1939 Werner Heisenberg suggested [24] that our poor understanding of strong interactions implied that quantum mechanics probably breaks down at an energy scale around 1 GeV. In the modern era string theory suggests that at least the first three items above are merely low energy approximations below the string scale.

STRING THEORY

Although string theory has not (yet) done a good job of matching to the Standard Model at low energies, it has proven to be a great exercise for liberating our thinking. If string theory is correct, the fundamental physical entities are not quarks and leptons, but perhaps a whole collection of particle-like, string-like, and membrane-like objects. Furthermore these objects propagate in a 10+1 dimensional spacetime. If string theory is correct, both general relativity and quantum field theory break down at some energy scale M_s. We don't know what this energy scale is!

The HEMC may operate above the threshold for stringy effects and/or quantum gravity. Any discoveries hinting at M_* or M_s in the multi-TeV range will dramatically change our expectations for what we will observe at the HEMC. Because of our poor understanding of string theory, it is hard to sketch detailed scenarios for life above the string scale or the effective Planck scale. Two generic expectations are the production of heavier vibrational modes of the string (Reggeization) and the production of mini black holes.

PRODUCTION OF BLACK HOLES AT COLLIDERS

If I collide two particles at an impact parameter less than their Schwarzschild radius, I expect to form a black hole. Normally this is not an issue, since the critical impact parameter goes like

$$b_{sch} \sim \frac{\sqrt{s}}{M_p^2} \sim (10^{35} \text{ GeV})^{-1} \qquad (1)$$

Suppose there are 2 large extra dimensions, with an effective Planck scale M_* in the TeV range. Then

$$b_{sch} \sim \frac{1}{M_*} \left(\frac{\sqrt{s}}{M_*}\right)^{1/3} \qquad (2)$$

Thus future colliders may produce lots of little black holes [25–28]. These black holes decay via Hawking radiation either into the bulk (i.e. the extra dimensions) or into Standard Model particles on our brane. The lifetime is of order $1/M_*$. Actually we have a poor understanding of such small black holes, and, in addition, quantum gravity and string theory involve lots of other messy junk.

A sensible approach might be to concentrate on diffractive quantum gravity with s/M_*^2 of order one –the same Regge limit that was studied fruitfully in the 1960s for the case of the strong interactions. Indeed traditional high p_T processes may be highly suppressed, since scatterings with small impact parameters go predominately into black hole production. So it may be crucial to be able to do physics in the forward region at the HEMC!

WHAT DO EVENTS LOOK LIKE AT HEMC?

There are significant challenges involved in reconstructing events at $\sqrt{s} = 10$ TeV or higher. Most worrisome are the detector backgrounds from muon decays in the beam and subsequent showering from the high energy decay electrons. For example Bethe-Heitler muons regenerated in these showers can penetrate into the detector and produce energy spikes in the calorimeter.

It is amusing at least to see that the Pythia event generator is capable of producing $\mu^+\mu^-$ events at 10 TeV. Here for example is a Drell-Yan event at 10 TeV:

```
========================================

PYTHIA will be initialized for
a mu+ on mu- collider
at   10000.000 GeV center-of-mass energy

========================================

    ISUB  Subprocess name          Maximum value

==============================================

     1    f + fbar -> gamma*/Z0      1.2603E-10

==============================================


Event listing (HEP format)

particle/jet PHEP(1,I)   PHEP(2,I)   PHEP(4,I)
mu+             0.00000     0.00000  5000.00000
mu-             0.00000     0.00000  5000.00000
mu+           203.57645    73.31965  4243.37126
mu-             0.00000     0.00000   757.03250
Z0            203.57645    73.31965  5000.40376
b              43.71009   -15.72983   658.60806
b~            159.86636    89.04947  4341.79570
Z0)           203.57645    73.31965  5000.40376
gamma        -203.57571   -73.30043  4969.25770
gamma          -0.00074    -0.01922    30.33854
```

Just for fun I ran this event through SHW, the fast CDF/D0 detector simulator:

There are   4 triggerable objects:

```
No  name       eta    phi     ET     clust
 1  jet       3.85   3.48   211.08   1.00
 2  jet      -3.86   0.56   159.82   2.00
 3  jet      -3.37   5.94    41.18   3.00
 4  met       0.00   6.13    11.47   2.00
```

There are 2 reconstructed objects:

object name	ET	E-EM	E-had	eta	phi
1 jet	159.82	1833.86	1965.71	-3.86	0.56
2 jet	41.18	288.10	310.22	-3.37	5.94

The 4.97 TeV initial state photon triggers as a jet then fails to reconstruct as anything.

Here is another dramatic event: a 10 TeV hard scattering that produces less than 20 MeV of visible energy!

Event listing (HEP format)

particle/jet	PHEP(1,I)	PHEP(2,I)	PHEP(4,I)
mu+	0.00000	0.00000	5000.000
mu-	0.00000	0.00000	5000.000
mu+	0.00000	0.00000	4999.989
mu-	0.01376	-0.00666	4999.994
Z0	0.01376	-0.00666	9999.983
nu-tau	-3256.56287	2305.08795	4999.984
nu-tau~	3256.57663	-2305.09461	4999.999
gamma	-0.01376	0.00666	0.016

"MODEL-INDEPENDENT" CONCLUSIONS

• There is *a whole new effective theory* waiting to be explored at the TeV scale. The new physics will be rich, surprising, confusing, and take a long time to untangle.

• For exploration you will want high energies, reasonable luminosities, and reasonable detectors. For detailed studies, you need excellent luminosities and excellent detectors. You will need detailed studies not only to unravel the new effective theory, but also *to give you hints about physics at even higher scales*.

• Higgs physics will be interesting for a long time! Thus we should keep in mind machine options such as a First Muon Collider (FMC) s-channel Higgs factory, a heavy Higgs factory, as well as the $\gamma\gamma$ option for a Linear Collider.

• LC + an FMC offer different sensitivities, polarization, reduced backgrounds, better contained events, more precise measurements. This will crucial for pointing the way to physics at even higher energy scales.

Examples:

  • Untangling the neutralino and slepton sectors in SUSY. What kind of SUSY is it?

- Deciphering virtual effects of extra dimensions. Is your Drell-Yan anomaly due to spin 2 Kaluza-Klein graviton exchange?

- LHC/LC data will be essential for making good decisions. E.g. LHC/LC may data indicate an effective Planck scale of 4 TeV! This affects your choices for $\sqrt{s}$, luminosity, and detector design!

- We don't yet know how to estimate the next interesting energy scale post-LHC. Will a 3 - 4 TeV lepton collider or an LHC upgrade be good enough, or do we need to push immediately to a 10 - 15 TeV muon collider or 100 - 200 TeV VLHC?

TASK LIST FOR THE PHYSICS WORKING GROUP

The basic task of the HEMC Physics Working Group was to classify possible new physics opportunities at 3 - 4 TeV, 10 - 15 TeV, and 100 TeV. We developed a template for detailing HEMC physics opportunities, which is reproduced below:

1. What is the new physics?

2. What are the theoretical motivations? What theoretical frameworks does this fit into?

3. How will the LHC + LC + other experiments provide evidence or hints for

    - the existence of this new physics, and
    - the energy scale of this new physics.

4. What are the experimental signatures of this new physics at an HEMC? What do you need to measure? Are there backgrounds? Do you need or take advantage of

    - small beam energy spread,
    - polarization,
    - radiative return,
    - forward coverage,
    - heavy flavor tagging,
    - good hadronic calorimetry.

5. What is the minimal useful luminosity for probing this physics?

6. What are the most likely relevant energy scales?

7. References to existing literature.

Members of the working group fleshed out a number of examples in conformity to this template. The results are documented individually in these proceedings.

REFERENCES

1. Draft report of the "Physics at Run II" Higgs Working Group, available at http://fnth37.fnal.gov/higgs.html.
2. Draft reports of the "Physics at Run II" Supersymmetry/Higgs Workshop, available at http://fnth37.fnal.gov/susy.html.
3. I. Antoniadis, Phys. Lett. **B246**, 377 (1990).
4. N. Arkani-Hamed, S. Dimopoulos, and G. Dvali, Phys. Rev. **D59** 086004 (1999).
5. L. Randall and R. Sundrum, Phys. Rev. Lett. **83**, 3370 (1999) [hep-ph/9905221].
6. L. Randall and R. Sundrum, Phys. Rev. Lett. **83**, 4690 (1999) [hep-th/9906064].
7. J. Lykken and L. Randall, hep-th/9908076.
8. J. Lykken, Phys. Rev. **D54** 3693 (1996).
9. E. Dudas and J. Mourad, hep-th/9911019.
10. E. Accomando, I. Antoniadis, and K. Benakli, "Looking for TeV Scale Strings and Extra Dimensions", hep-ph/9912287.
11. S. Cullen, M. Perelstein, and M. Peskin, "TeV Strings and Collider Probes of Large Extra Dimensions", hep-ph/0001166.
12. "Zeroth Order Design Report for the Next Linear Collider", SLAC-R-0474, May 1996.
13. B. H. Wiik, "The TESLA Project", Part. Accel. **62**, 43 (1998).
14. T. Tauchi, KEK-PREPRINT-98-180 *Invited talk at International Conference on Hadron Structure (HS 98), Stara Lesna, Slovakia, 7-13 Sep 1998.*
15. S. Geer, C. Johnstone, D. Neuffer, "Design Concepts for a Muon Storage Ring Neutrino Source", Fermilab-Pub-99-121.
16. FNAL Feasibility Study on a Neutrino Source Based on a Muon Storage Ring, http://www.fnal.gov/projects/muon_collider/nu-factory/nu-factory.html.
17. B. Autin, A. Blondel, and J. Ellis (eds), "Prospective Study of Muon Storage Rings at CERN", CERN-99-02, ECFA 99-197.
18. V. Barger, S. Geer, K. Whisnant, "Long Baseline Neutrino Physics with a Muon Storage Ring Neutrino Source", hep-ph/9906487.
19. P. McIntyre, E. Accomando, R. Arnowitt, B. Dutta, T. Kamon and A. Sattarov, hep-ex/9908052.
20. V. Barger, K. Cheung, T. Han, C. Kao, T. Plehn and R. Zhang, hep-ph/9910500.
21. C. Ankenbrandt et al, "Status of Muon Collider Research and Development and Future Plans", Fermilab-Pub-98-179, Phys. Rev.ST Accel.Beams **2** 081001 (1999).
22. J-P Delahaye et al, "CLIC, a 0.5 TeV to 5 TeV e^+e^- Compact Linear Collider", CERN/PS 99-005, CERN/PS 99-062, Acta Phys. Polon. **B30**, 2029 (1999).
23. G. Anderson et al, "Summary of the Very Large Hadron Collider Physics and Detector Workshop", hep-ph/9710254.
24. D. J. Gross, hep-th/9411233.
25. G. 't Hooft, Phys. Lett. **B198**, 61 (1987).
26. D. J. Gross and P. F. Mende, Nucl. Phys. **B303**, 407 (1988).
27. H. Verlinde and E. Verlinde, Nucl. Phys. **B371**, 246 (1992) [hep-th/9110017].
28. T. Banks and W. Fischler, hep-th/9906038.

Effects of Kinematic Correction on the Dynamics in Muon Rings

Kyoko Makino and Martin Berz

Department of Physics and Astronomy and
National Superconducting Cyclotron Laboratory
Michigan State University, East Lansing, MI 48824, USA

Abstract. Among many other challenges faced by muon accelerators and storage rings, the influence of nonlinear effects has to be studied carefully compared to conventional proton and electron machines. The short lifetime of muons as well as their production mechanism make the cooling of muon beams a necessity; however, in many scenarios still rather large transversal emittances persist, which makes nonlinear effects more pronounced than usual.

There are a variety of sources for nonlinear effects besides deliberate and random nonlinear multipoles; particularly, there are nonlinear effects due to fringing fields, which have been known in the design of high-resolution spectrographs and recently have also been studied for storage rings [1]. Details of some first estimates of their effects are given in an accompanying paper [2]. In this paper we address another nonlinear effect that is directly connected to the presence of large emittances, namely the so-called kinematic correction, and report first results of the effects on some current muon storage ring designs.

INTRODUCTION

The most general form of the Hamiltonian of a charged particle in electromagnetic fields in curvilinear coordinates $\{s, x, y\}$ can be written in terms of the generalized momentum $\vec{P}^G = (P_s^G, P_x^G, P_y^G)$ as [3]

$$H = q\Phi + c\sqrt{\frac{(P_s^G + P_x^G \tau_1 y - P_y^G \tau_1 x - \alpha q A_s)^2}{\alpha^2} + (P_x^G - qA_x)^2 + (P_y^G - qA_y)^2 + m^2 c^2}. \quad (1)$$

Here, τ_1 is the rate of rotation around the beam axis, τ_2 and τ_3 are curvatures in y-s and x-s planes, and α is defined as $\alpha = 1 - \tau_3 x + \tau_2 y$. The generalized momentum is expressed in terms of the kinematic momentum $\vec{p} = (p_s, p_x, p_y)$, where we usually expect $p_s^2 \gg p_x^2, p_y^2$, as

$$P_s^G = (p_s + qA_s)\alpha - (p_x + qA_x)\tau_1 y + (p_y + qA_y)\tau_1 x$$
$$P_x^G = p_x + qA_x, \quad P_y^G = p_y + qA_y.$$

Apparently the first term in the square root in the Hamiltonian (1) is the leading term, and the approximation to omit the second and the third terms, which entail nonlinear effects purely due to dynamics and independent of the fields, is widely used.

In case of muon accelerators and storage rings and their typically large transversal emittances, p_x and p_y are not safely negligible compared to p_s anymore. The restoration of those omitted terms is often referred to as kinematic correction, and the effects of the kinematic correction as well as the other nonlinear effects have to be studied and treated more seriously for muon machines than for conventional proton and electron machines.

COSY INFINITY

The code COSY INFINITY [4] is a transfer map based beam physics code working to arbitrary high order [5,6]. All nonlinear terms are included in the necessary equations from the outset, and the nonlinear effects can be taken care of up to any desired order.

As shown in [3], particles following x-s planar motion, in which case $(\tau_1, \tau_2, \tau_3) = (0, 0, -h)$, obey the following set of equations of motion [7]:

$$x' = a(1 + hx)\frac{p_0}{p_s} \tag{2a}$$

$$y' = b(1 + hx)\frac{p_0}{p_s} \tag{2b}$$

$$a' = \left\{(1+\delta_m)\frac{1+\eta}{1+\eta_0}\frac{p_0}{p_s}\frac{E_x}{\chi_{E0}} - \frac{B_y}{\chi_{M0}} + b\frac{p_0}{p_s}\frac{B_s}{\chi_{M0}}\right\}(1+hx)(1+\delta_z) + h\frac{p_s}{p_0} \tag{2c}$$

$$b' = \left\{(1+\delta_m)\frac{1+\eta}{1+\eta_0}\frac{p_0}{p_s}\frac{E_y}{\chi_{E0}} + \frac{B_x}{\chi_{M0}} - a\frac{p_0}{p_s}\frac{B_s}{\chi_{M0}}\right\}(1+hx)(1+\delta_z), \tag{2d}$$

Here $m = m_0(1 + \delta_m)$, $q = z_0 e(1 + \delta_z)$, with the index $_0$ expressing the respective quantities of the reference particle. We use the quantities $a = p_x/p_0$ and $b = p_y/p_0$. Furthermore,

$$\chi_{E0} = \frac{p_0 v_0}{z_0 e}, \quad \chi_{M0} = \frac{p_0}{z_0 e}, \quad \eta = \frac{K_0(1+\delta_k) - z_0 e(1+\delta_z)V(x,y,s)}{m_0 c^2 (1+\delta_m)}$$

and

$$\frac{p_s}{p_0} = \sqrt{(1+\delta_m)^2 \frac{\eta(2+\eta)}{\eta_0(2+\eta_0)} - a^2 - b^2}. \tag{3}$$

In this expression, the kinematic correction terms in the Hamiltonian, the second and the third terms in the square root in (1), correspond to the terms a^2 and b^2 in the square root in (3). Note that the kinematic correction appears from the second order in p_s/p_0 and hence in p_0/p_s. The expression p_s/p_0 appears in all equations (2a) to (2d). The B_y term in a' and the B_x term in b' are leading terms in (2c) and (2d), so the first kinematic correction effect may appear from the $h \cdot p_s/p_0$ term in a', and thus produce second order effects for x and a, and third order effects for y and b. The correctness of the appearing higher order kinematic correction terms was verified in [8].

COSY INFINITY includes kinematic correction routinely without any approximation as indicated in the equations (2a) to (2d) and (3). In order to study the effects of kinematic correction, which is frequently omitted in other codes, an option was created that ignores the terms a^2 and b^2 in the square root in (3). In the following, computational results were obtained by COSY INFINITY by using its default set of equations and by using a similar set of equations with kinematic correction turned off.

KINEMATIC CORRECTION IN MUON STORAGE RINGS

In the following we present observations related to the kinematic correction effects. We limit ourselves to the mere observation of the effects, without attempting to devise strategies for their correction through nonlinear elements, which of course should also include the influence of all other relevant nonlinear effects.

We studied the same muon storage rings as the ones used in [2]. This section reports on the effects on the 30 GeV neutrino factory ring and the 30 GeV Higgs factory ring.

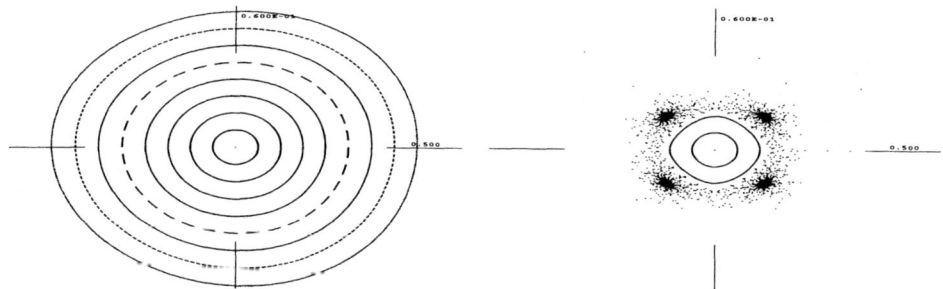

FIGURE 1. Tracking pictures for 1000 turns in a 30 GeV neutrino factory ring without (left) and with (right) kinematic correction in the same scale of 500mm×60mrad. With correction, only those particles up to 100mm×15mrad are stable.

TABLE 1. The beginning part of the Taylor transfer map of a 30 GeV neutrino factory ring without kinematic correction.

Expansion coefficients of x,a,y,b depending on the exponents of xayb				
(x,	(a,	(y,	(b,	xayb
-0.1936744	0.1416905	0	0	1000
-6.792904	-0.1936744	0	0	0100
0	0	0.1961760	-0.2456274E-01	0010
0	0	39.14526	0.1961760	0001
0.5590897E-01	-0.1884921E-02	0	0	2000
-0.2374424	-0.7032995E-01	0	0	1100
-0.9129287	-0.4467585	0	0	0200
0	0	0.1229122	-0.3076403E-03	1010
0	0	0.6994628	0.2591721E-02	0110
0	0	-0.4902813	-0.1229122	1001
0	0	4.130383	-0.6994626	0101
-0.8336143E-02	-0.1539706E-02	0	0	0020
0.2386714	0.1206977E-01	0	0	0011
13.28519	2.453805	0	0	0002
-0.3176306E-01	0.9418167E-03	0	0	3000
0.1598218	0.3223752E-01	0	0	2100
1.083413	-0.9744992E-01	0	0	1200
0.6192792	-1.809375	0	0	0300
0	0	0.4798076E-02	0.2628223E-03	2010
0	0	0.1013059	0.9227448E-02	1110
0	0	2.193448	0.1482136E-02	0210
0	0	0.5991117	0.5041080E-01	2001
0	0	21.11926	-0.4081118E-01	1101
0	0	2.237190	0.3706776	0201
-0.2023107E-02	-0.4458563E-03	0	0	1020
-0.2897106E-02	-0.8739436E-02	0	0	0120
-0.1355373	-0.6453340E-01	0	0	1011
-2.883894	-0.1806350E-01	0	0	0111
-10.94397	0.9968734E-01	0	0	1002
26.07000	-0.2402591	0	0	0102
0	0	0.8583696E-03	0.3180518E-05	0030
0	0	-0.1950716E-01	-0.9494305E-03	0021
0	0	-1.513093	0.1950715E-01	0012
0	0	-8.077975	1.367970	0003
0.1285578E-01	-0.1272996E-03	0	0	4000
....	....	....	....	....

TABLE 2. The beginning part of the Taylor transfer map of a 30 GeV neutrino factory ring with kinematic correction.

\multicolumn{5}{c}{Expansion coefficients of x,a,y,b depending on the exponents of xayb}
(x,	(a,	(y,	(b,	xayb
-0.1936744	0.1416905	0	0	1000
-6.792904	-0.1936744	0	0	0100
0	0	0.1961760	-0.2456274E-01	0010
0	0	39.14526	0.1961760	0001
0.9450713E-01	0.3330788E-02	0	0	2000
0.1377180	-0.5612056E-01	0	0	1100
-0.5801038	-0.6286416	0	0	0200
0	0	0.1229122	-0.3076403E-03	1010
0	0	0.6994628	0.2591721E-02	0110
0	0	-0.4902813	-0.1229122	1001
0	0	4.130383	-0.6994626	0101
-0.8336143E-02	-0.1539706E-02	0	0	0020
0.2386714	0.1206977E-01	0	0	0011
13.28519	2.453805	0	0	0002
11.55775	-0.1950424	0	0	3000
-81.95874	4.704898	0	0	2100
392.7585	-40.95482	0	0	1200
-1667.934	610.1026	0	0	0300
0	0	-0.7719322	-0.4832862E-01	2010
0	0	14.24336	0.4605147	1110
0	0	-417.3649	-2.944554	0210
0	0	-6.277132	-8.007910	2001
0	0	354.6553	34.10640	1101
0	0	-782.7162	-69.64528	0201
0.2344129	-0.2585728E-01	0	0	1020
-3.593587	1.490467	0	0	0120
26.50950	-1.496885	0	0	1011
-116.8393	6.467636	0	0	0111
2245.991	-29.01786	0	0	1002
-4789.016	244.2371	0	0	0102
0	0	-1.130680	-0.7756062E-02	0030
0	0	-6.403961	-0.4518258	0021
0	0	-430.3112	-29.87579	0012
0	0	-1116.550	-1705.361	0003
-1.322453	-0.9516536E-01	0	0	4000
....	....	....	....	....

TABLE 3. Amplitude dependent tune shifts.

Kinematic correction		off	on		Order	Exponents	
Fringe field effects		off	off	on		x	y
30 GeV neutrino factory ring	x motion	0.718979	0.718979	0.718979	0	0	0
		-0.0228772	12.0363	469.397	2	2	0
		-0.0305743	5.32077	741.941	2	0	2
	y motion	0.218573	0.218573	0.218573	0	0	0
		-0.0305743	5.32077	741.941	2	2	0
		-0.0071152	5.92760	472.150	2	0	2
30 GeV Higgs factory ring	x motion	0.864288	0.864288	0.864288	0	0	0
		-0.837791	35.8432	855.939	2	2	0
		-1.76978	42.0333	2226.02	2	0	2
	y motion	0.665356	0.665356	0.665356	0	0	0
		-1.76978	42.0333	2226.02	2	2	0
		-0.225010	422.116	4180.00	2	0	2

The neutrino factory ring consists of bending elements and quadrupoles, and the Higgs factory ring furthermore has sextupoles. Tables 1 and 2 show the beginning part of high order Taylor transfer maps of the neutrino factory ring. Table 1 shows the case without kinematic correction, and Table 2 the case with kinematic correction, which is obtained as the default of COSY INFINITY. As expected from the discussion in the previous section, the effects start to appear from the second order in x and a, and from the third order in y and b.

The size of the dynamic aperture was estimated by tracking particles for 1000 turns with transfer maps up to 7th order. In case of no kinematic correction, particles with x up to 1000mm and a up to 150mrad are stable in the neutrino factory ring. However when the kinematic correction is included, the particles with x up to 100mm and a up to 15mrad are stable. Figure 1 illustrates how the particles are preserved in case the kinematic correction is on in the right picture in comparison to the case without correction in the left picture, where the particles look very stable.

Based on normal form methods, amplitude dependent tune shifts were computed using COSY INFINITY [9,10]. Table 3 summarizes the results with and without kinematic correction; in addition, the effects of fringing fields are listed for comparison [2], where the full gap size of the magnets is assumed to be 10cm and the standard Enge function is used to describe the fall-off of the field at each fringing region [4].

The pictures in Figure 2 show tune footprints of the neutrino factory ring for particles up to 6mm in radius in both x and y directions. The horizontal axis shows the x tune and the vertical shows the y tune. The upper picture shows the tune footprint without kinematic correction. The lower picture shows the two footprints with and without correction in the same scale; the small dark spot at the lower left is the one without correction.

Altogether, for the design of the neutrino factory ring used in the study, the

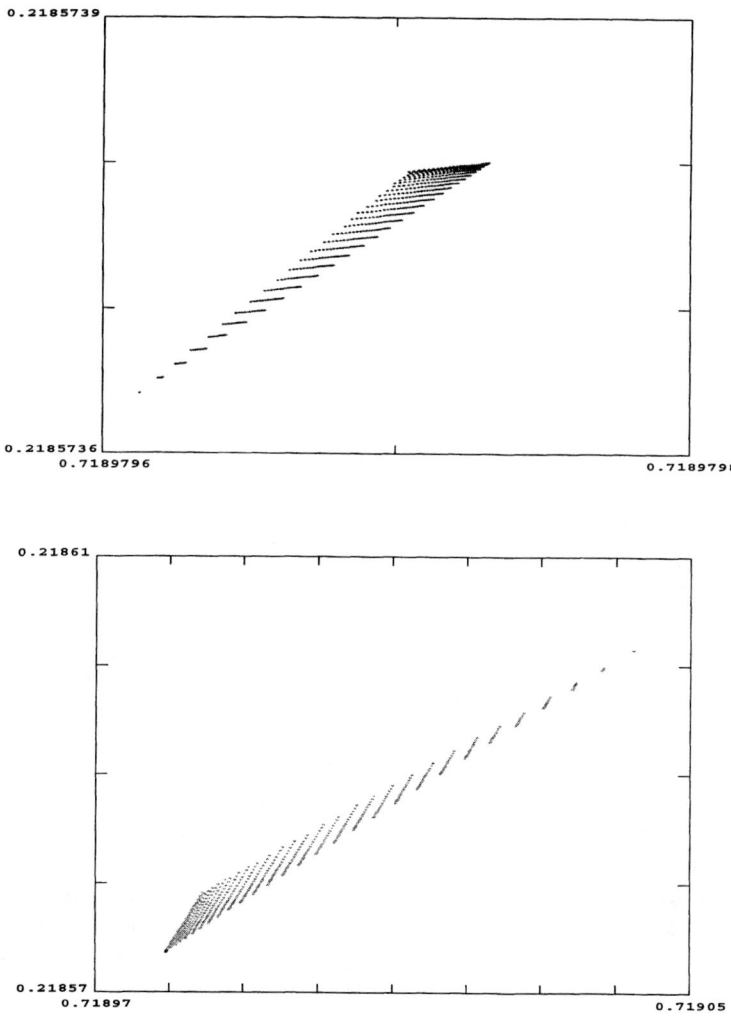

FIGURE 2. Tune footprints of a 30 GeV neutrino factory ring without (upper) and with kinematic correction. The lower picture shows both in the same scale, where the one without correction is the tiny dark spot at the lower left.

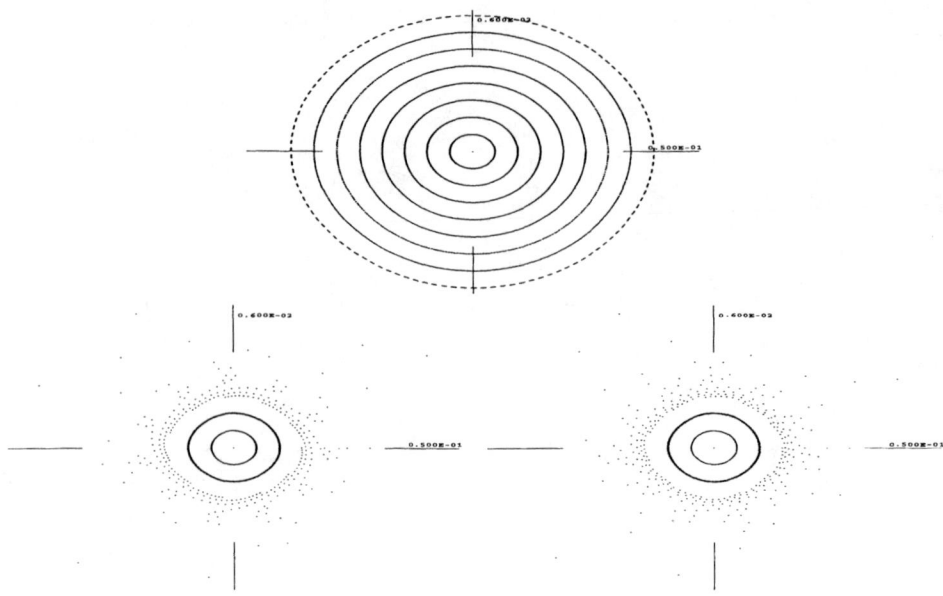

FIGURE 3. Tracking pictures for 1000 turns in a 30GeV neutrino factory ring without (upper) and with (lower) fringe field effects, and further without (left) and with (right) kinematic correction in the same scale of 50mm×6mrad. With fringe field effects, only those particles up to 10mm×1.5mrad survive.

dynamic aperture decreased by a factor of around 100 in x-a, and there is a corresponding large increase in the tune footprint area. To illustrate the other nonlinear effects, pictures in Figure 3 give a glimpse to the consequences of fringe field effects, which are reported in more detail in [2]. The setting of the system is the same to the previous tracking pictures, but with a smaller phase space region. The upper picture shows the situation with kinematic correction, and the particles are stable in this region without the fringe field effects. Now, the lower pictures show the cases with fringe field effects (left), and fringe field effects plus kinematic correction (right). With fringe field effects, the particles survive up to about 10mm×1.5mrad in x-a, which represents a further decrease of the phase space by a factor of 100, and at this scale, the effects of the kinematic correction are relatively suppressed.

The 30 GeV Higgs factory ring was studied in the similar way. Roughly speaking, the amount of effects by those nonlinear corrections is similar to that of the neutrino factory ring.

The tracking pictures in Figure 4 and 5 can be used to estimate the size of the dynamic aperture. Particles were tracked for 1000 turns with transfer maps up to 7th order, and those with x up to 40mm and a up to 300mrad are stable in

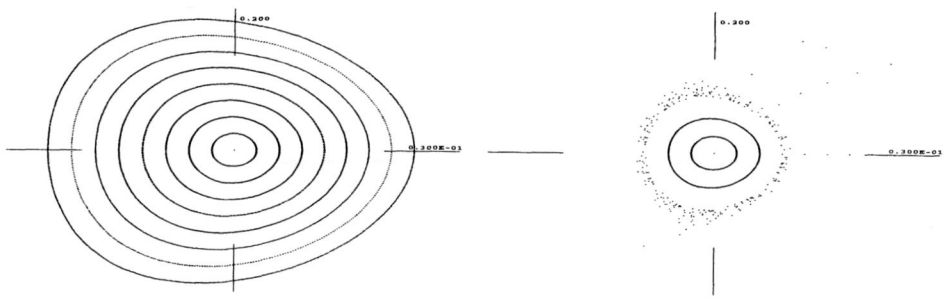

FIGURE 4. Tracking pictures for 1000 turns in a 30 GeV Higgs factory ring without (left) and with (right) kinematic correction in the same scale of 30mm×200mrad. The fringe field effects are not included. With correction, only those particles up to 6mm×50mrad are stable.

the Higgs factory ring when neither kinematic correction nor fringe field effects are included. By turning on only the kinematic correction, the stable size decreased to x up to 6mm and a up to 50mrad, which is a decrease by a factor of 40 in x-a. By including the fringe field effects, a further decrease by a factor of 100 in x-a is observed. However, under the influence of the fringe field effects, the difference between with and without kinematic correction is suppressed.

The amplitude dependent tune shifts are listed in Table 3. Figure 6 shows the tune footprints for particles up to 0.5mm in radius in both x and y directions. The pictures show the comparison with and without kinematic correction, but the fringe field effects are not included. The upper picture shows the tune footprint without the kinematic correction, and the lower picture shows the one with the correction. In the lower picture, the tiny horizontal bar at the lower left is the region of the footprint without the correction for comparison.

FIGURE 5. Tracking pictures for 1000 turns in a 30 GeV Higgs factory ring under the influence of the fringe field effects in the scale of 3mm×20mrad. Particles up to 0.6mm×5mrad are stable without (left) and with (right) kinematic correction.

Altogether a decrease of the stable region by up to a factor of 100 and an increase of the tune footprint area by up to a factor of 10^5 are observed by the kinematic correction when there is no influence of the fringe field effects.

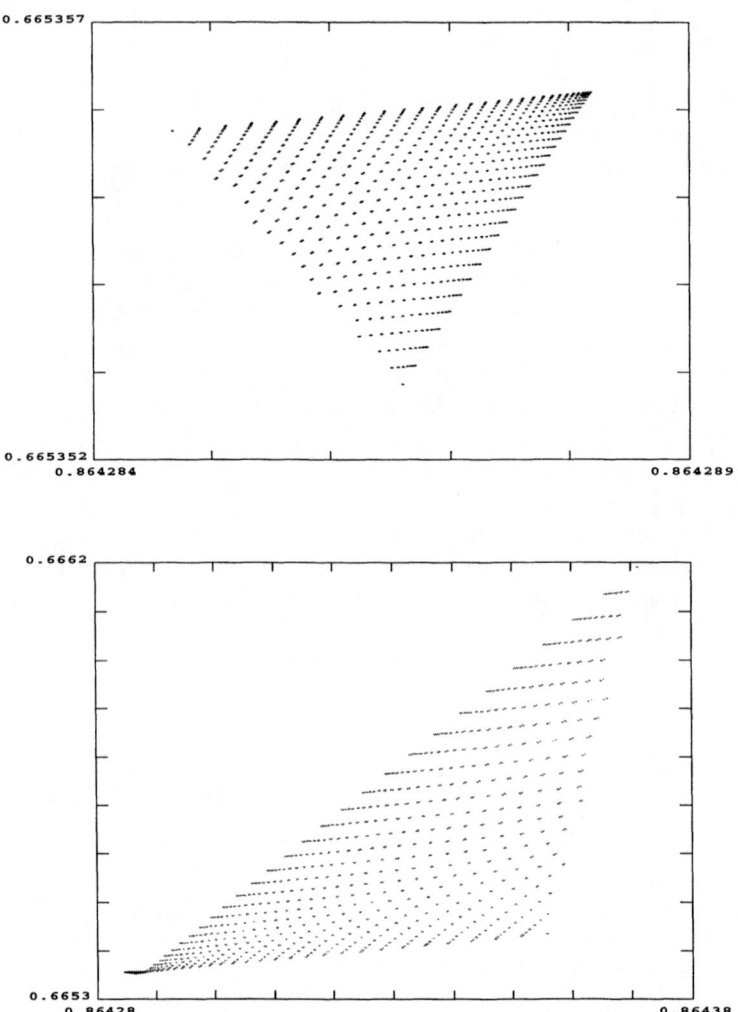

FIGURE 6. Tune footprints of a 30 GeV Higgs factory ring without (upper) and with kinematic correction. The lower picture shows both footprints in the same scale, where the one without correction is the tiny bar at the lower left.

ACKNOWLEDGMENTS

This study was motivated through a discussion with J. Gallardo. We thank C. Johnstone and W. Wan for providing the storage ring designs for the study. The work was supported by the US Department of Energy and the Alfred P. Sloan Foundation.

REFERENCES

1. W. Wan, C. Johnstone, J. Holt, M. Berz, K. Makino, and M. Lindemann. The influence of fringe fields on particle dynamics in the Large Hadron Collider. *Nuclear Instruments and Methods A*, in print.
2. M. Berz, K. Makino, and B. Erdélyi. Fringe field effects in muon rings. In these proceedings.
3. K. Makino. *Rigorous Analysis of Nonlinear Motion in Particle Accelerators*. PhD thesis, Michigan State University, East Lansing, Michigan, USA, 1998. Also MSUCL-1093.
4. M. Berz. COSY INFINITY Version 8 reference manual. Technical Report MSUCL-1088, National Superconducting Cyclotron Laboratory, Michigan State University, East Lansing, MI 48824, 1997. See also http://cosy.nscl.msu.edu.
5. K. Makino and M. Berz. COSY INFINITY version 8. *Nuclear Instruments and Methods*, A427:338, 1999.
6. K. Makino and M. Berz. COSY INFINITY Version 7. In *Fourth Computational Accelerator Physics Conference*, volume 391, page 253. AIP Conference Proceedings, 1996.
7. M. Berz. Computational aspects of design and simulation: COSY INFINITY. *Nuclear Instruments and Methods*, A298:473, 1990.
8. M. Berz, B. Erdélyi, W. Wan and K. Y. Ng. Differential algebraic determination of high-order off-energy closed orbits, chromaticities, and momentum compactions. *Nuclear Instruments and Methods*, A427:310, 1999.
9. M. Berz. *High-Order Computation and Normal Form Analysis of Repetitive Systems, in: M. Month (Ed), Physics of Particle Accelerators*, volume AIP 249, page 456. American Institute of Physics, 1991.
10. M. Berz. Differential algebraic formulation of normal form theory. In *M. Berz, S. Martin and K. Ziegler (Eds.), Proc. Nonlinear Effects in Accelerators*, page 77. IOP Publishing, 1992.

Very Low-Energy Cooling Possibilities Towards Muon Colliders and Neutrino Factory

By Kanetada Nagamine

Meson Science Laboratory, Institute of Materials Structure Science,
High Energy Accelerator Research Organization (KEK-MSL),
Tsukuba, Ibaraki, Japan

and

Muon Science Laboratory, Institute of Physical and Chemical Research (RIKEN-Muon),
Wakoh, Saitama, Japan

Abstract: Based upon recent success in ultra-slow positive muon production by employing laser resonant-ionization of the thermal muonium, a new scheme of muon colliders and a new production method of an intense and bright high-energy neutrino ($\bar{\nu}_\mu$, ν_e) beam are proposed. There, by using these muons as ion sources, one can accelerate and store to produce high-energy/high-intensity muon beams and high-quality neutrino beams.

§1. Introduction: Zero-Energy μ^+ Cooling

Recently, the concept of an ultra-slow μ^+ has been proposed [1, 2] and realized at KEK-MSL [3,4]. In addition, the possibility of using this ultra-slow μ^+ for an "ion-source" for further acceleration towards $\mu^+\mu^-$ colliders has been emphasized [5,6].

Let us explain the new concept of an ultra-slow μ^+. As described in Fig. 1, there are two processes to realize an ultra-slow μ^+ source: (1) thermal Mu production in vacuum by stopping the μ^+ at a rear-side of a selected metallic material, like hot tungsten, followed by μ^+ diffusion towards the surface of that foil as well as Mu evaporation from the surface; (2) efficient muonium ionization by e.g. laser resonant ionization via 1s → 2p → unbound excitation utilizing intense pulsed lasers in combination with a pulsed muon beam. In order to realize this ultra-slow μ^+ production idea, a new laboratory space with a dedicated pulsed (50 ns width and 20

Hz repetition rate) proton beam line from the 500 MeV booster synchrotron was constructed at KEK-MSL. Hot tungsten (W) was adopted as the thermal-Mu-producing target. All of the target area as well as the following ion optics of slow μ^+ transport were maintained at a pressure below 10^{-8} Torr under the conditions of proton-beam delivery and target heating. The actual target comprised 2 mm thick BN (boron nitride) for efficient π^+ production and 50 nm thick W. As the most powerful ionization method of a thermal muonium, we have adopted

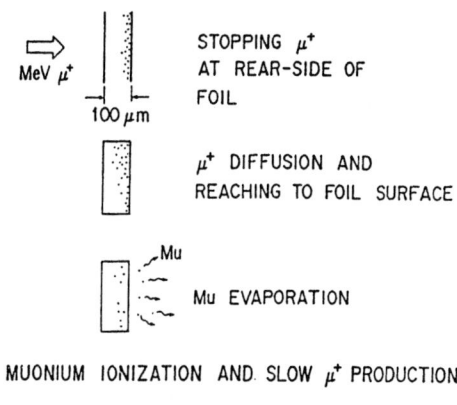

Fig. 1. Concept of ultra-slow μ^+ production by laser resonant-ionization of thermal muonium.

the method of laser resonant ionization by using pulsed lasers, namely, the single photon resonant transition of 1s → 2p (122 nm) followed by the photon-ionization transition 2p → unbound (<366 nm). For this purpose, a laser system was constructed for intense 122 nm VUV light with a 200 GHz frequency width (FWHM), which was matched to the Doppler broadening of the thermal Mu [4]. The basic structure of the ion-extraction optics for the ionized products of μ^+ comprised an SOA immersion lens for 9.2 kV acceleration as well as an ion transport by an electric bend and a magnetic bend with an axially focusing electric-quadrupole lens.

Recently, it has been noticed that, by combining with a large-solid angle MeV muon source, an intense ultra-slow μ^+ source with an intensity of 10^{12} μ^+/s and an emittance of better than 10^{-7} rad·m can be generated in the scheme shown in Fig. 2 [6]. There, some thick carbon target with a large solid-angle superconducting pion collector will be installed at the external beam line of medium energy, like a few GeV protons followed by the pion decay-section (super-super muon channel) coupled with the ultra-slow μ^+ generator mentioned above. Preliminary optics design work for a scheme of 90-degree muon-extraction a proton beam has been performed with a strong collaboration of K. Ishida [7]. As summarized in Table 1, by adopting a strong superconducting solenoid for pion collection and a reasonable length of the decay-solenoid installed at the existing strongest beam line of 0.8 GeV pulsed protons at ISIS-RAL, one can obtain 10^{10} μ^+/s. The momentum spread of the μ^+ obtained at the end of the decay solenoid is very large (nearly ±40%). However, it becomes possible to stop generated μ^+ inside a multi-layer of hot W, so that intense ultra-slow μ^+ beam

Table 1. Yield estimations for an advanced neutrino source

	Expected numbers (s^{-1})	Condition	Remarks
N_p	1.9×10^{15}	0.8 GeV × 300 µA	
$N_{\pi^+}^{tot}$	2.4×10^{13}	4 cm Carbon	$\sigma_{\pi^+}^{tot}$: 28 mb
N_{μ^+}	6.0×10^{15}	Pion Capture: 2.9 T, 20 cm bore × 1.5 m	Ref.(7)
		Muon Decay: 3.0 T, 25 cm bore × 10m	p_μ: 88(42) MeV
N_μ^{stop}	2.9×10^{11}	20 × 100 µm W	
$N_{th.Mu}$	1.2×10^{10}	$\varepsilon_{th.Mu}$: 0.04	2000 K hot W
$N_{u.s.\mu^+}$	1.0×10^{10}	ε_{ion}: 0.8	Laser Resonant Ionization
$N_{\mu^+}^{Acc.}$ (10 TeV)	10^{10}	ε_{cap}: 1.0 ε_{acc}: 1.0	
$N_{\bar{\nu}_\mu, \nu_e}$	0.5×10^{10}	Full conversion in Race-Track Strange Ring	$E_\nu \cong 3.3$ TeV, $\theta_\nu \cong 10^{-5}$

$N_{\pi^+}^{tot}$, total π^+ at production target; N_{π^+}, total μ^+ at the exist of decay solenoid; N_μ^{stop}, μ^+ stopping number at thermal Mu producing material; $N_{th.Mu}$, thermal Mu yield; $N_{u.s.\mu^+}$, ultra-slow μ^+ yield after thermal Mu ionization.

can be generated. Simulation calculations are in progress to confirm the characteristic features of the produced ultra-slow μ^+ after some optimizations.

§2. Application of Zero-Energy Cooling; Collider Option

Among many other important applications, we now consider how these slow μ^+ along with slow μ^- to be realized in the future can contribute to the concept of a $\mu^+\mu^-$ collider. We consider that the slow-muon production described here will be used as an ion source for further acceleration. The key-factors necessary to judge this possibility are the intensity and emittance. Let us consider the ultimate values of these parameters, and compare them to those for the ionization cooling method proposed for the conventional $\mu^+\mu^-$ collider.

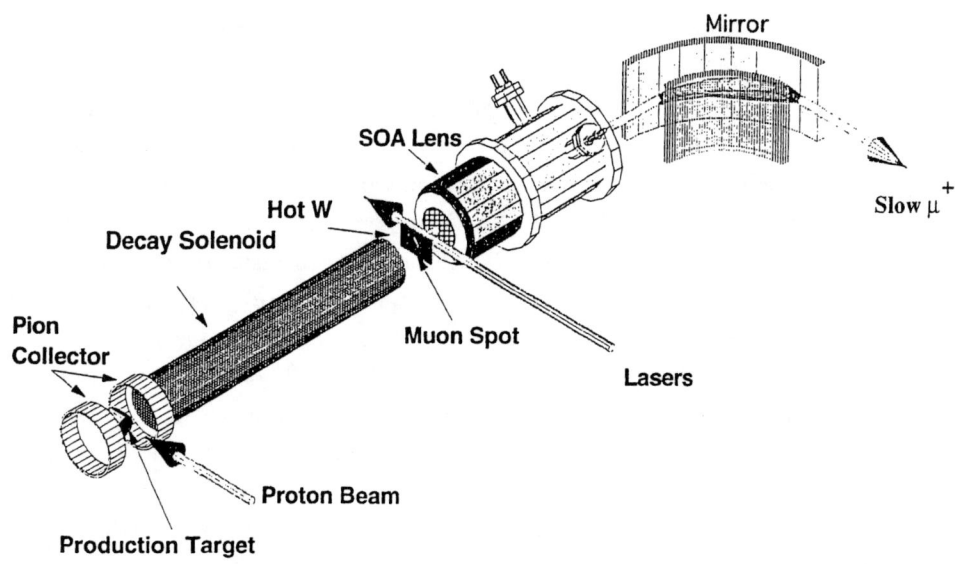

Fig. 2. Intense ultra-slow μ^+ production by employing the high-intensity muon production via a large acceptance pion collection.

As described in Table 1, at the end of μ^+ acceleration up to 10 TeV, an intensity of 10^{10} μ^+/s and a normalized emittance of 6×10^{-11} rad·m will be obtaiend. These values should be compared to 10^{11} and 10^{-8} in the ionization-cooling method.

The momentum spread of the μ^+ obtained at the end of decay solenoid is very large (~40%). In order to reduce the momentum spread of the produced μ^+, RF-rotation cooling by employing an induction-linac idea has been considered. By adopting 1000 cells with 50 kV voltage with a gradient of 1.5 MV/m, all to be inserted inside the decay solenoid, a model-calculation shows that one can reduce the momentum spread down to 24% [6]. Once the momentum bite after RF rotation cooling becomes small, it becomes possible to stop the generated μ^+ inside a few-layers of hot W. Simulation calculations are in progress to confirm the characteristic features of the produced ultra-slow μ^+ after some optimizations.

As for slow μ^-, by adopting a multi-layer of $H_2(T_2)$-DT target after establishing an intense MeV μ^- channel, we can expect 10^9 slow μ^+/s with an emittance of 10^{-9} rad·m at 1 TeV by using the concept of slow μ^- production via muon-catalyzed fusion [8].

§3. Application of Zero-Energy Cooling; Neutrino Source Option

Neutrinos (ν_e, $\bar{\nu}_e$, ν_μ, $\bar{\nu}_\mu$, ν_τ, $\bar{\nu}_\tau$), if generated in the form of an intense, and high-quality energetic-beam, can be useful not only for particle-physics experiments, like neutrino-oscillation studies, but also for probing the inside-nature of a really gigantic substance, such as the earth itself! Some proposals [9] exist to use neutrino to explore the inside-nature of the Earth. There, an intense neutrino beam is proposed to be generated by using an intense, high-energy (multi-TeV) proton accelerator. However, because of the limited quality of the neutrino beam, the idea should be considered to be unrealistic, or at least very difficult to be realized.

By using high-energy proton accelerators which produce energetic pions (π^+, π^-), one can expect two-types of methods for neutrino generation: 1) pion-decay neutrinos through $\pi^+ \to \mu^+ + \nu_\mu$ or $\pi^- \to \mu^- + \bar{\nu}_\mu$; 2) muon-decay neutrinos through $\mu^+ \to e^+ + \nu_e + \bar{\nu}_\mu$ or $\mu^- \to e^- + \bar{\nu}_e + \nu_\mu$. So far, most of the accelerator-based neutrino experiments, including the inner-earth studies described above, are considering to use an energetic neutrino beam obtained through the decay of energetic pions or muons based upon the decay-kinematics law. The quality of the neutrino beam can be drastically improved once one can improve the beam quality of the beam of the parent particle, pions or muons.

The high-intensity ultra-show μ^+ described here can be considered as an ion source for further acceleration, by installing on RFQ, DTL preaccelerator followed by e.g. a superconducting linac. Before the realization of $\mu^+\mu^-$ colliders, one of the most important applications of an intense low-energy μ^+ source would be for an advanced source of neutrinos ($\bar{\nu}_\mu$, ν_e) with accelerated μ^+. There, after prompt acceleration up to more than 100 MeV, with the help of special relativity, the life-time becomes longer, so that there is a small loss of the muons during the process of the further acceleration. Then, by installing a relevant decay-section for the accelerated μ^+, intense, and high-quality neutrinos could be produced via $\mu^+ \to \bar{\nu}_\mu + \nu_e + e^+$, employing the scheme as shown in Fig. 3.

The important key factors for the realization of this advanced neutrino beam can be summarized as follows:

i) Quality of accelerated muons

From the ultra-slow μ^+ of a laser resonant-ionization of thermal Mu from a hot metal surface, coupled with the super-super muon channel, one can expect more than a 10^{10}

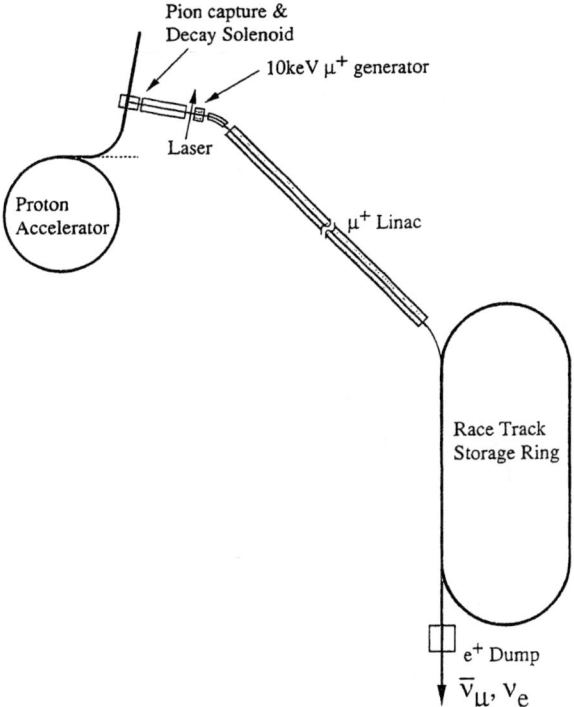

Fig. 3. Proposed scheme of the advanced generation of neutrino beam based upon a decay of the muons accelerated from the ultra-slow μ^* ion-source.

μ^+ source with an extremely small phase space (0.2 eV × (cm)^2). The limiting factor due to the capture efficiency to the adjacent linear accelerator is small.

ii) Muon-decay section

Since the decay length (L_μ) of the accelerated muons is fairly long, like $L_\mu(m) \approx 4.7 p_\mu$ (MeV/c), we should have an ingenious way to make the muon decay-section with a long distance and good confinement. The idea of a race-track type muon storage ring, as shown in Fig. 3, is one of the attractive ideas.

iii) Quality of the produced neutrino beam

A neutrino beam produced via the decay of accelerated muons should have properties due to the kinematics of three-body muon-decay. The energy spread as well as the decay cone are the limiting factors. Regarding the decay cone, the opening half-angle, θ_μ, becomes smaller at a higher muon energy (E_μ) in the relation m_μ/E_μ.

All of these properties for the case of the proposed system optimized for a future 0.8 GeV × 300μA proton synchrotron are summarized in Table 1. With the help of a future intense pulsed-proton source (with more than MW power) to also be available aa a spallation neutron source, a level of 10^{12} μ$^+$/s can be obtained. A rough estimate for other proton accelerators can be obtained by scaling to the proton beam power.

The arrival timing of the neutrino and the quality of the neutrino beam can be monitored via a charged-particle appearance reaction, like $\bar{\nu}_\mu \to \mu^+$. Such a reaction cross-section becomes larger at higher energy, so that the detection becomes easier for high-energy neutrinos.

There are several important applications for such an advanced neutrino beam. Some distinguished examples are neutrino oscillation, neutrino application to an exploration of the inner-structure of Earth, and neutrino tele-communications. Among them, the importance of exploring the inside nature of Earth is obvious. As predicted by various papers (e.g. [9]), the use of the neutrino is quite promising to explore the details of the inner-structure of Earth. The neutrino intensity transmitted through Earth for a fixed neutrino energy ($\bar{E}_\nu$) does depend upon the density distribution of the inner-earth structure. With the help of a variable value of the average neutrino energy ($\bar{E}_\nu \cong E_\mu/3$) by changing the muon energy (E_μ), more involved information can be obtained concerning the density as well as element distribution of the inner part of Earth.

By employing the advanced neutrino beam proposed here, a time-dependent change of the inner-Earth structure might be monitored, providing very important information for the earthquake prediction.

Helpful discussions with Drs K. Ishida and Y. Miyake are greatly acknowledged.

REFERENCES

1. K. Nagamine and A. P. Mills, Jr.: *Los Alamos Report* **LA10714-C,** 21-23 (1986).
2. K. Nagamine: *Atomic Phys.* **10**, 225-242 (1987).
3. K. Nagamine, Y. Miyake, K. Shimomura, P. Birrer, J. P. Marangos, M. Iwasaki, P. Strasser and T. Kuga: *Phys. Rev. Lett.* **74**, 4811-4814 (1995).
4. Y. Miyake, J. P. Marangos, K. Shimomura, P. Birrer, T. Kuga and K. Nagamine: *Nucl. Instruments* **B95**, 265-275 (1995).

5. K. Nagamine: *Nucl. Phys. B* (Proc. Suppl.) **51A**, 115-124 (1996).
6. K. Nagamine,: AIP Conf. Proceedings **441**, 295-309 (1998).
7. K. Ishida and K. Nagamine: KEK Proceedings 98-5 II & JHF-98-2 II (1998) p.12-18.
8. K. Nagamine: Proc. Japan Academy **65**, 225-228 (1989).
9. A.DeRújula, S. L. Glashow, R. R. Wilson and G. Charpak: *Physics Reports* **6,** 341-396 (1983).

The limitations of muon beam density at ionization cooling

V.V. Parkhomchuk

INP, Novosibirsk-90, Russia

I INTRODUCTION

The muon colliders with extremely high luminosity require high density muon beam at interaction points. The same parameters of this collider show at the table 1.

luminosity, L $[cm^{-2} \cdot s^{-1}]$	10^{36}
center of mass energy (TeV)	10
number muon at bunch N_μ	$3 \cdot 10^{12}$
6-dim. norm. emit $\epsilon_{6N} cm^3$	$8.5 \cdot 10^{-5}$
x,y norm. emit $\pi mm \cdot mrad$	38
long. emittance $eV \cdot s$	0.021
fract. momentum spread	$6 \cdot 10^{-4}$
long. norm. emitt. $\gamma * \beta * \delta p/p * \sigma_b (cm)$	6.2
spot size, $\sigma_{x,y}(\mu m)$	1.3
bunch length (mm)	2.2
$\gamma = E_\mu/m_\mu$	$4.73 \cdot 10^4$

The ionization cooling can be use only on the relatively small beam energy when $\gamma \cdot \beta \sim 1$. The strong scattering muon on the nuclear of the target generate the equilibrium angle at ionization cooling at order of magnitude $\theta \sim \sqrt{m_e/m_\mu} = 0.07$. For achieve emittanse $\epsilon_x = \gamma\beta\beta_\perp * \theta^2 = 3.810^{-3} cm$ ionization cooling chanel required the extremely strong focusing system with $\beta_\perp = 0.87 cm$. The lithium lenses with density current j=3.3 MA/cm^2 there is one of the discussed solution of this chanel $\beta_\perp = 1.6\sqrt{\gamma * \beta/j(MA/cm^2)}$ (cm). For so stronger focusing chanel we will have the muon beam size $\sigma_\perp = \sqrt{\epsilon_x/\gamma/\beta * \beta_\perp} = 0.057 cm$. At longitudinal direction scattering not so strong $\delta p/p \sim 0.03$ and longitudinal focusing can be made $\beta_\parallel = 6888$ cm and longitudinal length at cooling $\sigma_\parallel = \sqrt{\epsilon_\parallel * /\gamma/\beta * \beta_\parallel} = 206 cm$.

If we will redistribute the emittance from perpendicular at longitudinal direction condition on 6 dimensional emittanse give us limitation for the focusing system on cooling section:

$$(\gamma * \beta)^3 \theta_\perp^4 \cdot \theta_\parallel^2 \cdot \beta_\perp^2 \cdot \beta_\parallel = 8.5 \cdot 10^{-5} \tag{1}$$

that give condition:

$$\beta_\perp^2 * \beta_\parallel = 4000 cm^3 \tag{2}$$

At this transformation the density of muon beam:

$$n_\mu = \frac{N_\mu}{(2\pi)^{3/2} \sigma_\perp^2 \sigma_\parallel} = 2.8 10^{11} 1/cm^{-3} \tag{3}$$

will be constant and we should study ways to save so high density muon beam at processes of cooling.

II THE TRANSVERSE DEFOCUSING AT MUON BEAM

The forces of the space charge muon beam produce defocusing muon beam and equation of motion for single muon inside this beam we can write at form:

$$\frac{d^2 x}{ds^2} = \frac{2\pi r_\mu n_\mu}{\gamma^3 \beta^2} x = \frac{1}{\beta_d^2} x, \tag{4}$$

where $r_\mu = e^2/m_\mu c^2 \approx 1.4 10^{-15}$ cm. Defocusing length $\beta_d = 7.4$ cm present analog Laslet tune shift using for estimation the space charge effects at the storage ring. It seems reasonably to have the external focusing length $\beta_\perp < 0.1 \beta_d$. But for usual storage ring this interaction more or less smooth continue along the beam orbit. At the ionization cooling we should study question of compensation of this strong defocusing forces at moment come in the ionization target. The strong error for the motion of different muons near axis and far from axis produce dilution the beam emittance and effective heating particles. Calculation of this dilution should be made for real geometry the ionization cooling system. The muon beam will generate at the lithium target currents that not dissipated just after go out muon beam. This persistent currents connected with resistivity of the lithium target. For estimation we can take the decay of the current inside beam volume at form L/R model:

$$\frac{dI_c}{dt} = -\frac{R}{L} I = -\frac{\rho c^2}{\pi \sigma_\perp^2 (0.5 + \ln(b/a))} I \tag{5}$$

where $\rho = 4.5 10^{-5} om * cm = 5 10^{-17} (CGS)$- resistivity of liquid Li. Each portion of muon path thru lithium after leaving lithium generate persistent current and amplitude of this current is:

$$\frac{\delta I_c}{Ib} \approx \frac{\rho c^2}{\pi \sigma_\perp^2 (0.5 + ln(b/a))} \tau \qquad (6)$$

The gain connected with this wake fields is equal to:

$$g = \frac{L_{Li}\sigma_\parallel}{\beta_d} \frac{\rho c}{\pi \beta \sigma_\perp^2 (1 + 2 \cdot ln(b/a))} \qquad (7)$$

And for final stage of cooling close to 2. At this equation intensity of muon beam presented of the defocusing length β_d. As easy to see we need more careful study of this effect with calculation of resistivity for high temperature plasma that generated after patching fraction of muon beam.

III HOW CHANGES IONIZATIONS COOLING FOR HIGH DENSITY MUON BEAM?

After path the single muon at liquid Lithium exited plasma oscillation with plasma frequency $\omega_e = c\sqrt{4\pi n_e r_e}$. The Debay thransere size of this region is equal to $D_\perp = v/\omega_e = \beta/\sqrt{4\pi n_e r_e}$. The amplitude fields inside this fluctuation is equal to eL_c/D_{perp}^2. The time that muon move is equal to $\Delta t = D_\perp/\beta/c/\theta$. The probability to be inside this chanel is $w = (D_\perp/\sigma_\perp)^2 N_\mu$. For electron density at Li $n_e = 1.310^{23} cm^{-3}$, $\omega_e = 210^{16}$ and $D_\perp = 1.410^{-6} cm$. After path thru target $N_\mu = 3 \cdot 10^{12}$ muons w=1800. It means that at the all crosections lithium lengths the exited zone overlap many times and the muons move behind the head of bunch will have additional diffusion. The muon go ahead of next muon exite plasma wave with the amplitude $E \sim e/D$ and time of interaction with this chanel $\tau \approx D/\theta/v$. The diffusion by this effect is equal to:

$$\frac{dp^2}{dt} = \frac{4\pi e^4 N_\mu L n_c}{v D \sigma_\perp^2 \theta^2} \qquad (8)$$

For estimation role of this diffusion we should made comparison with the scattering on the electrons at the lithium target:

$$\frac{dp^2}{dt} = \frac{4\pi e^4 n_e L n_c}{v} \qquad (9)$$

The gain at diffusion is equal to:

$$g = \frac{N_\mu}{n_e D \sigma_\perp^2 \theta^2} \approx \frac{3 \cdot 10^{12}}{1.3 \cdot 10^{23} 1.4 \cdot 10^{-6} 0.057^2 0.07^2} = 1 \qquad (10)$$

As easy to see this effect limits to achieve so high density at muon beam $N_\mu/\epsilon_\perp^2$ and should be taken at account for optimization cooling system.

HIGH ENERGY PHYSICS POTENTIAL AT MUON COLLIDERS

Z. Parsa[*]

*Brookhaven National Laboratory,
Physics Department 510 A, Upton, NY 11973-5000, USA*

Abstract.
In this paper, high energy physics possibilities and future colliders are discussed. The $\mu^+\mu^-$ collider and experiments with high intensity muon beams as the stepping phase towards building Higher Energy Muon Colliders (HEMC) are briefly reviewed and encouraged.

I INTRODUCTION

The high energy physics community is interested in the potential of colliders beyond the e^+e^- colliders, and LHC, finding Higgs bosons (thus understanding the origin of electroweak symmetry breaking), and supesymmetric (SUSY) particles, etc. In devising a strategy for technical accelerator possibilities we can look at the historical record as represented by the Livingston Plot in Figure 1. There, accelerator energy is plotted as function of calendar time for various accelerators. It shows where we come from, where we are and where we are going.

Experiments over the last two decades have convincingly shown that the strong, electromagnetic, and weak forces are all closely related and are simply described by the "Standard Model." In particular the anticipated sixth quark, top, has been found at Fermilab, and the predicted properties of the Z boson, one of the carriers of the weak force, have been tested to better than 0.1%. Although there is now little doubt that the Standard Model is a very good description of the basic forces responsible for all atomic and nuclear physics, there remain many open questions [1,2].

Perhaps the most urgent is to understand how masses of the elementary particles originate. To that end, new physics beyond what has been observed is required. The simplest possibility, the "Higgs Mechanism" predicts the existence of a fundamental Higgs Boson. Finding that elusive particle or whatever new physics is actually responsible for mass generation motivated the Superconducting Supercollider (SSC)

---

[*] Supported by US Department of Energy contract DE-AC02-98CH10886
[†] E-mail: parsa@bnl.gov

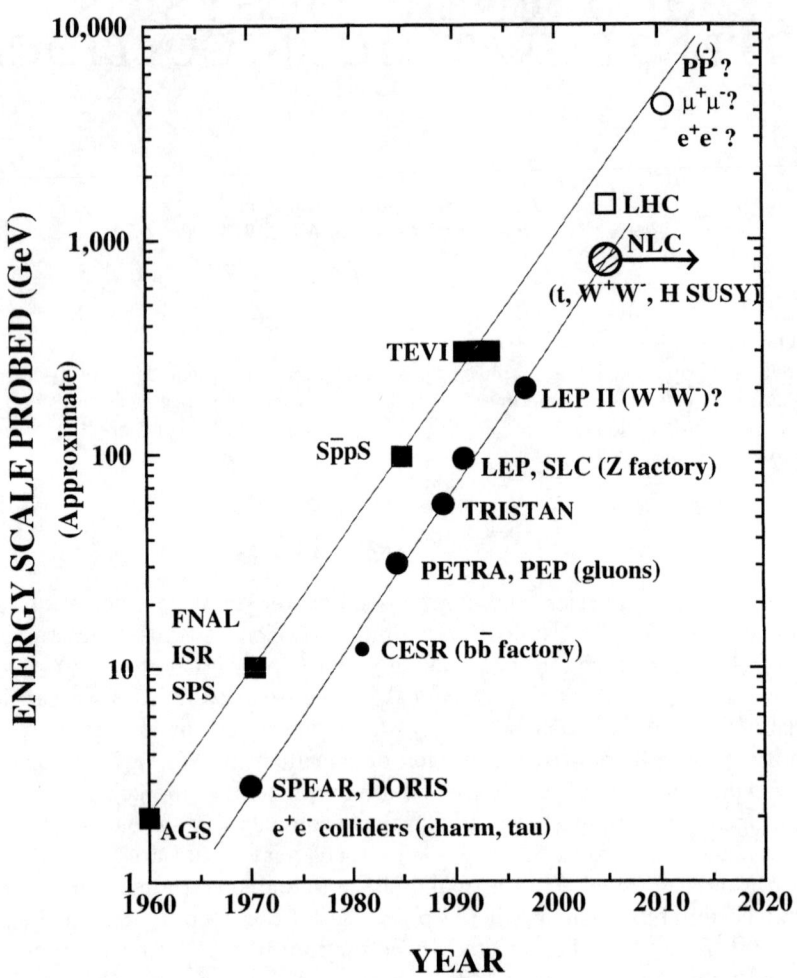

FIGURE 1. Livingston Plot - High Energy Frontiers (Accelerators) and their physics programs.

and remains the primary goal of the next generation of colliders. A number of other interesting and more elaborate models have been proposed, but there is as yet no direct experimental evidence supporting any of them.

Table 1 illustrates (a summary of the standard model), minimal spectrum of particles along with some of their basic properties. The fermions are grouped into three generations of spin 1/2 leptons and quarks which span an enormous mass range.

TABLE 1. Elementary Particles and Their Properties

Symbol	Spin	Charge	Color	Mass (GeV)	
ν_e	1/2	0	0	$< 4.5 \times 10^{-9}$	
e	1/2	-1	0	0.51×10^{-3}	First
u	1/2	2/3	3	5×10^{-3}	Generation
d	1/2	-1/3	3	9×10^{-3}	
ν_μ	1/2	0	0	$< .16 \times 10^{-3}$	
μ	1/2	-1	0	0.106	Second
c	1/2	2/3	3	1.35	Generation
s	1/2	-1/3	3	0.175	
ν_τ	1/2	0	0	$< 2.4 \times 10^{-3}$	
τ	1/2	-1	0	1.777	Third
t	1/2	2/3	3	174.3 ± 5.1	Generation
b	1/2	-1/3	3	4.5	
γ	1	0	0	0	
$W^\pm$	1	± 1	0	80.39 ± 0.04	
Z	1	0	0	91.187 ± 0.002	
g	1	0	8	0	
H	0	0	0	$106 < m_H < 235(800)$	

II PHYSICS QUESTIONS AT HEMC

Following is a partial list of questions generated in the HEMC physics working group, (some of which we briefly address in this presentation. For additional questions and discussions see [11,12]):

What is the new physics?

What are the theoretical motivations?

How do the LHC + LC + other experiments provide evidence/hints of this new physics? of the energy scale of this new physics?

What are the experimental signatures at a HEMC? What do you need to measure? Are there back grounds? ....

What is the minimal useful luminosity?

What is the most likely relevant scale/scales?

III FUTURE COLLIDERS – "NEW PHYSICS"

Particle beam colliders are the primary tools for performing high energy physics research. Collisions of high energy particles produce events in which much of the energy of the beams can be converted into the masses of new heavy particles not normally found in nature. By studying the production and decay of these new particles, the underlying structure of the universe and the laws that govern it are unveiled.

High energy accelerators take us to new domains where top, Higgs, and "New Physics" can be directly produced and studied. The LHC, scheduled for 2005 will take us to 14 TeV with very high luminosity $\simeq 10^{34}$ cm^{-2}s^{-1}. Besides finding the Higgs, it will be capable of uncovering supersymmetry, Z' bosons, technicolor or many other scenarios with "new physics" $\lesssim 1$ TeV. Beyond those facilities, new ideas and technologies are required. The Next Linear Collider (e^+e^-) offers an exciting viable possibility. The Recent, growing enthusiasm for a $\mu^+\mu^-$ collider with high energy $\gtrsim 3$ TeV and luminosity $> 10^{35}$ cm^{-2}s^{-1}, if feasible, would be a significant technological leap forward.

High energy physicists are anxiously waiting for the next dramatic experimental discovery. Fortunately, anticipated future collider facilities offer broad discovery potential. The Fermilab main injector upgrade will allow the $p\bar{p}$ Tevatron to operate at $\sqrt{s} \lesssim 2$ TeV and luminosity $\sim 2 \times 10^{32}$. Those improvements broaden the discovery potential while allowing precision measurements and searches for rare B and τ decays. The Higgs mass region of 110 ~ 130 GeV may be explored via $W^{\pm}H$ and ZH associated production if the $H \to b\bar{b}$ mode is resolvable, [2] - [8].

Asymmetric B factories provide new ways to explore CP violation. LEPII has achieved e^+e^- center-of-mass energy of about 202 GeV and will push its energy to $\sqrt{s} \simeq 204$ GeV or higher. If a standard model or SUSY Higgs with mass $\lesssim 110$ GeV exists, it should be found. Perhaps, they will also get a first glimpse of SUSY. Furthermore, the $W^{\pm}$ mass has been measured to $\leq \pm 45$ MeV at LEPII, and Fermilab, providing an interesting constraint on the Higgs mass via quantum loop relations.

In some longer term (~ 2005), the LHC pp collider with $\sqrt{s} = 14$ TeV should find the Higgs scalar or tell us it doesn't exist. If SUSY exists $\lesssim 1$ TeV, it will be discovered. Hopefully, completely unexpected revelations will also be made.

Beyond the LHC, various collider options are possible. The Next Linear Collider (NLC) would start e^+e^- collisions at $\sqrt{s} = 500$ GeV and be upgradeable to 1–1.5 TeV. It would have high luminosity $> 5 \times 10^{33}$ and polarization. The NLC also offers $\gamma\gamma$, e^-e^-, and $e^-\gamma$ collider options which expand its physics potential. There has been also some discussion of possible future e^+e^- colliders with $\sqrt{s} \simeq 5$ TeV, a major step, if achievable. The NLC will be a superb tool for studying the Higgs, SUSY, Technicolor etc., [9,10]. Other possibilities include a $\mu^+\mu^-$ collider and Very Large Hadron Collider (pp with $\sqrt{s} \simeq 100$ TeV or more) which are less advanced. The muon collider concept is very interesting, but require series studies and technology demonstrations. An effort at BNL will aim to produce very intense muon beams and use them to do physics (such as $\mu^- N \to e^- N$). Such hands on efforts combined with a vigorous R&D program could lead to the First Muon Collider (FMC). Various machine energies have been considered including 100 GeV, 500 GeV, 3 TeV etc., [14]- [19]. In the next section the concept of muon collider and parameter sets will be addressed. Recently, the concept of muon storage ring based Neutrino Source has generated considerable interest in the High Energy Physics community. Beside providing the first phase toward a muon collider, it would generate more intense and well collimated neutrino beams than currently available.

The Very Large Hadron Collider(VLHC) with $\sqrt{s} \simeq 100$ TeV and $\mathcal{L} \simeq 10^{35}$ looks technically feasible but is very expensive. Does a Very Large Hadron Collider with $\sqrt{s} \simeq 100$ TeV have viability? Our SSC experience suggests a prohibitive cost and difficult construction issues because of its size. However, new ideas about inexpensive magnets and tunnels and/or a new technology could offer hope for the needed significant reduction in cost.

IV MUON COLLIDER

Figure 2 shows a schematic of a high energy muon collider components [14,15]. A high intensity proton source is bunch compressed and focused on a heavy metal target. The pions generated are captured by a high field solenoid and transferred to solenoidal decay channel within a low frequency linac. The linac reduces, by phase rotation the momentum spread of the pions and of the muons into which they decay. Subsequently, the muons are cooled by a sequence of ionization cooling stages, and must be rapidly accelerated to avoid decay. This can be done in recirculating accelerators (as at CEBAF) or in fast pulsed synchrotrons. Muon collisions occur in a separate high field collider storage ring with a single very low beta insertion.

A muon collider with center of mass energy less than about 10 TeV can be circular and relative to NLC (a Next Linear Collider) of the same energy, it could be far smaller in size. For the same luminosity a muon collider can tolerate a far larger spot size than an electron linear collider since the muons make about 1000 crossings. Muon ($m_\mu/m_e = 207$) have the same advantage in energy reach as electron but has little beamstrahlung, thus very small energy spread is obtainable. In addition the direct coupling of lepton-lepton system to Higgs boson has a cross section proportional to the square of the lepton mass. thus the cross section for direct Higgs production from a $\mu^+\mu^-$ collider is about 40,000 times that from an e^+e^- collider system.

A large effort has been devoted to design and assessing the feasibility of building a high energy muon collider at a 4-3 TeV, .5-.4 TeV and .1 TeV [14,15]. Figure 2 illustrates concept of a 4 TeV muon collider complex. Machines with energies higher than $3 \sim 4$ TeV, have a significant beam current constraints from the neutrino radiation limits. Thus to reach the required high luminosities without unacceptable radiation hazards, a significant improvements in the muon emittance to the current base-line values are required. Although muon colliders remain a promising

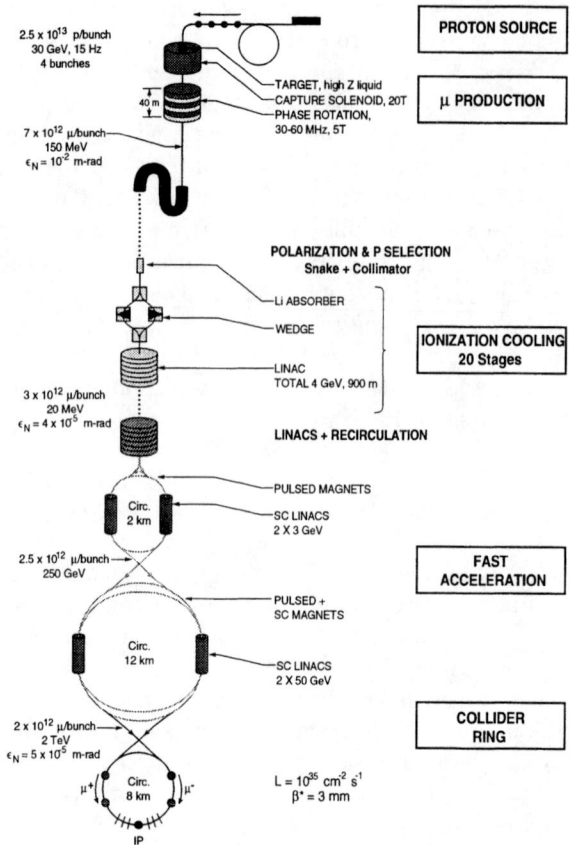

FIGURE 2. Schematic of a 4 TeV Muon Collider.

complement, or alternative to e^+e^- colliders, much work is still needed, including demonstration of μ production and cooling, detector, and neutrino radiation.

A 100 TeV muon collider provides a unique opportunity for exploring high energy physics, and mysteries of the elementary particles and their role in the universe. Table 2 [13] illustrates a 10 and 100 TeV parameter sets distributed by workshop organizers. These parameters are speculative and require technical extrapolation much beyond present limits and may not be possible for many decades to come? Unfortunately, the estimates of the size, cost and radition problems seems to prohibit muon colliders at such energies, without some drastic new developments? For energies below 3 TeV, and for a fixed muon current, the neutrino radiation falls proportional to the energy cubed, which may be less of a problem?

V PHYSICS POTENTIALS

Muon Colliders have unique physics and technical advantages and disadvantages as compared to e^+e^- and hadron colliders and are to be considered as complementary. For the same energy and integrated luminosity, anything that can be done at e^+e^-, should be possible at $\mu^+\mu^-$ collider, and more. E.g., possibilities for s-channel Higgs production, and a higher center-of mass energy with reduced backgrounds. Both of which are due to the large muon mass as compared to the electron mass. Higher energy may be crucial e.g., in improving signals for WW scattering, and the kinematical reach for pair production of SUSY particles.

The figure of merit in physics searches at an e^+e^- or $\mu^+\mu^-$ collider, is expressed by the QED point cross section for $e^+e^- \to \mu^+\mu^-$:

$$\sigma_{QED}(\sqrt{s}) = \left(\frac{100[fb]}{s[TeV^2]}\right) \times \left(\frac{\alpha(s)}{\alpha(M_z^2)}\right) \qquad (1)$$

As before [16], we will neglect the factor $\left(\frac{\alpha(s)}{\alpha(M_z^2)}\right)$ as it varies slowly with s. Also, if the integrated luminosity needed for studying the new physics signals is:

$$\left(\int Ldt\right)\sigma_{QED} \gtrsim 1000 \; events, \qquad (2)$$

then the $\mu^+\mu^-$ collider design should be able to deliver an integrated luminosity of

$$\left(\int Ldt\right) \gtrsim 10s[fb^{-1}]. \qquad (3)$$

If this is to be accumulated for one year of running time, the required luminosity estimate is

$$L[fb^{-1}] \gtrsim 10^{33}s[cm^{-2}][sec^{-1}]. \qquad (4)$$

E.g., The estimates of the luminosity requirements for the collider energies of interest (in parameter Table 2, distributed by organizers [13]) in this workshop are given below, (some of these energies and high luminosities may not be possible in practice?):

- For $\sqrt{s} \simeq 100[TeV]$, $L[fb^{-1}] \gtrsim 10^{37}[cm^{-2}][sec^{-1}]$
- For $\sqrt{s} \simeq 10[TeV]$, $L[fb^{-1}] \gtrsim 10^{35}[cm^{-2}][sec^{-1}]$
- For $\sqrt{s} \simeq 4[TeV]$, $L[fb^{-1}] \gtrsim 10^{34}[cm^{-2}][sec^{-1}]$
- For $\sqrt{s} \simeq 3[TeV]$, $L[fb^{-1}] \gtrsim 10^{33}[cm^{-2}][sec^{-1}]$
- For $\sqrt{s} \simeq 0.1[TeV]$, $L[fb^{-1}] \gtrsim 10^{31}[cm^{-2}][sec^{-1}]$.

VI PHYSICS WITH INTENSE MUON BEAMS

Using intense muon beams, forefront low energy research may be possible, including possibilities of: Precision measurements (e.g., muon decay $\tau_\mu \to G_F$, and Michel parameters); Neutrino source; Muon scattering; Muon capture $\mu^- p \to \nu_\mu n$. More interesting are the possibilities of Anomalus magnetic moment, Parity violation in muonic atoms (better than 1%); T violation; and ν_μ mass. Most interesting and compelling possibilities include the Muon number non-conservation - (rare or forbidden processes - Discovery would revolutionize physics.) such as

$\mu^+ \to e\gamma$;

$\mu^+ \to e^+ e^- e^+$;

$\mu^- N \to e^- N$.

Other processes could include

$\mu^- N \to e^+ N'$;

$\mu^- N \to \mu^+ N'$;

$\mu^+ e^- \to \mu^- e^+$;

$\mu^- e^- \to e^- e^-$;

$\mu^+ e^- \to e^+ e^-$.

VII DISCUSSION

The concept of muon collider once entirely speculative, now promises to extend the high energy frontier to an unprecedented domain, with center of mass energies of 3 TeV or beyond as its goal. Considerable effort has already gone into the conceptual design of muon colliders, but much more work and study is needed. The muon Collider Collaboration represents a dedicated effort to address those issues and bring to realm the possibility of a future muon collider complex.

A 100 TeV muon collider (energy of interest at this workshop) would provide a unique opportunity for exploring high energy physics, and mysteries of the elementary particles and their role in the universe. Unfortunately, the estimates of the size, cost and radiation problems seems to prohibit muon colliders at such energies, for many decades to come, without some drastic new technology developments?

Although a full high energy muon collider may take a considerable time to realize, intermediate steps in its direction are possible and could help facilitate the process. Employing an intense muon source to carry out forefront low energy research, such as the search for muon - number non - conservation, represents one interesting possibility. For example, the MECO proposal at BNL aims for 2×10^{-17} sensitivity in their search for coherent muon - electron conversion in the field of a nucleus.

TABLE 2. HEMC99 Parameter Sets For High Energy Muon Colliders

center of mass energy, E_{CoM}	0.1 to 3 TeV	10 TeV	100 TeV	100 TeV
additional description	MCC status report	evol. extrap.	evol. extrap.	ultracold beam
collider physics parameters:				
luminosity, $\mathcal{L}$ [10^{35} cm^{-2}.s^{-1}]	$8 \times 10^{-5} \to 0.5$	10	10	1000
$\int \mathcal{L} dt$ [fb^{-1}/year]	$0.08 \to 540$	10 000	10 000	1.0×10^6
No. of $\mu\mu \to ee$ events/det/year	$650 \to 10\,000$	8700	87	8700
No. of 100 GeV SM Higgs/year	$4000 \to 600\,000$	1.4×10^7	2.1×10^7	2.1×10^9
CoM energy spread, σ_E/E [10^{-3}]	$0.02 \to 1.1$	0.42	0.080	0.071
collider ring parameters:				
circumference, C [km]	$0.35 \to 6.0$	15	100	100
ave. bending B field [T]	$3.0 \to 5.2$	7.0	10.5	10.5
beam parameters:				
(μ^- or) μ^+/bunch, N_0[10^{12}]	$2.0 \to 4.0$	3.0	0.80	0.19
(μ^- or) μ^+ bunch rep. rate, f_b [Hz]	$15 \to 30$	27	7.9	65
6-dim. norm. emit., ϵ_{6N}[10^{-12}m^3]	$170 \to 170$	85	10	1.0×10^{-3}
ϵ_{6N}[10^{-4}m^3.MeV/c^3]	$2.0 \to 2.0$	1.0	0.12	1.2×10^{-5}
P.S. density, N_0/ϵ_{6N} [10^{22}m^{-3}]	$1.2 \to 2.4$	3.5	8.0	19 000
x,y emit. (unnorm.) [$\pi.\mu$m.mrad]	$3.5 \to 620$	0.81	0.018	4.4×10^{-4}
x,y normalized emit. [π.mm.mrad]	$50 \to 290$	38	8.7	0.21
long. emittance [10^{-3}eV.s]	$0.81 \to 24$	21	47	8.1
fract. mom. spread, δ [10^{-3}]	$0.030 \to 1.6$	0.60	0.113	0.100
relativistic γ factor, E_μ/m_μ	$473 \to 14\,200$	47 300	473 000	473 000
time to beam dump, $t_D[\gamma\tau_\mu]$	no dump	no dump	1.0	1.0
effective turns/bunch	$450 \to 780$	1040	1350	1350
ave. current [mA]	$17 \to 30$	55	4.0	7.8
beam power [MW]	$1.0 \to 29$	131	100	198
synch. rad. critical E [MeV]	$5 \times 10^{-7} \to 8 \times 10^{-4}$	0.012	1.75	1.75
synch. rad. E loss/turn [GeV]	$7 \times 10^{-9} \to 3 \times 10^{-4}$	0.017	25	25
synch. rad. power [MW]	$1 \times 10^{-7} \to 0.010$	0.91	99	195
beam + synch. power [MW]	$1.0 \to 29$	130	200	390
power density into magnet liner [kW/m]	$1.0 \to 1.7$	4.3	1.2	2.4
interaction point parameters:				
spot size, $\sigma_{x,y}$ [μm]	$3.3 \to 290$	1.3	0.21	0.015
bunch length, σ_z [mm]	$3.0 \to 140$	2.2	2.5	0.49
$\beta^*_{x,y}$ [mm]	$3.0 \to 140$	2.1	2.5	0.49
ang. divergence, σ_θ [mrad]	$1.1 \to 2.1$	0.63	0.086	0.030
ip compensation factor: $N_0/N_{0,eff.}$	1	1	1	10
beam-beam tune disruption, $\Delta\nu$	$0.015 \to 0.051$	0.085	0.100	0.100
pinch enhancement factor, H_B	$1.00 \to 1.01$	1.08	1.11	1.11
beamstrahlung frac. E loss/collision	negligible	6.8×10^{-8}	1.5×10^{-6}	9.0×10^{-7}
final focus lattice parameters:				
max. poletip field of quads., $B_{5\sigma}$ [T]	$6 \to 12$	15	20	20
max. full aper. of quad., $A_{\pm 5\sigma}$ [cm]	$14 \to 24$	22	19	6.6
quad. gradient, $2B_{5\sigma}/A_{\pm 5\sigma}$ [T/m]	$50 \to 90$	140	210	610
β_{max} [km]	$1.5 \to 150$	580	19 000	64 000
ff demag., $M \equiv \sqrt{\beta_{max}/\beta^*}$	$220 \to 7100$	17 000	89 000	360 000
chrom. quality factor, $Q \equiv M \cdot \delta$	$0.007 \to 11$	10	10	45
neutrino radiation parameters:				
collider reference depth, D[m]	$10 \to 300$	100	100	100
ave. rad. dose in plane [mSv/yr]	$3 \times 10^{-5} \to 0.03$	3.8	2.5	4.9
str. sec. len. for 10x ave. rad. [m]	$1.3 \to 2.2$	1.0	4.2	4.2
ν beam distance to surface [km]	$11 \to 62$	36	36	36
ν beam radius at surface [m]	$4.4 \to 24$	0.8	0.08	0.08

To reach that goal requires the production, capture and stopping of muon at an unprecedented $10^{11} \frac{\mu}{sec}$. If successful, such an effort would significantly advance the state of muon technology. More ambitious ideas for utilizing high intensity muon sources are also being explored. Indeed, if very high intensities, $\sim 10^{21} \frac{\nu}{year}$, are attained and nature has been kind in her neutrino mass and mixing parameters, one could envision a complete exploration of the 3 × 3 neutrino mixing matrix and even the detection of CP violation in the oscillation phenomena.

High intensity muon experiments, neutrino factories, and other intermediate steps toward the muon collider are extremely important. They will greatly expand our abilities and build confidence in the credibility of high energy muon colliders.

REFERENCES

1. Z. Parsa (Editor), *Future High Energy Colliders* AIP-Press CP 397 (1997).
2. W.J. Marciano, Keynote Address in *Proc. of Snowmass 1996* and *Proc. of Santa Barbara symposium on "Future High Energy Colliders" 1996* Ed. Z. Parsa, AIP CP 397, pp 11-25 (1997); *ibid*, private communication 1999.
3. C-N. Yang, Oskar Klein Memorial Lecture; *Phys. Today* **33**,42 (1980); A. Zee, "Fearful Symmetry", Macmillan 1986.
4. S. Weinberg, *Phys. Rev. Lett.* **19** (1967) 1264; A. Salam, in *Elementary Particle Theory*, ed N. Svartholm (Almquist & Wiksells, 1968) p. 367.
5. See A. Sirlin, *Comments on Nucl. and Part. Phys.*, **21**, 287 (1994).
6. D. Gross and F. Wilczek, *Phys. Rev. Lett.* **30**, 1323 (1973); H. D. Politzer, *Phys. Rev. Lett.* **30**, 1346 (1973).
7. W. Marciano, *Phys. Rev.* **D29**, 580 (1984).
8. A. Stange, W. Marciano, and S. Willenbrock, Phys. Rev. D**50**, 4491 (1994).
9. S. Kuhlman *et al.*, *Physics Goals of the Next Linear Collider*, BNL report 63158.
10. *ZDR Report for NLC*, SLAC, May 1996; updates (1999).
11. See Physics working group summary and other papers in the HEMC'99 proceedings.
12. Proceedings of HEMC'99 "Studies on Colliders and Collider Physics at the Highest Energies: Muon Colliders at 10 TeV and 100 TeV", 27 Sept.-1 Oct., 1999, Montauk, NY, USA.
13. B. King, Private comm.: provided a latex file for Table 2.
14. Muon Collider Collaboration; J. Norem private comm.: provided file for Fig. 2.
15. C.M. Ankenbrandt etal, *Status of muon collider research and development and future plans*, Phys. Rev. ST Accel. Beams **2**, 081001 (1999), and references therein; $\mu^+\mu^-$ *Collider, A Feasibility Study*, BNL-52503, FERMILAB-Conf-96/092, LBNL-38946 (July 1996), Proceedings of the 1996 DPF Summer Study on High Energy Physics, Snowmass'96; Collaboration web page http://www.cap.bnl.gov/mumu.
16. V. Barger, M. Berger, K. Fujii, J. Gunion, T. Han, C. Heusch, W. Hong, S. Oh, Z. Parsa, S. Rajpoot, R. Thun, W. Willis, BNL-61593 (1995).
17. Z. Parsa, New High Intensity Muon sources and Flavor Changing Neutral Currents, World scientific Publishing, pp 147-153 (1998).
18. Z. Parsa, Polarization and Luminosity requirements for the First Muon Collider, in AIP Conf. Proc. 472, pp. 251-259, (1998).
19. Z. Parsa, *Muon Storage Rings - Neutrino Factories*, in Proceedings of NNN'99, SUNY Stony Brook, NY. AIP-Press (2000); References therein.

Focusing and Acceleration of Bunched Beams

Z. Parsa[1]

Brookhaven National Laboratory, Physics Department 510 A, Upton, NY 11973
parsa@bnl.gov

V. Zadorozhny

Institute of Cybernetic, National Academy of Sciences of Ukraine
zvf@umex.istrada.net.ua

Abstract. A new approach to solving the kinetic equation for the beam distribution function, (very useful from the practical point of view), is discussed. In which we also obtain a complement to the Skrinsky's condition for the self-focused bunched beam. This problem belongs to the theory of nonlinear systems in which both regular and chaotic motion is possible. The kinetic approach, based on Vlasov-Poisson equations, are used to investigate the focusing and acceleration of bunched beam. Special attention is given to the studies of stability in a bunched beam by means of the two norm, which may be used to describe the motion of high - energy particles.

I INTRODUCTION

The localized physical structure such as bunched beam, halo beam, vortex, etc.; can be represented in the form of soliton-like solution of some evolution equations, by hydrodynamic or kinetic approach. In accordance with contemporary studies the soliton-like solution arises due to a sufficiently large influence of external forces and also because of nonlinear effect of self-influence. Objects, which are formed in this case, have unusual properties. The selfconsistent Vlasov equation [1] is one of the most frequently used equations for the time dependent description of many- particle systems. By varying the selfconsistent Vlasov equation we can find different analytical solutions by means of which, its phase portraits can be studied.The selfconsistent Vlasov equation is a partial differential equation of a first order and the time evolution of its solutions are entirely determined by the initial distribution function. This property will be used below to organize the localized structure. It may be noted that the localized structures in a finite region

---

[1] Supported by US Department of Energy contract DE-AC02-98CH10886

are integrable models. The construction of such solutions will have physical sense if and only if they will be stable (in some sense), and obviously, every asymptotically stable solution is a self - focused solution. As well known the notation stable may vary as defined in various works (papers), for clarity in the case under consideration we will appeal to the notation of Lyapunov (stable) and its modifications only. That is, we consider 1) the stable solution of Vlasov equation having relation with an initial value of a position and velocity of some particles. 2) The stable solution of Vlasov equation as stable initial-value problem for the initial perturbs of an initial value for the beam distribution function. 3) The stable solution of Vlasov-Poisson equation with respect to action of the corresponding magnetic field. The exact sense of these notations will be established below.

Always there exist an electro-magnetic field satisfying the Maxwell equation for a given arbitrary motion, i.e. any motion of the beam may be realized by an electro-magnetic field which satisfy the Maxwell equation.

In contemporary physical World there are many interesting and challenging problems that may be resolved by solutions of Vlasov equation. For example, the motion of the muon bunches into a collider.

In a muon collider complex, a rapid acceleration to the collider beam energy is needed to avoid expensive particle loss from the muon decays. It can be achieved, initially in a linear accelerator and later in recirculating linear accelerators, rapid-cycling synchrotron, or fixed - field - alternating - gradient accelerators. Positive and negative muon bunches are then injected in opposite directions into a collider storage ring and brought into collision at the interaction point. The bunches circulate and collide for many revolutions before decay has depleted the beam intensities to an interesting level. Useful luminosity can be delivered for about 1000 revolutions for a high-energy (e.g. $\sim 4\ TeV$) collider and about 450 revolutions for a low-energy (e.g. $\sim 100\ GeV$) one [4].

Since Vlasov equation has high degree of confidence for sufficiently small time of Beam motion, given the limited muon life time ($\tau_\mu = 2.2_{\mu s}$ at rest), the Vlasov equation well describes the muon beam dynamics. In this approach, design concepts for acceleration and strong focusing of the muon beam are being developed. Now, we can perform study of the nonlinear motion of the muon bunches by Vlasov-Maxwell system.

II PRELIMINARIES

We are going to consider the motion of many particles in the following standard form

$$\dot{x} = v, \quad \dot{p} = qE + \frac{q}{c}(v \cdot H) \tag{1}$$

here $p = mv$ is impulse of the particle, $v = (v_1, v_2, v_3)$ is the velocity, $x = (x_1, x_2, x_3)$ is the vector which characterizes the position of this particle in (Euclid) space $\{x_1, x_2, x_3\}$, i.e. (x_1, x_2, x_3) is the coordinate system, q is the charge of it, m

is the mass, E is the electric field and H denotes the magnetic field. The fields E and H are generated by a charged bunch and current, E_j, H_j (here $j = 1,...k$) are quantity of the beams. If there are some outward fields E_0 and H_0 then

$$E = \sum_{j=0}^{k} E_j, \quad H = \sum_{j=0}^{k} H_j.$$

Let Ω_{t_0} be any set of the phase space $E_0 = \{x, p\}$. If $\Omega_{t_0} \in E_0$ and $\{x_t(x_0), p_t(p_0)\}$ some solution of equation (1) for $(x_0, p_0) \in \Omega_{t_0}$ then obviously $\Omega_t = \{x_t(x_0), p_t(p_0); (x_0, p_0) \in \Omega_{t_0}\}$. Let $\rho(t, x, p)$ be some function such that it yields a simple equation

$$\int_{\Omega_{t_0}} \rho(t_0, x_0, p_0) dx_0 dp_0 = \int_{\Omega_t} \rho(t, x_t, p_t) dx_t dp_t$$

for any domain Ω_{t_0}, which we will call density (of the integral invariant). As is well known, if system (1) have some integral invariant with the density ρ then the function $\rho(t, x, v)$ satisfies the following equation

$$\partial_t \rho + \partial_x \rho v + \partial_v \rho [qE + \frac{q}{c}(v \cdot H)] + \rho \cdot div(v, qE + \frac{q}{c}(v \cdot H)) = 0$$

but is easy to see that, $div(v, qE + \frac{q}{c}(v \cdot H)) \equiv 0$, consequently

$$\partial_t \rho + v \partial_x \rho + [qE - \frac{q}{c}(v \cdot H)] \partial_v \rho = 0. \quad (2)$$

Further we denote ρ by f (for tradition), thus the following equation

$$\partial_t f + v \partial_x f + [qE - \frac{q}{c}(v \cdot H)] \partial_v f = 0 \quad (3)$$

is Vlasov equation which we will study.

Let us now find the solution $f(t, x, v)$ of Vlasov equation in the form

$$f = f_0(x, v) e^{-i\omega t} \quad (4)$$

Substituting (4) in (3), we get

$$v \partial_x f_0 + [qE - \frac{q}{c}(v \cdot H)] \cdot \partial_v f_0 = i\omega f_0 \quad (5)$$

and an operator form of the formula (5) yields

$$L f_0 = i\omega f_0 \quad (6)$$

for unknown f_0, where L is a linear operator in a given Hilbert space $L^2(\bar{\Omega}_{t_0})$ which is generated by an operator

$$L_0 \bullet = v \partial_x \bullet + [qE - \frac{q}{c}(v \cdot H)] \partial_v \bullet$$

The domain $D(L_0)$ of the operator L_0 is dense in the space $L_0^2(\Omega_{t_0})$. Thus the divergence of the phase space $\{x, v\}$ is identically equal to zero then dynamical system for the operator L retain some measure. Consequently [16] there is the solution (strong or weak) and $f_0 \in L^2(\Omega_{t_0})$.

III BUNCHED BEAM IN ELECTRIC FIELD

The Maxwell equation reduce to the form rotE=0 if magnetic field vanishes, and $E = -\nabla U$, where U is an electro-potential. An ultra-relativistic driving beam can be represented by the following equations

$$\dot{x} = v \qquad (7)$$

$$(m\,\dot{v}) = qE.$$

Here $x = (x_1, x_2, x_3)$ is a position vector, $v = (v_1, v_2, v_3)$ is a velocity vector of q is the beam particle charge, respectively. We suppose that a velocity field is given in the position space such that

$$\dot{x} = \eta(x) \qquad (8)$$

Next we consider problem of existence of some field E such that the motion of the beam will be identical to the motion given by (8), for case when initial values coincide. The statement becomes apparent when it is considered as the following. The equation (7) have an integral manifold

$$v - \eta(x) = 0 \qquad (9)$$

Introducing into consideration the vector function

$$\Theta = v - \eta(x)$$

and calculating its time derivative with respect to the given equation (7) we get

$$\dot{\Theta} = F(x, \Theta). \qquad (10)$$

Obviously $F(x, \Theta) \equiv 0$ for $\Theta = 0$. Let us consider (10) together with the equation (7). In accordance with existence and uniqueness Theorem for ordinary differential equations any solution with initial data $x = x_0, v = v_0, \Theta = \Theta_0$ for $t = 0$ has the trivial solution $\Theta \equiv 0$ if $\Theta_0 = 0$. This means that equation (7) has integral manifold (9). In other words, if $v_0 = \eta(x_0)$ as $t = 0$ then $v = \eta(x)$ for all t.

Thus, the solution of the equation (7) satisfies also the equation (8) and consequently it is coinciding with its solution. Now let us consider the following problem: Suppose the integral manifold Θ and the distribution of velocity $\eta(v)$ on it is given. To find the field E which is focusing and accelerate the beam for all $\{x_0, v_0\} \notin \Theta$, let us consider equation (10) and corresponding to it the Cauchy problem for equation

$$\frac{\partial V}{\partial \Theta} F = w, \quad V(0) = 0 \qquad (11)$$

Thus $\frac{dV}{dt} \equiv \frac{\partial V}{\partial \Theta} F$ with respect to the given system (10) and the function w is to be such that the following holds

$$\int_0^\infty w(\Theta_t(\Theta_0))dt < \infty,$$

Then $V(\Theta_t) = V(\Theta_0) - \int_0^\infty w(\Theta_t(\Theta_0))dt$. Here $Q_0 = v - \eta(x)$ for $t = t_0$ and $Q_t(\Theta_0)$ is solution to equation (10). Although, we have neither the solution $Q_t(Q_0)$ nor a criterion of such choice for the function w, evidently, the function w, (the solution of equation(11)), may be represented by the following equation

$$V(\zeta) = \int_\Omega k(\zeta,\xi)w(\xi)d\zeta \tag{12}$$

where $\zeta, \xi \in \Omega \subset R^n$. Differentiating formula (12), the left-hand and right-hand sides with respect to the (9) we obtain

$$w(\xi) = \int_\Omega k(\zeta,\xi)w(\xi)d\zeta. \tag{13}$$

Here $\tilde{k} = \frac{\partial w}{\partial \xi} F(\xi)$, and $\xi \equiv \Theta$. The conclusion from this is that, the equation (11) will be resolved if its right side is a solution of the Fredholm's integral equation (13). For detailed proof see e.g. [15]. Here kernel $\omega(\zeta, \xi)$ is given by a sum of series, i.e.

$$k(\zeta, \xi) = \sum \lambda_i \varphi_i(\zeta) \overline{\varphi_i(\xi)}$$

where $\{\varphi_k(\xi)\}$ is a complete orthonormal set in $L^2(\bar{\Omega})$ and $\{\lambda_k\}$ is a set of numbers $\lambda_k = \mu_k + \bar{\mu}_k$ and μ_k is eigenvalue for the kernel $\tilde{\omega}$. A common way of regarding this fact is to suppose that the domain $\bar{\Omega}$ is the attractor for the singular point $\Theta = 0$ of the equation (11).

In view of the solution for the perturbation equation (9) it is clear that in some neighborhood Ω of the singular point $\Theta = 0$ condition for all $\lambda_k < 0$ holds. In this case the neighborhood Ω is an attraction region. From equation (13) we deduce some results about boundary of the domain Ω.

Obviously, the boundary is a closed surface $S = \bar{\Omega} - \Omega$ such that if $\xi \in S$ then $g(\xi) = 0$, i.e., $\frac{\partial v(\xi)}{\partial \xi} F(\xi) = 0$, $\xi \in S$. Also assume that on the surface the following conditions are satisfied: (i) it is surface in phase space such that the particle beam can not go through it i.e., it plays the roll of the impenetrable wall; (ii) it contains whole trajectory of dynamical system (11).

To be more clear we consider the following example.

EXAMPLE: Let $(x_2 = x_3 = 0, \quad v_2 = v_3 = 0)$ be an integral manifold. Denote $x_1 = x$, $v_1 = v$ and $H = (H_1 0, 0)$, thus

$$\dot{x} = v$$

$$\dot{v} = qE$$

$$\Delta U = -4\pi q \int_{-1}^{1} f(t, x, v) dv$$

This reasoning yields a simple equation:

$$\partial_t f + v \partial_x f - 4\pi q \int_{-1}^{1} f(t, x, v) dv \cdot \partial_v f = 0$$

Now taking the solution f in the form $f(t, x, v) = f_0(x, v) e^{i\omega t}$ we get

$$v \partial_x f - 4\pi q \int_{-1}^{1} f(t, x, v) dv \cdot \partial_v f_0 = i\omega f_0 \qquad (14)$$

Remark 1 *Taking* $U = U_0(x) e^{i\omega t}$, *we see that* $U_0 = -4\pi q \int_{-1}^{1} f_0(x, v) dv$.

The equation (14) is quasi-linear equation, therefore it will be reduced to a linear equation

$$v \partial_x \Phi - 4\pi q \int_{-1}^{1} f_0 dv \cdot \partial_v \Phi - i\omega f_0 \partial_{f_0} \Phi = 0 \qquad (15)$$

where $\Phi = \Phi(x, v, f_0)$ *is an integral of the linear equation (15) As known, the equation (15) yields*

$$\frac{dx}{v} = -\frac{dv}{8\pi f_0} = -\frac{df_0}{i\omega f_0} \qquad (16)$$

Remark 2 *The function* f_0 *play roll of an independent variable in the equation (5). Now it is easy to establish the existence of an integral basis [20] for equation (15):*

$$(8\pi f_0 z^2 + v^2, vx - \frac{i\omega}{2} f_0^2)$$

Remark 3 *Without loss of generality it can be assumed that* $v \in [-1, 1]$, *as was assumed in the above.*

The solution of Cauchy problem (initial-value problem) is called the solution of equation (14), $f_0(x, v)$ satisfying some initial conditions: $f_0^0 = \eta(v)$ for $x = x_0$. Here the function η is a distribution of the velocity v for some fixed point x_0.

Example 4 Let $x = x_0 = 0$ and $\eta(v) = \exp(-\frac{v^2}{\varepsilon})$ where ε is some constant. Using the technique of Cauchy problem solvability, we will easily see that the solution f_0 on the integral manifold have the following form

$$f_0 = \frac{i\omega x}{v} - \left\{\frac{m}{2\pi\Theta}\right\}^{\frac{3}{2}} \exp\left\{-\frac{m}{\Theta}\left[8\pi L f_0 x - \frac{1}{2}v^2\right]\right\}$$

where $0 \le x \le L$, $v \in [-1, 1]$.

Corollary 5 *We have arrived at the following result: The function $f_0(x, v)$ possesses the property and the form which it succeeded from the function φ. Consequently, there exist situations such that the solution $f_0 e^{i\omega t}$ have the soliton-like form. For this we choose the function $\varphi(v, x_0)$ in accordance with this conditions. E.g., see [17], [18]*

Example 6

$$\varphi(x_0, v) = 4\tan^{-1}\left\{\exp[(x_0 - vt)/(1 - v^2)^{1/2}]\right\}$$

The "classical" problem of the soliton solution can be posed as follows: identify the Lorentz force $F_l(E, H)$, such that the motion on the integral manifold will be an asymptotically stable. We are going to consider the problem of the stability in the above mentioned solutions. Assume conditions $x_1(t_0) = x_{10}$, $x_2(t_0) = x_{20}$, $x_3(t_0) = x_{30}$; and $v_1(t_0) = v_{10}$, $v_2(t_0) = v_{20}$, $v_3(t_0) = v_{30}$ are satisfied. Next we suppose that $(0, x_{20}, x_{30}) \in \Omega_x$, $(0, v_{20}, v_{30}) \in \Omega_v$, where $\Omega_x = \left\{x_{i0} : \sum_2^3 x_{i0}^2 \le 1\right\}$, $\Omega_v = \left\{v_{i0} : \sum_2^3 v_{i0}^2 \le 1\right\}$ without loss of generality it can be assumed. Thus the perturbed has place in the planes $(\{x_2, x_3\} \times \{v_2, v_3\})$. Let $D = \Omega_x \cdot \Omega_v$ be a topological multiplying the domains Ω_x and Ω_v.

Let $\{\varphi_k(x_2, x_3)\}$ and $\{\varphi_k(v_2, v_3)\}$ be complete orthonormal sets in Ω_x and Ω_v respectively, then $\{\varphi_k \Psi_r\}$ be a complete orthonormal set in D.

Now we can construct the kernel

$$k = \sum_i \sum_j \lambda_{ij} \varphi_i(x) \Psi_j(v) \overline{\varphi_i(y) \Psi_j(u)},$$

where $x, y, \in \Omega_x, v, u \in \Omega_v$. Now by using (12) and (13) we can find the solution of the Cauchy problem (11).

Note that, present techniques are just arriving at the point where this approach can be proved if the phase space of the physical system is limited and is closed, i.e. the phase space is a compact with bound.

IV TWO-DIMENSIONAL PROBLEM

In view of the above and numerical algorithms of the solutions, it is clear that simplification of the physical model is necessary to yield practical solutions.

The perturbation motion is described in the plane which is orthogonal to the main motion. Thus our approach can be improved by simplicity of the two-dimensional problem. In this section we further develop the technique which was proposed above. We shall use z-complex plane to express the disturbance. As before we consider the motion of a beam in a plane $\{x_2, x_3\}$ with a velocity $\{v_2, v_3\}$. Now we introduce complex variable z: $\quad z = x_2 + ix_3$ and $\dot{z} = \dot{x}_2 + i\dot{x}_3 = V$, then we have the following system

$$\dot{z} = V$$
$$\dot{V} = q\bar{E} + \frac{q}{c}[\overline{VH}] = \bar{F}_l \qquad (17)$$

where $\bar{E} = E_2 + iE_3$, and $[\overline{VH}] = [VH]_{x_2} + i[VH]_{x_3}$.
From these definitions, we immediately obtain the following relation

$$\frac{\partial f}{\partial t} + V\partial_z f + \bar{F}_l \partial_V f = 0.$$

By the use of the WKB approximation, that is, the wave packet form, we have

$$f(t, z, V) = f_0(z, V)e^{-i\omega t}. \qquad (18)$$

Substituting (18) in (17), we get

$$V\partial_z f_0 + \bar{F}_l \partial_V f_0 = i\omega f_0. \qquad (19)$$

Now have to resolve this equation. If we consider the variable $z \in C = \{z : |z| < 1\}$, then

$$f_0 = \sum c_k \varphi_k(z) \qquad (20)$$

the set $\{\varphi_i\}$ is some complete orthonormal system in the circle C. Let $\{\varphi_k(z) = re^{ik\theta}, r = (x_2^2 + x_3^2)^{1/2}\}$ be one. For the orthonormal system we obtain

$$c_k = \int_C \varphi_k(z) f_0(z, V) dz,$$

thus $c_k = c_k(V)$.
By substituting the value series (20) in (12) we obtain

$$f_0(z, V) = i\omega \int_C k(z, \bar{\xi}) f_0^*(\xi, V) d\xi, \qquad (21)$$

$z, \xi \in C$, $\omega(z, \xi) = \sum \varphi_i(z)\varphi_i^*(\xi)$.
Using the values of the coefficients $\{c_k\}$ we get

$$k(z,\bar{\xi}) = \frac{1}{\pi}\sum_{n=0}^{\infty}(n+1)(z\bar{\xi}) = \frac{1}{\pi}\frac{1}{(1-z\bar{\xi})^2}. \tag{22}$$

Substituting (22) in (21) we get

$$f_0(z,V) = \frac{1}{\pi}\int_C \frac{1}{(1-z\bar{\xi})^2}f_0^*(\xi,V)d\xi. \tag{23}$$

It may be noted that there exists an infinite number of different orthonormal systems the circle C, but the functions k for all systems coincide.

It is easy to see that the system (21) can be written as

$$\dot{z} = V, \quad \ddot{z} = F_l. \tag{24}$$

Using this equation we immediately obtain the important equation

$$i\omega f_0(z,V) = \int_C \widetilde{\widetilde{k}}(z,\bar{\xi})f_0^*(\xi,V)d\xi, \tag{25}$$

where

$$\widetilde{\widetilde{k}} = \frac{d^2k}{dt^2}\left(\frac{1}{\pi}\frac{1}{(1-z\xi)^2}\right). \tag{26}$$

We recall, the function $\widetilde{\widetilde{k}}$ is the second time derivative of k with respect to the given system (24), i.e.

$$\widetilde{\widetilde{k}}(z,\bar{\xi}) = \frac{6}{\pi}\frac{\xi^2}{(1-z\xi)^4}V^2 + \frac{2}{\pi}\frac{\xi}{(1-z\xi)^3}F_l.$$

Then here exist function $f_0(z,V)$ such that the integral equation (25) hold. The equation (25) is called the Fredholm integral equation with kernel $\widetilde{\widetilde{k}}(z,\xi)$. Here the set $\{i\omega\}$ is the spectrum σ of kernel $\widetilde{\widetilde{k}}$. Examining the differential equations

$$\frac{df_0}{dt} = \dot{f}_0$$

$$\frac{d\dot{f}_0}{dt} = -\omega^2 f_0.$$

We recognize that its characteristic numbers are $\pm i\omega$. Consequently the general solution

$$f(t,z,V) = c_1\sum c_k(V)r^k e^{i(k\Theta-\omega t)} + c_2\sum c_k(V)r^k e^{i(k\Theta+\omega t)}$$

where c_1 and c_2 are some constants which depend on initial condition. Therefore if it is chosen $c_2 = 0$ and $\omega = i\lambda$, where $\lambda > 0$ then the perturbed function

$f(t, z, v) \to 0$ as $t \to \infty$. Thus the bunched beam survive the full energy so that for this condition we have self-focusing and accelerating system.

It can be shown in the usual way, that if $i\omega \in \sigma(\widetilde{\widetilde{k}})$ be such that

$$\operatorname{Im} \sigma(\widetilde{\widetilde{k}}) \succeq \alpha > 0 \tag{27}$$

then the integral manifold $\{x_1, x_2 = 0, x_3 = 0\}$ is self-focused and acceleration.

Indeed, under condition (27) we can construct the function $\varsigma = f_0 \bar{f_0} \geq 0$ such that its time derivative with respect to the given system (21), (i.e. the expression $\frac{d\varsigma}{dt} = -2|\omega| f_0 \bar{f_0}$), is positive everywhere except at the singular point where it vanishes. Under this condition, consequently $\varsigma(z, V) \to 0$ as $t \to \infty$.

Furthermore, the boundary of asymptotically stable is defined by a condition

$$\int_C \widetilde{\widetilde{k}}(z, \bar{\xi}) f_0 d\xi = 0. \tag{28}$$

Thus the equation (28) is a common way of looking for the electro-magnetic field resolving the 2-D problem 1.

V THE STABILITY OF THE SOLUTION OF VLASOV-POISSON EQUATIONS IN THE SENSE OF TWO METRIC

As well known, the stability of local structures or solitons by two metric have important physical sense. Thus the solution of the Vlasov-Maxwell system is the stability for the metric ρ_0 and ρ if for all $\varepsilon > 0$, there exists $\delta(\varepsilon) > 0$ such that from $\rho_0(x_0, v_0) < \delta$ follows $\rho(f(t, x, v)) < \varepsilon$ for all $t > 0$. If in addition $\rho(f(t, x, v)) \to 0$ as $t \to \infty$ then we have the asymptotically stable. Let $\rho_t = \int_{D_0} |f(t, x, v)|^2 d\mu = \|f\|_{l_2}$ be the norm of the function in a space $L^2(D)$, then $\rho_0 = \int_{D_t} |f(0, x, v)|^2 d\mu$, where μ is some measure in $L_2(D)$. From the above mentioned result about ρ_t we deduce

$$\rho_t = \int_{D_t} |f(t, x, v)|^2 d\mu = \int_{D_t} |f(0, x, v)|^2 \exp \int_0^t div X d\tau d\mu$$

here $divX$ is a divergent of the vector field $X = \{v, F_l + G(t, x, v)\}$ and $G(t, x, v)$ is some disturbance of the Lorentz force (e.g., it may be collision of the particles). Notice there exists some constant B such that an inequality

$$\exp \int_0^t div X dt \leq B e^{-\alpha t} \tag{29}$$

iff $\text{Re}\,\sigma(k) \leq -\alpha < 0$. The inequality (29) yields: $\rho_t \leq B\rho_0 e^{-\alpha t}$ and $\rho_t \to 0$, as $t \to \infty$; as only $\text{Re}\,\sigma(k) \leq -\alpha$ if $\rho_0 < \delta$.

It is easily seen, that norm ρ in the Hilbert space $L^2(D)$ have to deal with volume and is determined up to measure zero. This volume may be m−dimensional chaotic attractor in some dissipative phase space.

REFERENCES

1. A.A. Vlasov, Many-Particle Theory and its Application to Plasma Physics, (1961).
2. E.g. see [3] – [14] and references therein. Also see presentations in this workshop.
3. C.M. Ankenbrandt etal., *Status of muon collider research and development and future plans*, Phys. Rev. ST Accel. Beams **2**, 081001 (1999), and references therein.
4. A.M. Skrinsky and V.V. Parkhomchuk, particles, Sov.J.Part.Nucl.,12,p.223(1981).
5. Z. Parsa, *Muon Storage Rings - Neutrino Factories*, in Procds. of the Workshop on the Next generation Nucleon decay and Neutrino detector AIP-Press (2000).
6. Z. Parsa, ed., Future High Energy Colliders, AIP CP **397**, AIP-Press, NY (1997).
7. Z. Parsa, *Ionization cooling and Muon Dynamics*, AIP CP **441**, 289-294 (1997).
8. Z. Parsa, *New High Intensity Muon sources and Flavor Changing Neutral Currents*, World scientific Publishing, pp 147-153 (1998).
9. Z. Parsa, ed., Beam Stability and Nonlinear Dynamics, AIP CP **405** (1997).
10. Z. Parsa, *Lasers and Future High Energy Colliders*, STS-Press, pp 823-830 (1997).
11. B. Kamal, W. Marciano, Z. Parsa, Resonant Higgs enhancement at the first muon collider, AIP CP**441**, pp174- (1997); ibid, AIP **435** pp567-662 (1997).
12. Z. Parsa, $\mu^+\mu^-$ Collider and Physics Possibilities (1993) (unpublished).
13. Z. Parsa, *Polarization and Luminosity requirements for the First Muon Collider*, in 8th Advanced Accelerator Concepts AIP CP **472**, pp. 251 - 259 (1999).
14. Z. Parsa, *Muon Dynamics and Ionization Cooling at Muon Colliders*, Procd. of EPAC98, Stockholm, Sweden, Vol **2**, pp.1055-.
15. V. Zadorozhny,*On the arise of the deterministic chaos in dynamical systems.* Proceedings of ICNPAA-98, v.**2**, p.797-803 (1999).
16. A.V. Balakrishnan, Introduction to optimization theory in a Hilbert space, Springer-Verlag (1971).
17. Z. Parsa, *Topological solitons in physics*, Am.J.Phys. Vol.**47**. N.1, p.56-62 (1979).
18. Z. Parsa, *Nonlinear Dynamics – Maps, Integrators and Solitons*, Procds. of 2nd Int'l Conf. on Nonlinear Problems in Aviation and Aerospace, Vol. **2**, pp. 589-600 (1999).
19. V.F. Zadorozhny,*On the stability volume of measure zero in to the dissipative dynamical systems*, J. Nonlinear studies,vol.**5**, n.1, p.115-122, (1998).
20. E. Kamke, Partielle differential gleichungen erster ordnung, Leipzig, (1959).

Detector Challenges for $\mu^+\mu^-$ Colliders in the 10–100 TeV Range

presented by Pavel Rehak* for D. Cline**, E. Gatti[†], C. Heusch[⊕],
S. Kahn*, B. King*, T. Kirk*, P. Norton[‡], V. Radeka*, N. Samios*,
V. Tcherniatine* and W. Willis[°]

*Brookhaven National Lab., Upton NY 1973-5000[1]
**Univ. of California, Dept. of Physics, 405 Hilgard Ave, LA, CA 92717
[†]Politecnico di Milano, P.za L.da Vinci 32, 20133 Milano, Italy
[⊕]Stanford Linear Accelerator Center, P.O. Box 4349, Stanford, CA 94309
[‡]Rutherford Appleton Lab., Chilton, Didcot, Oxon. OX11 0QX, England
[°]Nevis Labs, Columbia Univ., P.O.Box 137, Irvington-on Hudson, NY 10533

Abstract. The challenges to design, construct and operate a detector system in an interaction region at a high energy $\mu^+\mu^-$ collider are briefly summarized. A new solution, based on an extensive use of liquefied rare gases, is proposed for the vertex detector and for the inner tracker. The proposed solution takes a full advantage of the small size of the collision region. The region has radial dimensions below the μm range, while its extent in the beam direction is of the order of $1mm$ depending on the energy of the machine and on the details of the design. In the energy range considered in this workshop, the momenta of charged particles are high enough to be bent only slightly in the solenoidal magnetic field. The tracks are basically straight lines emerging from a known common point. The proposed solution is based on a projective geometry matched to the desired tracks. The aim of the design is to achieve a background blind radiation hard detector.

The second novel feature of the design is the calorimetric instrumentation of two large $20°$ tungsten cones used to shield the detector from the machine induced background. Here again the full advantage of the known directionality of particles initiating the shower process is incorporated into the design.

INTRODUCTION

The main challenge for the detector system around an intersection region of a $\mu^+\mu^-$ collider is to eliminate background particles reaching the detector practically at the same time as the particles of interest. There are three sources of background particles:

1. muon halo

---

[1]) Supported by Department of Energy contract DE–AC02–98CH10886

TABLE 1. Radial Fluences at $2 \times 2TeV$. Numbers of particles crossing a radial surface of $1 cm^2$ for two bunches of $10^{12} \mu's$ each

Radius (cm)	photons	neutrons	protons	pions	electrons	muons
5	2700	120	0.05	1.8	2.3	1.7
10	750	110	0.20	0.8		0.7
15	350	100	0.13	0.8		0.4
20	210	100	0.13	0.6		0.1
50	70	120	0.08	0.1		0.02
100	31	50	0.04	0.006		.008
calor.						.003
muon s.						.0003
cut–off threshold	25 keV	40 keV	10 MeV	10 MeV		

2. decays of the muon beams producing high energy electrons

3. beam–beam interactions

Detailed studies of individual background sources and the design of the shielding of the intersection region were carried for center-of-mass (COM) energies of 0.1, 0.5 and 4 TeV [1,2]. The shielding for the highest COM energy of 4 TeV was designed already in 1996. [1] Since then there has been a better understanding of the shielding geometry and an improved configuration was designed for lower energy $\mu^+\mu^-$ colliders. The decrease in fluences of background particles is about a factor of two. It was also found out that the background fluences decrease with the increase of the beam energy for the same number of muons in the bunch. In this paper, however, we will take the background fluences from studies [1], listed in Table 1, to be more conservative.

The most direct impact of the shielding on the design of the detector are two conical tungsten shields with angles up to $20°$, called noses, coming very close to the intersection region. During the discussions within the working group it become apparent that it is advantageous to instrument both cones (noses). The most important new technology proposed in this workshop is the "Background blind, radiation hard" vertex detector and tracker. This technology takes full advantage of i) the small size of the interaction point of the $\mu^+\mu^-$ collider and ii) the time structure of the signal and background. The view of an elementary cell is shown in Figure 1. The detection medium is liquid argon or neon. This medium is radiation hard and allows to fill almost any geometry. The electric field applied to the cell is such that signal electrons move across the short cell dimension ($2mm$). The collection time is about $1\mu s$, that is, short enough relative to the bunch crossing time. The shown cell points to the intersection region and the aspect ratio of the cell depth to the transverse dimensions leads to a large background suppression already from the rough ionization measurement.

Most of the paper will be dedicated to the analysis of the performance of the detector built from close packages of these cells. With prior knowledge of the direction of electromagnetic showers were electrons and γ's originated at the intersection point, the segmentation of the calorimeter was also optimized to take full advantage of the shower directionality. Driven by physics one can even think about an almost fully absorbing liquid xenon electromagnetic calorimeter.

This contribution starts with an overall description of the detector. Most considerations are for the $10TeV$ option. The main section presents a detail analysis of the new vertex detector and the new tracker. All possible limitations of the vertex detector are analyzed. The section about calorimetry emphasizes a possible construction with a high degree of directional selectivity. The Appendix shows a very good position resolution attainable with the cells of the proposed vertexer and tracker.

DETECTOR SYSTEMS

A cross section of the "Strawman Detector" (from [1]) is shown in Figure 2. The dimensions shown on the Figure 2 are only rough indications, as they would be different for different beam energies in order to fit between the two last quadrupoles. The detector has a solenoidal magnetic field and consists of following 6 components:

1. Vertex Detector

2. Inner Tracker Detector of charged particles

3. Instrumented Tungsten shield as a Calorimeter

4. Electromagnetic Calorimeter

5. Hadron Calorimeter

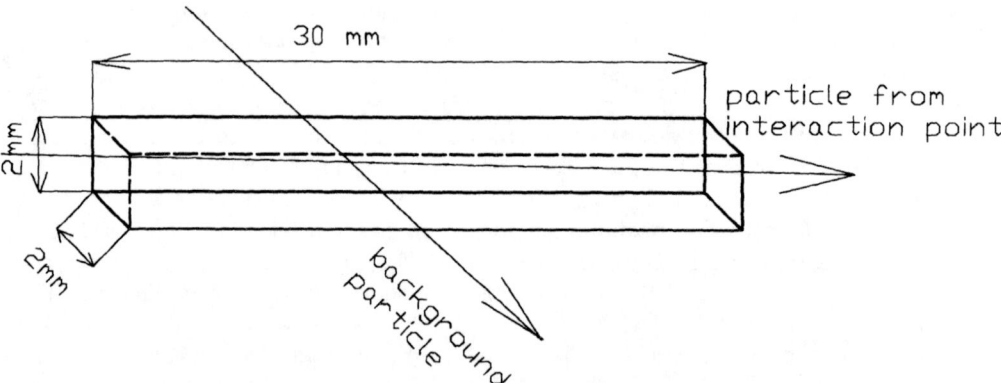

FIGURE 1. Basic cell of the vertex detector and tracker. Particles from the interaction point traverse the cell along the long direction producing a large signal charge. Ionization within the cell due to background particles is much smaller.

6. Muon Spectrometer

The first attempt to specify criteria for the detector in the high muon energy range is summarized in Table 2. A solenoid is the natural choice for interaction with almost isotropic distribution of track of interest. The strength of the field may be increased for a higher energy machine and the actual field depends on the state of the magnet technology at the time of the construction. The coil placement behind the hadron calorimeter is not fixed. The need of a higher magnetic field in the inner tracker may move the coil right behind the electromagnetic calorimeter.

A new vertex detector and a new inner tracker are being proposed here. These systems take full advantage of the small extent of the intersection region and the time structure of the bunch crossings.

The instrumentation of the nose of the tungsten shield is the other novel feature of this report. It is felt that the physics potential of the $\mu^+\mu^-$ detector can be greatly enhanced by it. Most importantly, even a coarse measurement of electromagnetic

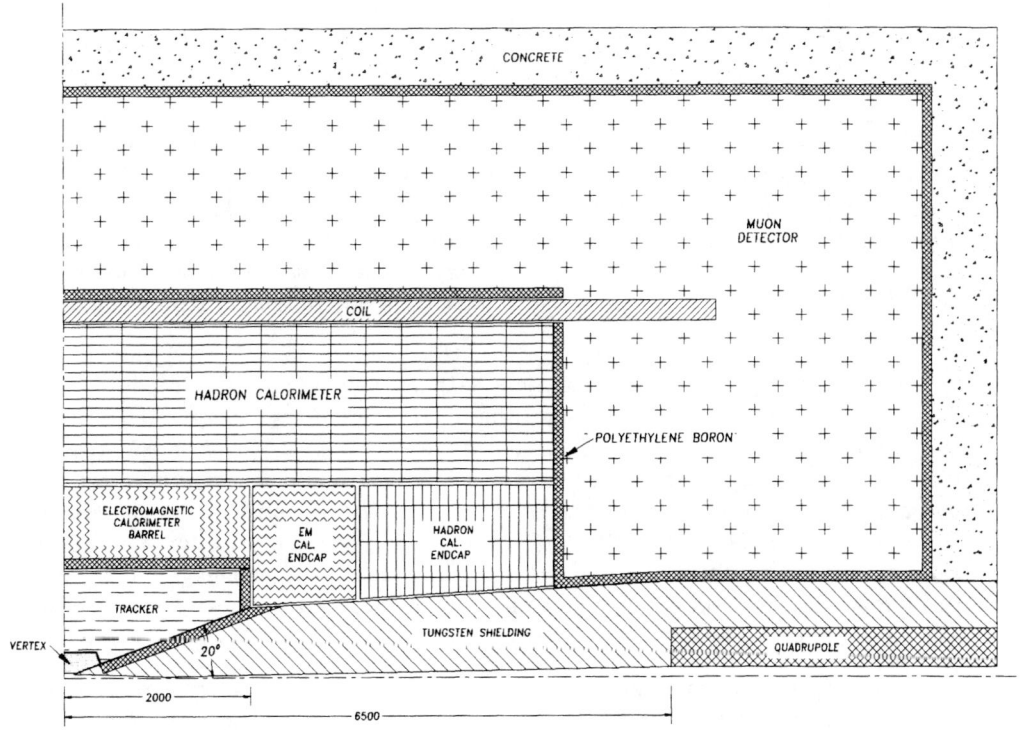

FIGURE 2. Strawman detector

TABLE 2. Detector Performance Requirements. (10 TeV)

Detector Component	Minimum Resolution/Characteristics
Magnetic field	Solenoid; $B > 3T$
Vertex Detector	b-tagging; $\sigma \lesssim 10\mu m$; small cells
Inner Tracker	$\Delta p/p^2 \lesssim 2 \cdot 10^{-4}(GeV)^{-1}$ at large p High granularity background blind
Nose calorimeter	$\Delta E/E \sim 35\%/\sqrt{E} \oplus 4\%$ for EM showers $\Delta E/E \sim 70\%/\sqrt{E} \oplus 8\%$ for hadron showers
EM Calorimeter	$\Delta E/E \sim 10\%/\sqrt{E} \oplus 0.7\%$ (baseline) $\Delta E/E \sim 5\%/\sqrt{E} \oplus 0.4\%$ maybe required projective geometry. Depth $\sim 25X_0$
Hadron Calorimeter	$\Delta E/E \sim 60\%/\sqrt{E} \oplus 3\%$ Granularity: projection cells and transverse Total depth (EM + HAD)$\gtrsim 7\lambda$
Muon Spectrometer	$\Delta p/p \lesssim 15\%$ at 1TeV Combining with the Inner Tracker

and hadron energies will improve considerably the detection of missing transverse energy. The importance of identifying and of measuring the missing energy is reported by the theory working group of this workshop.

There are two versions of the Electromagnetic calorimeter considered. Both versions identify electrons, photons and the core of jets. The projection geometry and the granularity are crucial to reduce the background. The first, baseline version is a sampling calorimeter. The second, that is, the high performance version of the calorimeter may be required by physics. A possible realization of such a high resolution would be an almost fully absorbing liquid xenon Electromagnetic calorimeter.

The hadron calorimeter measures jets with enough precision to separate W's from Z^0s and provide an indication of the missing transverse energy. It is a very large and expensive detector and whose not many details are given in this report. It is located in a region well shielded from the radial flux of background particles originating in the nose of the tungsten shield. Due to its large transverse dimensions it intercepts high flux of longitudinal muons. The segmentation of the hadron calorimeter may be different from the segmentation of the electromagnetic one due to different background profiles and also due to different shapes of hadron and electromagnetic showers.

The muon spectrometer, located outside the hadron calorimeter, and an additional shielding, identifies and measures muons. The momentum measurement of muons is the combination of the momentum measurement in the inner tracker and in the muon system itself. Due to the background flux of the muons moving mainly parallel to the beam directions a high degree of redundancy is required.

VERTEX DETECTOR

Main Design

The vertex detector is based on liquid noble gases rather than on silicon as the detection medium for the following reasons:

1. radiation hardness

2. ability to deliver position resolution down to a few μm range

3. formation of long detection cells pointing to the interaction region leading to superior background rejection.

The vertex detector is the nearest detector to the beams and is most exposed to radial backgrounds. At the same time it has the most demanding performance specifications. The task of the vertex detector is to identify secondary vertices created mainly B and τ decays and measure the points on the trajectories of charge tracks close to the collision point.

The proposed vertex detector is shown in Figure 3. It is based on long and narrow individual cells, closely packed, pointing in both z and ϕ coordinates directly to the intersection point. The high aspect ratio between the depth and the transverse dimension of the cell is possible thanks to the small extent of the intersection region. The size of the intersection region is only a few μm in the transverse dimensions and about a mm along the direction of the beams. The tracks of interest, having fairly high momentum have a very small curvature in the scale of the Figure 3. Individual cells have a typical ratio of depth to transverse dimension of about 15.

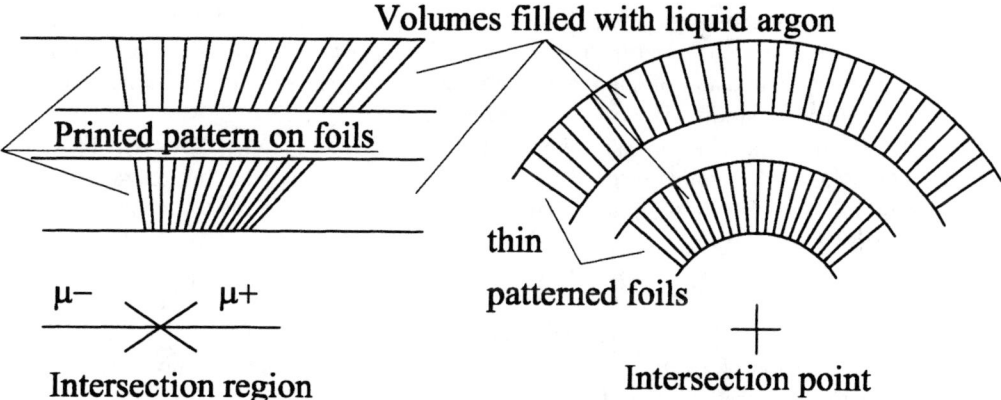

FIGURE 3. Cross sections through the vertex detector. The first layer starts at a distance of about 4cm. A typical transverse dimension of a cell is 2mm, the length of the cell 3cm Electric field is applied perpendicular to the radial direction leading to a short collection path for signal electrons.

TABLE 3. Effective thresholds for different detection media

Detection medium and detector	Typical energy threshold
Gas proportional counters	$\sim 100 eV$
Silicon planar detectors	$\sim 10 keV$
Long liquid argon cells	$\sim 1 MeV$

The energy deposited within a cell by a charge particle originating at or close to the interaction region is thus substantially larger than background tracks leading to the background rejection (see below) already at the single hit level before the additional constraint of combining hits into segments of a track.

The proposed approach can be seen as an extension of the concept of pixel micro-telescopes [2,3]. Here the detection medium, being liquid argon or liquid neon, allows perfectly projective shape pointing into the interaction region and the extension of the detection depth by two orders of magnitude from $300 \mu m$ for a typical silicon layer to a few cm. A minimum ionizing particle deposits about $6 Mev$ in $3 cm$ of liquid argon. The effective thresholds for charged particles in gas, silicon and liquid argon are given in Table 3. We can see that the effective threshold in the proposed long liquid argon cell is 2 orders of magnitude higher than in a typical silicon detector and 4 orders of magnitude higher than in a gas detector. In the large background environment of the $\mu^+\mu^-$ collider this feature contributes to the early background rejection.

The electric field is applied in a direction perpendicular to the direction of the particle produced in the intersection region. The charge carriers, that is, electrons and positive ions produced by the ionization, move a relatively short distance ($\sim 1 mm$) before being collected by electrodes. The maximum collection time for electrons is less than $1 \mu s$. This time is much shorter than the time before the next muon bunch crossings. This time is used to process the signal waveforms due to the moving electrons. Detailed analysis in the Appendix shows that the precision of the determination of the particle crossing is about $10 \mu m$ in both directions perpendicular to the direction of the charged particle.

It is stressed that this precision is attainable with liquid noble gases which are well suited as high resolution position detectors. The density of the ionization is three orders of magnitude higher in the liquid phase than in gases at atmospheric pressure and comparable to semiconductors. The liquids provide enough signal charge without the need for an avalanche amplification. More importantly, the range of delta rays is much reduced in liquids. The ionization is confined within a μm diameter column, roughly the same as in semiconductor detectors. We believe that the resolution of $10 \mu m$ obtainable from the signal processing is not degraded by ionization processes in liquid noble gases.

The number of read out channels of the three layers (cylinders) is several times 10^5 with less than 10^5 channels in the closest layer to the intersection. This is a modest number of channels and connections. One can replace the electronics of the closest detection layer in case of radiation damage. The layers further away from the interaction point receive less radiation not only because of the geometry but also due to the absorption of photons in the previous layers.

Background rejection

The first idea about the background rejection of the cell can be obtained from a simple calculation from Figure 1. It is assumed that the angular distribution of background particles is uniform in a hemisphere away from the intersection point. A background particle, crossing the entrance surface of the cell will be detected if it crosses, let us say, one half of the length of the cell. The probability of being accepted is thus $p_{acc} \approx (2mm)^2/(2\pi \cdot (15mm)^2) \approx 1\%$. The rejection can be tuned by changes of the cell geometry.

The simulated fluences of various background particles are reported in Table 1. The first layer of the vertex detector at about $5cm$ from the interaction point receives the highest density of background particles. The total density of charged background particles, that is, protons, pions, electrons and muons is about $6/cm^2$. The largest cell has the entrance surface of $2mm \times 2mm = .04cm^2$. It means that the probability for a charged background particle to enter a cell is about 0.2. The probability to produce the signal above the threshold and result in a hit in the largest cell is only 0.2%.

The flux of neutron does not represent a problem for the proposed vertex detector and tracker. There is very little energy transfered to liquid argon, or liquid neon in the interaction with a neutron. Moreover, due to the saturation effects in these liquids, the dense ionization of a slowly moving atom is not visible.

The most important background is the flux of photons. The energy spectrum of photons in the vertex region is shown in Figure 4. A Monte Carlo program was developed to study the sensitivity of the proposed vertex detector to this photon flux . The background photons were incident on the layer of the vertex detector where they interacted, mostly by Compton scattering. Trajectories of the resulting electrons were followed in the magnetic field within the cells.

The histogram of length of electron trajectories for three different cell dimension is shown in Figure 5. The resulting integral spectrum of energy deposited in a cell, normalized to 1 is shown in Figure 6 for the same geometries of the detector cells. The energy deposited by a minimum ionizing particle originating from the interaction point in 3 cm of liquid argon is about $6.6 MeV$. One assumes that the threshold for individual cells is about $4 MeV$. The acceptance of the photonic background in a case of $0.5mm \times 2mm$ cell is only 0.3%.

The probability of a photon conversion in the energy range of a few MeV in $1cm$ of liquid argon is $6.2 \cdot 10^{-2} cm^{-1}$ or about 20% in 3cm of the cell's length. Let us concentrate on the first measuring layer (cylinder) of the vertex detector about $5cm$ from the interaction region. Within one cell with an entrance area of $0.5mm \times 2mm$ there are $2700 cm^{-2} \times 0.01 cm^2 \times 0.2 = 5.4$ photon conversions per bunch crossing. Taking the above energy acceptance (suppression) factor of 0.3%, the mean cell occupancy of the closest layer is about 1.6%. The occupancy decreases rapidly with the distance from the beams. Already the cells of the second layer (cylinder) can have the entrance surface $1mm \times 2mm$ with the same mean occupancy. The dimensions of the entrance areas of the subsequent layers (cylinders) is determined by requirements of the position resolution rather than background occupancy.

The mean cell occupancy can be decreased by replacing liquid argon with liquid neon as the detection medium. The photon conversion probability in this energy range is about a factor of two lower in neon than in argon. The change to neon

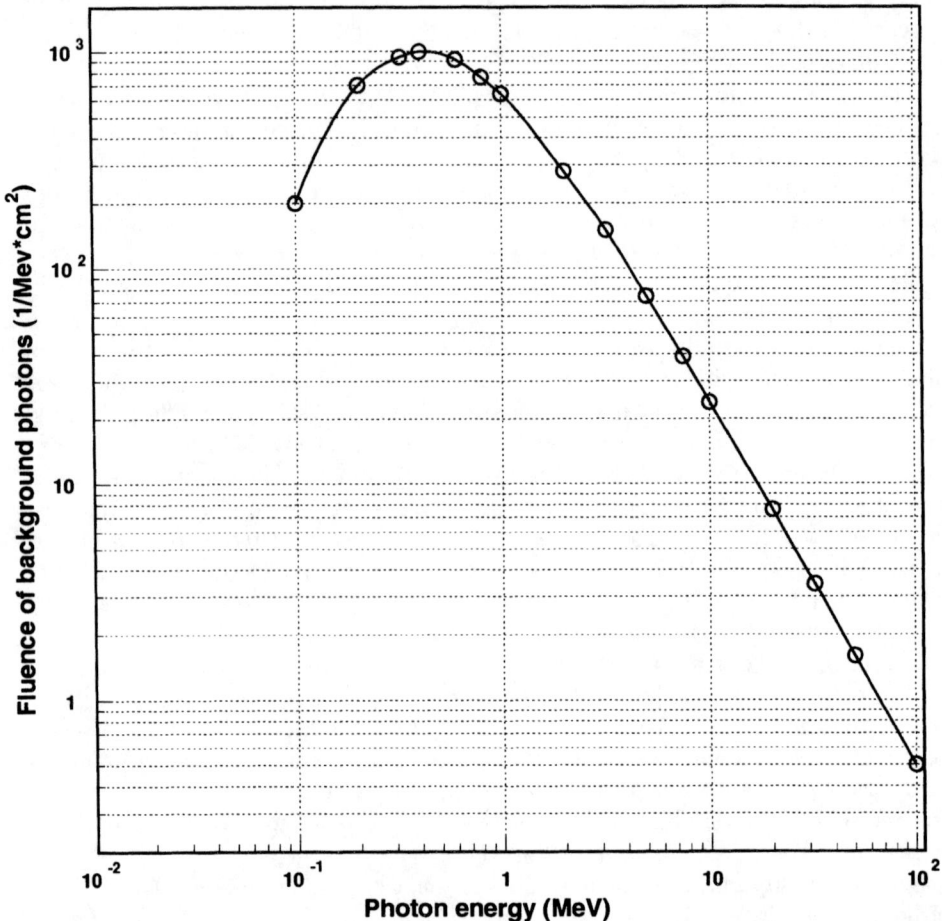

FIGURE 4. Simulated spectrum of photons in the vertex region. The cut–off energy was 25 keV. The vertex detector and the inner tracker were assumed to be massless during the simulation. Inclusion of realistic vertex and tracker detectors may shift the peak toward lower values.

will bring the occupancy of the first layer (cylinder) down to 0.8% for the same cell geometry. A threefold increase in radiation length X_0 of neon is an additional advantage over liquid argon. Multiple scattering in the system based on neon rather than argon is a factor of $\sqrt{3}$ smaller. A better momentum resolution can be

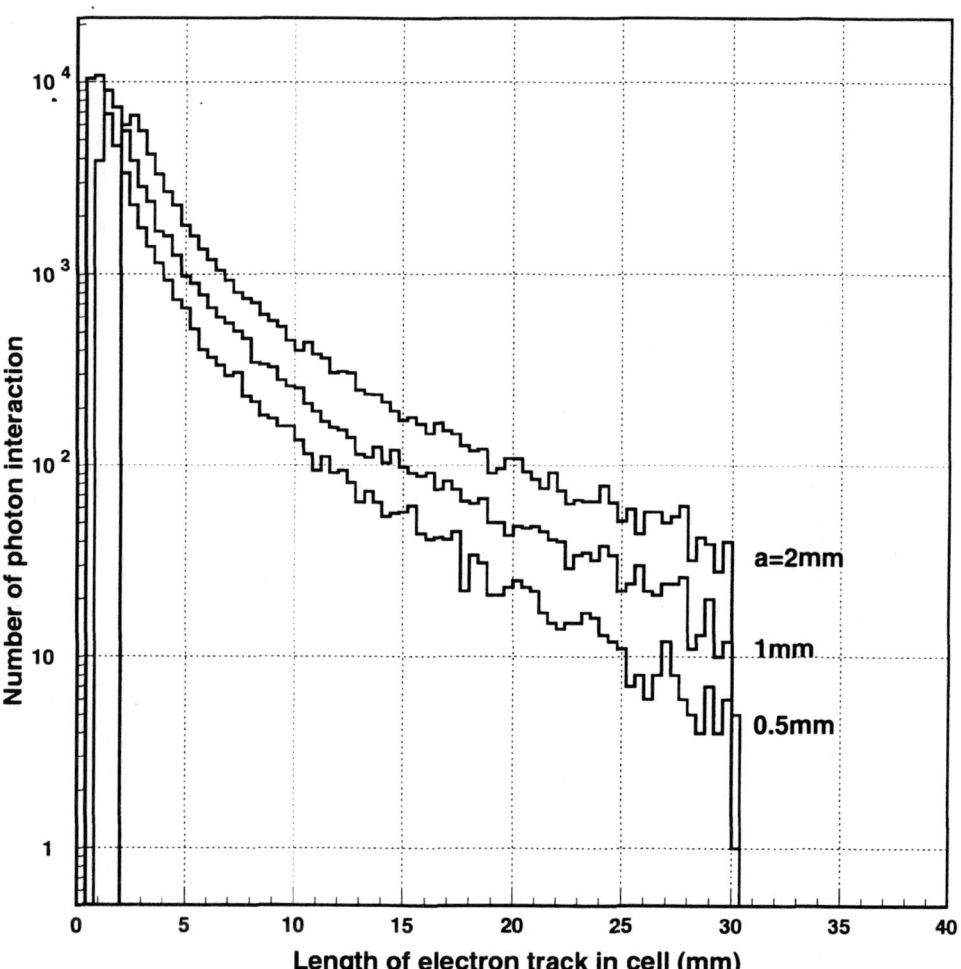

FIGURE 5. Distribution of lengths of electron trajectories resulting from the interaction of background photons in liquid argon of a single cell. Three different areas of the entrance surface of the cell were considered. The one dimension was fixed at 2mm while the second was 0.2, 1 or 2mm.

achieved for a given detector with liquid neon as a detection medium. Alternatively, we can build the detector with a smaller volume of the inner tracker or lower the magnetic field to obtain the same momentum resolution.

There is, however, a principal problem with the transport of electrons in pure liquid neon. The atomic polarizability is so small that the electron is trapped in a bubble in the liquid. Electrons thus have a very small mobility. It is known [4]

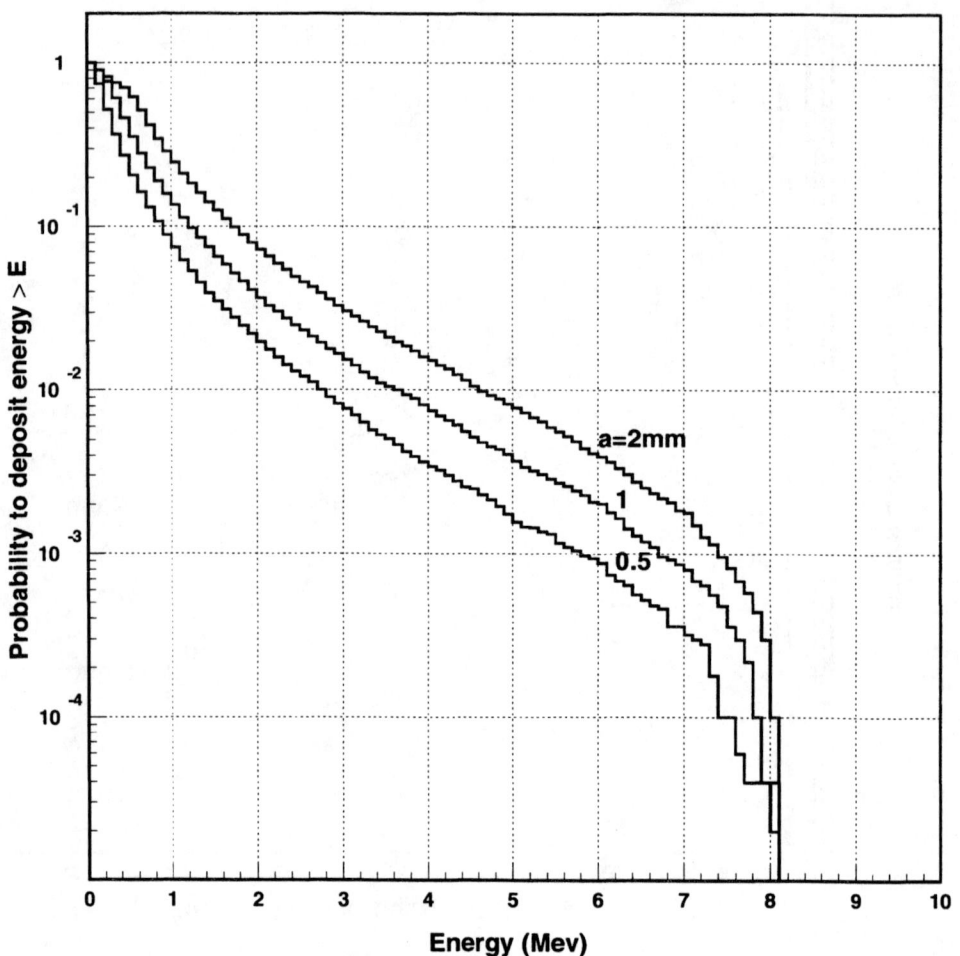

FIGURE 6. Integral spectrum, normalized to one, of the energy deposited in an individual cell. The simulations were done for three different cell geometries.

that this bubble is only marginally stable and we can suppose it is not present when a small amount of liquid argon or liquid methane is mixed with neon. More investigation is needed to find the best liquid.

Position Resolution

The cell dimensions of typically 1mm are too large to use the cell location as the only information about the position of the crossing fast particle. The vertex detector requires a precision of about $10\mu m$ and the interpolation of factors of about 100 is required in both measured directions. To obtain this high degree of interpolation the electric signal due to the motion of electron is treated in the optimal way. The optimal processing of the signals is presented in the Appendix. Principal ideas only are outlined here.

The sketch on the left hand side in Figure 7 represents the top view of the cell, that is, with a particle of interest coming perpendicular to the plane of paper. The cell dimension not visible in this projection is the depth traversed by the particle. The trapezoidal shape of the cell has been neglected and surfaces are assumed to be perfectly parallel. In the y direction the size of the cell is defined by a high voltage plane on one side ($y = 0$) and the electrons collecting electrodes at $y = d$ connected to preamplifiers. In the x direction the size of the cell is defined by the pitch of the electrode called a. The ionization produced by a fast particle originating in the intersection point is projected into a point (x_0, y_0). It is reminded here that liquid argon is a single carrier medium. Electrons move inside under the influence

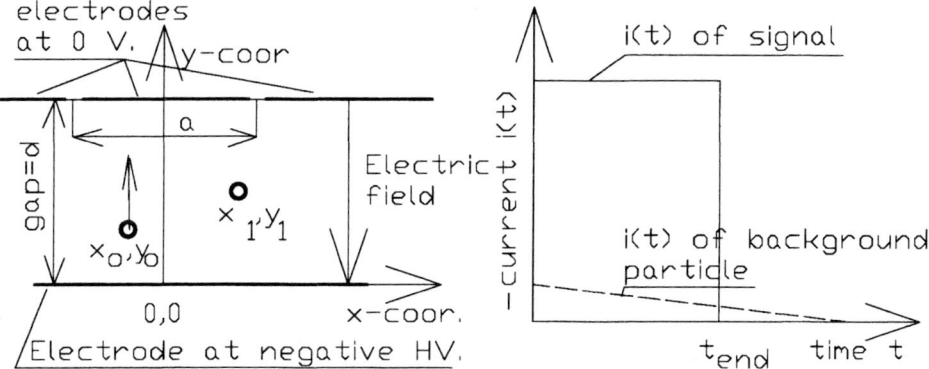

FIGURE 7. (LHS) The geometry and the electric field of several consecutive cell. The projection is such that the trajectories of particles of interest are seen as individual points. The movement of electrons produced at $(x_0; y_0)$ toward the electrode is indicated by an arrow. (RHS) the negative current waveform induced by i) signal particle–full line and ii) by a background particle–dotted line.

of the the applied electric field, while the velocity of positive ions is about 4 orders of magnitude smaller and they will be considered fixed during the signal processing time.

The plot on the right on Figure 7 shows the negative induced current due to the point–like ionization at x_0, y_0 at time 0 under the assumption of one dimensional geometry, that is, $a \gg d$. Right after the ionization was created electrons move to the collecting electrode with a constant velocity in a uniform electric field. This movement of electron induces a constant current in the electrode connected to a preamplifier. The current stops when signal electrons arrive at the electrode. The duration of the current pulse is the distance traveled by signal electrons divided by the drift velocity of electrons. In the coordinate system of Figure 7 $(d - y_0) = t_{end} \cdot v_{drift}$. Simply put, the duration of the signal current measures the y–coordinate of the particle.

The plot on the right on Figure 7 shows a dotted line indicating the current waveform produced by a background charged particle which crossed the cell along the y axis. The ionization was produced almost uniformly along the y axis. Signal electrons start to reach the collecting electrode close to it already at time zero and continue to arrive with a constant rate during the drift time of electrons across the full gap. The resulting current waveform is triangular and the area under the dotted line which is related to the total ionization of the background track is much smaller than the area under the signal current waveform due to the geometry of the cell. It is intuitive to see that we can measure the end point of the signal current waveform in the presence of background particles.

The position of the second coordinate, called x in Figure 7, of the particle within the cell is determined by the signal division among the consecutive electrodes of width a. (See the Appendix)

Other Limitations

Less sensitive regions

When the fast charged particle passes too close to the collecting electrodes, $(d - y_0) \ll d$, the signal electrons move only a short distance and the induced charge is small. One can arrive to the same conclusion from an equivalent point of view. The remaining positive charge at the production point x_0, y_0 induces an opposite polarity signal at the close–by electrode as the collected charge of negative electrons. The net signal is therefore small. To avoid the loss of sensitivity in one part of the cell, Figure 8 shows a perspective view of an array of cells where the direction of the electric field reverses along the cell. The least sensitive region in the lower half of the cell corresponds to the most sensitive region in its upper part. There is always a backup region behind a less sensitive region of a cell. The obvious drawback is the doubling the number of read out channels.

At this point it is noted that in order to avoid the dead layer of high voltage and of collecting electrodes it may be beneficial to introduce a small deviation (twist) from a perfect projection geometry. The background rejection and the resolution would be effected only slightly, while the small dead regions would be completely eliminated.

$\vec{E} \times \vec{B}$ Effect

The proposed cell provides the most precise position measurement along the direction of the motion of electrons (y coordinate in Figure 7 or ϕ coordinate in Figure 3). The deflection of a charged particle in a magnetic field is also measured along this direction. The achievable momentum resolution is thus optimized, however, we have to deal with the case where the electric field is perpendicular to the magnetic field leading to the maximal $\vec{E} \times \vec{B}$ effect. In the uniform electric field of the cells the effect causes signal electrons to drift along lines witch are at an angle α from the direction of the applied electric field and at a slightly reduced velocity.

The angle α is called a Lorentz angle in gas detectors and Hall angle in semiconductor detectors. It will be referred to as a Hall angle since rare gas liquids are closer to semiconductors than gases. Moreover, it will be assumed that some of the physics of Hall angle from semiconductors can be applied to liquids. The vector of average velocity of an electron under the influence of the electric field $\vec{E}$ and the magnetic field $\vec{B}$ equals (Equation 4 in Reference [5]):

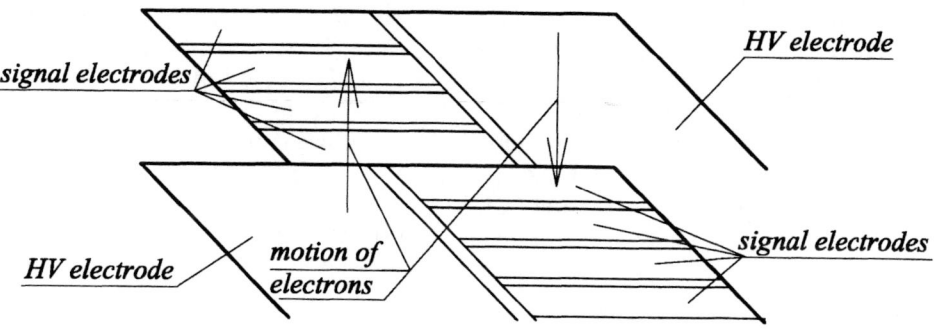

FIGURE 8. Perspective view of an array of cells where signal electrodes and HV electrode are swapping position along the trajectory of a fast particle. The direction of the electron motion is reversed in two parts of the cell.

$$\begin{pmatrix} dx/dt \\ dy/dt \\ dz/dt \end{pmatrix} = -\mu_r \begin{pmatrix} 1+\mu_H^2 B_x^2 & \mu_H^2 B_x B_y - \mu_H B_z & \mu_H^2 B_x B_z + \mu_H B_y \\ \mu_H^2 B_x B_y + \mu_H B_z & 1+\mu_H^2 B_y^2 & \mu_H^2 B_z B_y - \mu_H B_x \\ \mu_H^2 B_x B_z - \mu_H B_y & \mu_H^2 B_z B_y + \mu_H B_x & 1+\mu_H^2 B_z^2 \end{pmatrix} \begin{pmatrix} E_x \\ E_y \\ E_z \end{pmatrix} \quad (1)$$

where the reduced electron mobility μ_r, due to the presence of $\vec{B}$, is given by

$$\mu_r = \frac{\mu}{1+\mu_H^2 B^2} \quad (2)$$

where $dx/dt, dy/dt$ and dz/dt are the components of the average velocity, μ is the electron mobility with all limitation due to the velocity saturation in liquids and μ_H is the Hall mobility.

The above system of 3 equations completely describes the electron transport within the liquid of the cells. Using the coordinate system of Figure 7 in which $E_x = E_z = 0$, $E_y = E$ and $B_y = B_z = 0$, $B_x = B$. This orientation of the magnetic field is exact for a cell located at $90°$ from the intersecting beams. At other angles there is a non zero component of the magnetic field B_z (perpendicular to the plane of the paper). A substitution of these components of $\vec{E}$ and $\vec{B}$ into the tensor equation leads to:

$$\begin{pmatrix} dx/dt \\ dy/dt \\ dz/dt \end{pmatrix} = -\mu_r \begin{pmatrix} 0 \\ E \\ \mu_H B \cdot E \end{pmatrix} \quad (3)$$

The ratio $\frac{dz/dt}{dy/dt} = \mu_H B = \tan \alpha$ is the definition of the Hall angle α and the Hall mobility μ_H. The Hall mobility differs from the electron mobility by fine details of the average time between scattering events with and without applied magnetic field. The ratio of the Hall mobility to the electron mobility is a weak function of the magnetic field. We will assume that this ratio μ_H/μ is the same in liquids as in silicon where it is about 1.2 [5].

The effective value of the electron mobility in liquid argon is small. By dividing the electron velocity of $3mm/\mu s$ by the applied electric field of $5kV/cm$ the effective value of μ is $0.6 \times 10^{-2} m^2/Vs$. Assuming conservatively that the ratio μ_H/μ is between 1 and 2, the Hall mobility μ_H value is between $0.006 m^2/Vs$ and $0.012 m^2/Vs$. In a field of $3 Tesla = 3Vs/m^2$ the tangent of the Hall angle is between 2% and 4%.

For a vertex cell at $90°$ the trajectory of electrons deviates from the direction of the electric field only in the z coordinate which is not measured. In this location the only measurable effect of the magnetic field is the decrease of the electron drift velocity by a second order factor $1 + \mu_H^2 B^2 \approx 1.0001$, too small to consider. At other locations of the cell we will see a projection of the Hall angle directly into the measured coordinate x. The required precision of the interpolation is about 1% while $\tan \alpha$ can be as large as 4%. To bring the $\vec{E} \times \vec{B}$ correction down to a percent level it is sufficient to understand the value of the Hall mobility down to a 25% level.

Positive Space Charge

The mobility of positive ions in liquid argon is about four orders of magnitude smaller that the mobility of electrons. Positive ions drift very slowly to the HV electrode of the cell with a typical time of $3ms$ to be swept away. The presence of positive charges accumulated during this time influences the electric field within the cell and may modify the simple proportionality between the y position of the particle and the electron drift time. The induced signal due to the slow motion of accumulated positive charges can produce a noise like signal at the input of the preamplifiers.

Let us estimate the effect of the space charge in the first layer of the vertex detector at the distance $5cm$ from the interaction region. It was seen in the subsection describing the main background due to the conversion of photons that there are about 5.4 photon converting within one $0.5mm \times 2mm \times 30mm$ cell per bunch crossing. The average energy deposited by a photon in that cell, as can be seen form Figure 6, is slightly less than $0.5MeV$. Photon background deposits about $2.5MeV$ in a single cell per one bunch crossing. $25eV$ of energy is needed to produce 1 electron–ion pair, thus there are 10^5 pair produced in the volume of the cell per 1 crossing. The number of bunch crossing per second depends on the parameters of the machine and it is assumed here that there are $2 \cdot 10^4$ crossings per second. The rate of the ion production in the first cell is $2 \cdot 10^9 s^{-1}$ per volume of the cell ($= .5 \times 2 \times 30 mm^3$) or the volume intensity of the source s is $0.7 \cdot 10^8 mm^{-3} s^{-1}$.

It is also assumed that the ions are created uniformly in time and in the volume of the cell and from the moment of the creation move uniformly along the $-y$ axis of Figure 7 with velocity $v_{ion} = 300mm/s$ toward the HV electrode. Using the equation of continuity in one dimension with the source term s, one can find that the density of positive ions along the axis y has a triangular shape given by the expression $\rho(y) = (d-y) \cdot s/v_{ion}$. This form can be understood intuitively. Positive ions are swept away from the signal electrode region $y \leq d$ which is supplied only by the source term s. The HV electrode region $y \geq 0$ contains not only sourced ions but ions arriving from the whole length d. The mean density of positive ions in the cell is $2 \cdot 10^5 mm^{-3}$.

The presence of positive charges in the volume of the cell perturbs the uniform electric field of a charge free cell $E = U/d$. The size of the perturbation can be obtained by solving the Poisson equation

$$\frac{\partial U_p(y)}{\partial y^2} = -\frac{q \cdot \rho(y)}{\epsilon_0 \cdot \epsilon_r} \left(= \frac{q \cdot s \cdot (y-d)}{\epsilon_0 \cdot \epsilon_r \cdot v_{ion}}\right), \tag{4}$$

where U_p is the potential of the perturbation, q is the absolute value of the electron charge, ϵ_0 permittivity of vacuum and ϵ_r is the relative dielectric constant of liquid argon. The boundary condition are $U_p(0) = U_p(d) = 0$ because the value of the potential on the electrodes is fixed. The Poisson equation with the ion density $\rho(y)$ can be solved directly by double integration. The solution, which satisfies the boundary condition is

$$U_p = \frac{q \cdot s}{\epsilon_0 \cdot \epsilon_r \cdot v_{ion}} \cdot \left(\frac{y^3}{6} - \frac{y^2 \cdot d}{2} + \frac{y \cdot d^2}{3}\right). \tag{5}$$

The electric field $E_p = -\partial U_p/\partial y$ is:

$$E_p = -\frac{q \cdot s}{\epsilon_0 \cdot \epsilon_r \cdot v_{ion}} \cdot \left(\frac{y^2}{2} - y \cdot d + \frac{d^2}{3}\right) \tag{6}$$

The electric field generated by the space charge of positive ions E_p is negative close to the HV electrode, that is, $y \geq 0$, then its value goes through zero and it is positive for $y \leq d$. The maximum absolute value of E_p at $y = 0$ is about $2V/mm$, which is 0.4% of the applied field of $500V/mm$. It seems that the perturbation of the electric field due to the positive ions does not represent a problem even within the cells positioned nearest to the interaction point.

The slow movement of positive ions is seen as a current by the preamplifier connected to the electrode of the cell. The average value of this current is just 1/2 of the rate of the ion production in the cell times the charge of individual ion or $\bar{I} = 1/2 \cdot 2 \cdot 10^9 \cdot qs^{-1} = 160pA$. This is a very small current when compare to the leakage currents of position sensing silicon detectors and can be easily accommodated by the preamplifier.

The spectral density of this current may be a more important issue. The current power spectral density of this current of $160pA$ viewed as being composed of individual uncorrelated electrons is of little importance for the signal processing time of interest, that is, about $1\mu s$. To see it, the "leakage current" of $160pA$ corresponds to the flow of 10^9 electron charges per second. In average an electrode collects 1000 charges within $1\mu s$. Poisson fluctuations of this number of charges is ± 30. This is negligible relative to a signal due to a particle produced at the interaction point and crossing the cell which is about 10^5 electrons.

The current due to the motion of positive ions is not composed from individual elementary charges. The average energy deposited in the cell by a background photon is $0.5Mev$ which produces $2 \cdot 10^4$ electrons. The uncorrelated charges are therefore much higher. However, due to the slower velocity of positive ions as compared to signal electrons (four orders of magnitude) the net effect is practically unchanged. We can conclude that the fluctuations in the current due to motion of positive charges does not degrade the performance of the vertex detector.

Now, all numbers needed to estimate the heating of liquid argon due to the drift of electrons and positive ions within a cell with $1kV$ applied are available. The sum of electron and ion currents is $320pA$ per one $0.5 \times 2mm^2$ entrance area cell of the innermost layer. Ohmic heating is $0.32\mu W$ per cell. With number of cells slightly less than 10^5 in this layer the total ohmic heating is only about $30mW$. this is much smaller than the expected heat dissipation of the front end electronics.

Inner Tracker

There is no reason to change the technology of the inner tracker from the proposed technology of the vertex detector. To keep the number of channels at reasonable level, the transverse dimensions of the cells can be increased. The background here is not a problem and individual cells can be made large enough so that the position resolution matches the mechanical tolerances of rather large layers of the tracker. The number of layers and the depth of individual layers is a compromise which leads to the best momentum resolution of the charged particles. If neon is used, we may even think of a tracker volume filled completely with the liquid neon.

This inner detector can be considered as a first class pre–shower detector. It may have just the right number of radiation length in front of the last position sensitive layer to convert a large enough percentage of high energy gammas. Clearly, it provides a very good position information for the shower localization.

The details of the momentum measurement for the charged particles will not be addressed here. The outer radius of the inner tracker has to be larger for the $100 TeV$ collider than for $10 TeV$ collider. Taking the outer radius of the tracker to be $1.2m$ at $90°$, the momentum resolution can be slightly better than $\delta P/P \lesssim 10^{-4}/GeV/c$. Thus $1 TeV/c$ particles will be measured with 10% resolution and it will be possible to identify the charge of a particle with 3σ confidence level up to the momentum of $4 TeV/c$.

Electromagnetic Calorimeter and Nose Calorimeter

In the previous studies [1] the choice of technology for the Electromagnetic Calorimeter was an accordion liquid argon calorimeter similar to one being under construction for the ATLAS Experiment at the CERN LHC. This is still a valuable option. To instrument two tungsten shield cones (noses) reaching close to the interaction point a different approach to the geometry of the calorimeter is needed. Once we have a valid technology for the nose calorimeters, we may as well apply it to the Electromagnetic calorimeter if it simplifies the system and provides an equivalent or superior performance.

The proposed technique is a refinement of the ATLAS liquid argon forward calorimeter [6] which was originally proposed for GEM for SSC [7]. The ATLAS and GEM unit cell of the calorimeter (see Figure 1 in Ref. [6]) consists of a solid brass rod of $4.5mm$ diameter inserted in a brass tube of inner diameter of $5mm$, resulting in a cylindrical gap of $250\mu m$ around the rod. The rod is fixed at the both ends and the gap spacing is maintained by an insulating quartz fiber wound around the rod, having the same diameter of $250\mu m$ as the gap filled with liquid argon.

An electric field is applied across the gap to collect the ionization electrons in the argon. The signal is then fed out, at the rear end of the rod through a low impedance cable to a preamplifier. The forward calorimeter is a matrix of such cells with tubes mutually parallel in order to have the same sampling fraction through the depth of the calorimeter.

The design of the instrumented nose calls for a projective geometry, with finer sampling and smaller gap to maintain the shielding properties of the cone. The granularity should be very high to improve the discrimination against background. The density of the detector should be at least 95% of the density of the tungsten. To keep the constant energy sampling ratio independent of the radius in the projective geometry the gap has to change. We may change the radius of the tube or the radius of the rod.

More discrimination power relative to background can be gained by the longitudinal division of the cell along the shower direction. We can segment the outer surface of the rod and buried cables within the rod to connect all longitudinal segments out through the back.

The Electromagnetic Calorimeter can be constructed with similar cells. The material will not be the tungsten, but most likely brass or copper. Being further away form the intersection, the tube diameter may be larger. The main possible advantage of the projective geometry of these cells as compared to accordion (also in a projective geometry) is a superior pointing to the interaction region by rod cells. The design will require more complete simulations.

During the discussion with the theory group it was pointed out, that for some broad class of physics an ultimate energy and position resolution of the Electromagnetic calorimeter may be needed. An natural candidate would be a fully absorbing liquid xenon calorimeter. The cells of the calorimeter are practically identical to the cells of the vertex detector or tracker. The electronics will be much simpler, sensing only the total charge deposited in individual cells. Clearly, to work with liquid xenon requires a higher degree of a chemical cleanliness than when working with liquid argon.

The background situation and the trigger formation are well described on Page 64 of Reference [2].

Hadron Calorimeter

None of the energy generated by background photons is expected to penetrate into the hadronic calorimeter. The main background is due to the neutrons and Bethe–Heitler muons passing practically parallel to the beam direction. Calculations [2] shows that the background levels of neutrons and muons do not prevent a missing energy trigger down to 20 to 50 GeV level. Two options for the hadron calorimeter are considered. i) Liquid argon which has the advantage of being blind to neutrons and ii) Scintillator tile calorimeter which seems to be more economical. Both options are still valid. There was very little discussion about the hadron calorimeter options during the workshop.

Muon System

All 3 options mentioned in [1,2] are still under consideration. These are: i) Cathode strip chambers, ii) Threshold Cerenkov counter and iii) Long drift pad chambers with pad read out. The new strategy is to move the main precision muon detectors to a larger radius, to benefit of the rapid fall off of the background. A new muon absorber, in addition to calorimeters is introduced. In the forward directions, this strategy may include a Cerenkov detector measuring the angle of the muon. The strategy can be combined with a magnet supplying a solenoidal field for the inner tracker and the muon detector. To avoid wasted, expensive, magnetic field volume an additional bending takes place in the absorber. The momentum information from the muon system is combined with the measurement of the momentum from the inner tracker to improve the resolution and to reject muons originating from pion decays.

Appendix. Position resolution in a proposed cell

We will use the coordinate system shown on the left hand side of Figure 7. The ionization produced by a particle created in the interaction point is seen as a point ionization in this view. Let us concentrate on the particle which produced a unit of ionization charge at the point $(x_1, y_1) = (0.9, 1)$ in the coordinate system, where we have fixed the cell dimension at $a = d = 2mm$.

Electrons created by the ionizing particle move along a straight line $x = x_1 = 0.9$ toward the electrode while positive ions move so slowly that we will consider them being fixed at point x_1, y_1 during the signal processing time. While electrons are moving the signal current is induced on all electrodes "visible" from the electron trajectory. While the concept of the induced current is intuitive, here we will use more often a closely related concept of the induced charge on an electrode. The induced charge is simply the time integral of the induced current. Our preference to use the integral of the current rather than the current itself is due to the different shape of noise spectral density in both cases.

We have seen that the effective leakage current of a cell is very small. The parallel noise at the input of the preamplifier is negligible and the only relevant noise source is the series noise of the preamplifier which is proportional to the frequency. We can transform the series noise into a white noise, that is, the noise which has the uniform density independent of frequency, by integrating the input signal. This pre-whitening filter transforms induced current into induced charge. When processing signals from the first layer of the vertex detector (where the leakage current is not completely negligible due to the cumulative current from motion of positive ions) to obtain white noise, we have to integrate with an integration constant equal to the noise time corner τ. Here we will assume that $\tau \gg 1\mu s$, $1\mu s$ being the processing time for electron signals.

The signal induced at any electrode by the presence of charge Q located at position x, y within the gap can be obtained by solving the direct electrostatic problem, that is, by solving the Laplace equation of electrostatics with the charge Q at x, y and all electrodes (boundaries) at zero potential. A charge induced at any particular electrode is, according to Gauss's theorem, the surface integral of the electric field ending on the surface times dielectric constant ϵ_0. When the charge is moving within the gap, in order to know the induced charges at the electrode produced by the charge Q in the next position we have to find a complete solution for a different electrostatic field, possibly a time consuming approach.

We can apply Green's Reciprocation Theorem [8] to be able to know the field induced by a charge Q located at any arbitrary point x, y on a given electrode with only one solution of the electrostatic field problem. The final result of the application of the Green's Reciprocation Theorem can be stated: the charge induced at a given electrode by a charge Q located at x, y equals to the product of the charge Q times the value of the potential of an induction (auxiliary) field U at the point x, y. The induction (auxiliary) field is the solution of an electrostatic problem, when the value of 1 is applied at the electrode on which we want to know the induced charge and 0 on all other electrodes of the problem and the space between is free of any charges.

The induction (auxiliary) field for the central electrode of Figure 7, ($y = d$, $-a/2 < x < a/2$) can be found in a closed form the by method of conformal mapping. In the volume of the cell, that is, for $0 < y < d$ one can write:

$$U(x,y) = \frac{1}{\pi} \left\{ \arctan\left[\tanh(\pi \frac{x+a/2}{2d}) \cdot \tan(\pi \frac{y}{2d})\right] \right.$$
$$\left. - \arctan\left[\tanh(\pi \frac{x-a/2}{2d}) \cdot \tan(\pi \frac{y}{2d})\right] \right\}, \qquad (7)$$

or:

$$U(x,y) = \frac{1}{\pi} \left\{ \arctan\left[\tanh(\beta) \cdot \tan(\gamma)\right] - \arctan\left[\tanh(\alpha) \cdot \tan(\gamma)\right] \right\}, \qquad (8)$$

where α (no relation to the Hall angle), β and γ have the following dependence on the x, y coordinates:

$$\alpha = \pi \frac{x-a/2}{2d}; \quad \beta = \pi \frac{x+a/2}{2d}; \quad \gamma = \pi \frac{y}{2d} \qquad (9)$$

The charge induced on the central electrode by the moving electrons is $Q_-(t) = -Q \cdot U(x_1, y = y_1 + v \cdot t)$ during their drift time $\tau = (d - y_1)/v$ where v is the drift velocity of electrons in the liquid. The negative sign corresponds to a standard convention of electron having a negative charge. When electrons arrive onto the collected electrode, U on the electrode equals 1, and they induce the full charge on the electrode. This full induction is equivalent to an actual collection of the electrons by the electrode. The charge induced by positive ions is $Q_+ = Q \cdot U(x_1, y_1)$

and does not change with time. The net induced charge is the sum of the charges induced by electrons and positive ions: $Q(t) = Q_-(t) + Q_+$. This time dependence can be seen from the left hand side of Figure 9. At time zero the electrons and positive ions are at the same location and the net induced charge is zero. After collecting all electrons the net negative charge on the electrode is less than 1 due to the induction of opposite polarity charges of ions remaining within the cell's volume.

The charge induced on the neighboring electrodes can be found from the translational symmetry of the geometry along the x-axis. The charge induced on an electrode centered at $x_c = 0 + k \times a$, (k being an integer), by a charge located at x, y is the same as the charge induced on the central electrode by the charge located at point $(x - k \times a, y)$, which is given by Eg. 7 as $U(x - k \times a, y)$. The right hand side of Figure 9 shows the induced charge on the two electrodes adjacent to the central electrode. The continuous line corresponds to the charge induced on the electrode with its center at $x = a$, that is, right from the central electrode. The particle crossed the cell at $x_1 = 0.9$, not too far from the boundary between these two cells at $x = 1$. During the first part of the electron drift time, the charge

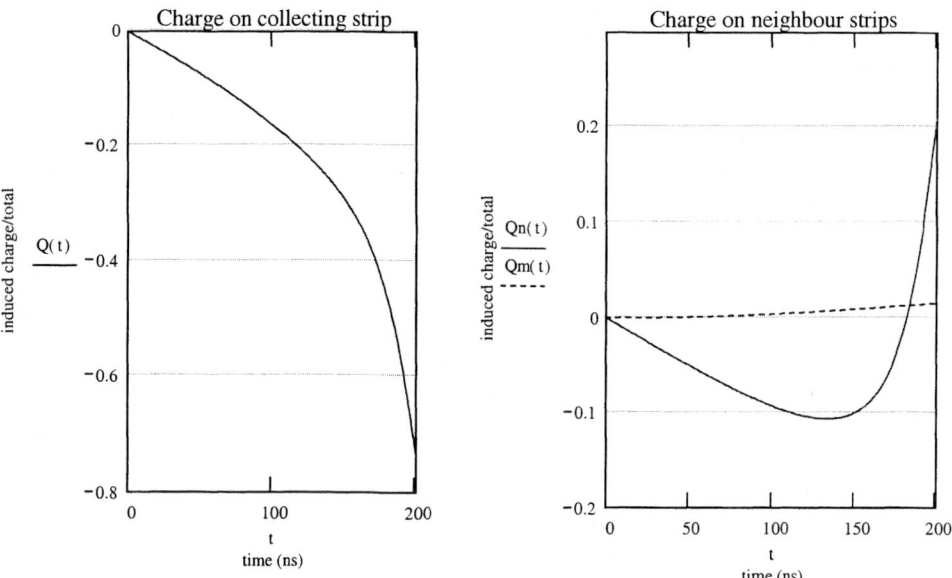

FIGURE 9. Time dependence of induced charge normalized to 1 by ionization at point $(x_1, y_1) = (0.9, 1)$ seen: i) on signal collected electrode (left hand side) and ii) on the neighbor electrodes (right hand side). The drift velocity of electrons was $5 \mu m/ns$. After 200 ns of electron collection time, all waveforms remain constant.

induced on this neighbor cell is only slightly smaller than the charge induced on the central electrode. However, during the final moments of the electron drift, electrons are headed to the central electrode and the amount of negative charge induced by electrons decreases. When electrons arrive to the central electrode the only charge induced on the neighbor electrode is the positive charge due to remaining ions.

The shapes of the induced charge signals on each electrode depend only on 3 parameters and on the noise in the read outs. The 3 parameters are: x and y coordinates of the crossing particle and the amount of charge Q produced by the particle. It is assumed that the time of the particle crossing is known from the timing of the bunch crossing. The problem is to find these 3 parameters from the measurable waveforms of the induced charge on the electrodes of interest. Noise present in all read out channels is a stochastic process and we will use statistical methods to infer the best estimates of these 3 parameters from observed waveforms for each event.

The noise superposed on the ideal waveforms is white as proven above. In the time domain (that is, when we are inspecting a measured noisy waveform as function of time as presented on an oscilloscope) the white noise means that the fluctuations at different times are all the same and that there is no correlation between fluctuations at any two different times. We can think to sample induced charges (and noise) from all electrodes and to find x, y and Q to "fit" shapes of all recorded waveforms. We will use the most powerful statistical tool, the maximum likelihood method for our analysis. The analysis presented here is similar to that of [9].

The logarithm of the likelihood function L is

$$\ln(L) = \sum_k \sum_n \left\{ \ln\left[\frac{1}{\sqrt{2\pi} \cdot \sigma_k(t_n)}\right] - \frac{[Q_k(x,y,Q,t_n) - m_k(t_n)]^2}{2\sigma_k^2(t_n)} \right\}, \quad (10)$$

where $\sum_k$ denotes the summation through the waveform produced by different electrodes k, $\sum_n$ is the summation through different sampling times t_n, $m_k(t_n)$ is the measured (sampled) waveform from the kth electrode at the sampling time t_n and $Q_k(x,y,Q,t_n)$ is the value of the noiseless waveform from k^{th} electrode at time t_n when a particle crossed the cell at the point x, y and produced the total ionization charge Q, and $\sigma_k(t_n)$ is the square root of the variance of the kth waveform at time t_n.

The sampled waveform has to be passed through a bandwidth limited or Nyquist filter before being sampled. This filter limits also the white noise of individual samples. For our case $2\sigma_k^2(t_n) = C_{tot}^2 \cdot e_s^2/\Delta t$ where C_{tot} is the total input capacitance of individual electrodes including the capacitance of the input transistor and the capacitance of connections, e_s^2 is the voltage power spectral density of the preamplifier and Δt is the sampling interval.

The noiseless waveform from the kth electrode can be written as:

$$Q_k(x,y,Q,t_n) = Q \cdot [U(x-ka,y) - U(x-ka, y+v\cdot t_n)] \quad 0 < t_n < \tau \quad (11a)$$

$$Q_k(x,y,Q,t_n) = Q \cdot [U(x-ka,y) - \delta_k] \quad \tau < t_n < T, \quad (11b)$$

where τ is the drift time of electrons from the creation point x, y to the electrode at $x = d$.

$$\tau = (d - x)/v \tag{12}$$

and we have introduced a total processing time T. This time must be shorter than the time between two consecutive bunch crossings. Practically there is very little improvement beyond $T = 3\tau_{max} = 3d/v$. Eq. 10 can be now be rewritten as

$$\ln(L) = N - \sum_k \sum_n \frac{[Q_k(x, y, Q, t_n) - m_k(t_n)]^2}{C_{tot}^2 e_s^2 / \Delta t}, \tag{13}$$

where $Q_k(x, y, Q, t_n)$ is given by Eqs. 11 and N is a normalization factor. Following the method of [9] we will linearize $Q_k(x, y, Q, t_n)$ as functions of the unknown parameters x, y and Q at a starting point $x^{(0)}, y^{(0)}$ and $Q^{(0)}$ of an iteration process.

$$Q_k(x, y, Q, t_n) = Q_k(x^{(0)}, y^{(0)}, Q^{(0)}, t_n) + \frac{\partial Q_k(x, y, Q, t_n)}{\partial x} \cdot (x - x^{(0)})$$

$$+ \frac{\partial Q_k(x, y, Q, t_n)}{\partial y} \cdot (y - y^{(0)}) + \frac{\partial Q_k(x, y, Q, t_n)}{\partial Q} \cdot (Q - Q^{(0)}) \tag{14}$$

All partial derivatives in Eq. 14 are meant to be taken at $x^{(0)}, y^{(0)}$ and $Q^{(0)}$ and do not depend on x, y and Q. We substitute Q from Eq. 14 into Eq. 13 and obtain a linearized expression for the logarithm of likelihood to be maximized relative to 3 parameters x, y and Q.

$$\ln(L) = N - \sum_k \sum_n \frac{\Delta t}{C_{tot}^2 e_s^2} \cdot \left[Q_k(x^{(0)}, y^{(0)}, Q^{(0)}, t_n) + \frac{\partial Q_k(x, y, Q, t_n)}{\partial x} \cdot (x - x^{(0)}) \right.$$

$$\left. + \frac{\partial Q_k(x, y, Q, t_n)}{\partial y} \cdot (y - y^{(0)}) + \frac{\partial Q_k(x, y, Q, t_n)}{\partial Q} \cdot (Q - Q^{(0)}) - m_k(t_n) \right]^2 \tag{15}$$

Partial derivatives of $\ln(L)$ with respect to x, y and Q must equal to zero leading to a system of 3 linear equations:

$$(x - x^{(0)}) \sum_{k,n} \left[\frac{\partial Q_k(t_n)}{\partial x}\right]^2 + (y - y^{(0)}) \sum_{k,n} \left[\frac{\partial Q_k(t_n)}{\partial x}\right] \cdot \left[\frac{\partial Q_k(t_n)}{\partial y}\right] +$$

$$(Q - Q^{(0)}) \sum_{k,n} \left[\frac{\partial Q_k(t_n)}{\partial x}\right] \cdot \left[\frac{\partial Q_k(t_n)}{\partial Q}\right] = -\sum_{k,n} [Q_k(t_n) - m_k(t_n)] \frac{\partial Q_k(t_n)}{\partial x} \tag{16a}$$

$$(x - x^{(0)}) \sum_{k,n} \left[\frac{\partial Q_k(t_n)}{\partial x}\right] \cdot \left[\frac{\partial Q_k(t_n)}{\partial y}\right] + (y - y^{(0)}) \sum_{k,n} \left[\frac{\partial Q_k(t_n)}{\partial y}\right]^2 +$$

$$(Q - Q^{(0)}) \sum_{k,n} \left[\frac{\partial Q_k(t_n)}{\partial y}\right] \cdot \left[\frac{\partial Q_k(t_n)}{\partial Q}\right] = -\sum_{k,n} [Q_k(t_n) - m_k(t_n)] \frac{\partial Q_k(t_n)}{\partial y} \tag{16b}$$

$$(x - x^{(0)}) \sum_{k,n} [\frac{\partial Q_k(t_n)}{\partial x}] \cdot [\frac{\partial Q_k(t_n)}{\partial Q}] + (y - y^{(0)}) \sum_{k,n} [\frac{\partial Q_k(t_n)}{\partial y}] \cdot [\frac{\partial Q_k(t_n)}{\partial Q}] +$$

$$(Q - Q^{(0)}) \sum_{k,n} [\frac{\partial Q_k(t_n)}{\partial Q}]^2 = - \sum_{k,n} [Q_k(t_n) - m_k(t_n)] \frac{\partial Q_k(t_n)}{\partial Q} \qquad (16c)$$

All partial derivatives in Eqs. 15 and 16 have an explicit analytical form. Here, just for completeness, are the derivatives of $U(x,y)$

$$\frac{\partial U(x,y)}{\partial x} = \frac{\sin \gamma \cos \gamma}{2d} \left(\frac{1}{\cos^2 \gamma + \sinh^2 \beta} - \frac{1}{\cos^2 \gamma + \sinh^2 \alpha} \right) \qquad (17a)$$

$$\frac{\partial U(x,y)}{\partial y} = \frac{1}{2d} \left(\frac{\sinh \beta \cdot \cosh \beta}{\sinh^2 \beta + \cos^2 \gamma} - \frac{\sinh \alpha \cdot \cosh \alpha}{\sinh^2 \alpha + \cos^2 \gamma} \right), \qquad (17b)$$

where α, β and γ were defined in Eq. 9.

The first solution of linear system of equations 16 gives values for x, y and Q. These value become $x^{(0)}, y^{(0)}$ and $Q^{(0)}$, that is, initial values for the next iteration. All derivatives have to be recalculated at $x^{(0)}, y^{(0)}$ and $Q^{(0)}$ and the system of Eqs. 16 have to be solved again. This iteration process has to be repeated until the respective differences between $x^{(0)}, y^{(0)}$ and $Q^{(0)}$ and x, y and Q are sufficiently small. We will not attempt to prove that the process converges in general to the physical solution.

Let us study the statistical properties of the solution thus obtained. The likelihood function in Eq. 15 can be formally regarded as a probability density function for the parameters x, y and Q viewed as random variables while the final values $x^{(0)}, y^{(0)}$ and $Q^{(0)}$ are the expected values. The logarithm of the probability density function $p(x, y, Q)$ for jointly normal random variables x, y and Q is written in its canonical form as:

$$p(x,y,Q) = N_{cf} - \frac{1}{2} \overline{X}^T \cdot \mathcal{M}^{-1} \cdot \overline{X}, \qquad (18)$$

where N_{cf} is the normalization constant, $\overline{X}$ is the column vector composed of $(x - x^{(0)})$; $(y - y^{(0)})$; $(Q - Q^{(0)})$, $\overline{X}^T$ is its transposed (row) vector and $\mathcal{M}^{-1}$ is the inverse of the symmetric, positive–definite variance matrix [10]. $\overline{X}^T \cdot \mathcal{M}^{-1} \cdot \overline{X}$ is the quadratic form of this three-variate normal distribution. We have to carry on the square of the polynomial terms in brackets of Eq. 15 and try to express it in the form of Eq. 18. After some manipulations the inverse of the variance matrix $\mathcal{M}^{-1}$ can be written

$$\frac{2\Delta t}{C_{tot}^2 e_s^2} \sum_{k,n} \begin{pmatrix} [\frac{\partial Q_k(t_n)}{\partial x}]^2 & [\frac{\partial Q_k(t_n)}{\partial x}] \cdot [\frac{\partial Q_k(t_n)}{\partial y}] & [\frac{\partial Q_k(t_n)}{\partial x}] \cdot [\frac{\partial Q_k(t_n)}{\partial Q}] \\ [\frac{\partial Q_k(t_n)}{\partial x}] \cdot [\frac{\partial Q_k(t_n)}{\partial y}] & [\frac{\partial Q_k(t_n)}{\partial y}]^2 & [\frac{\partial Q_k(t_n)}{\partial y}] \cdot [\frac{\partial Q_k(t_n)}{\partial Q}] \\ [\frac{\partial Q_k(t_n)}{\partial x}] \cdot [\frac{\partial Q_k(t_n)}{\partial Q}] & [\frac{\partial Q_k(t_n)}{\partial y}] \cdot [\frac{\partial Q_k(t_n)}{\partial Q}] & [\frac{\partial Q_k(t_n)}{\partial Q}]^2 \end{pmatrix}, \qquad (19)$$

because the absolute term $\sum_{k,n} \left[Q_k(x^{(0)}, y^{(0)}, Q^{(0)}, t_n) - m_k(t_n)\right]^2$, which gives the information about the statistical quality of the solution, is absorbed in the normalization term N_{cf} and the coefficients of all linear terms sum up to zero. This can be understood heuristically from the condition of maximum of likelihood function L at the solution point. More directly, it can be noticed that the factors multiplying the vector $(x - x^{(0)})$, $(y - y^{(0)})$ and $(Q - Q^{(0)})$ form (apart from a multiplication constant) the right hand side vector of the system of Eqs. 16. At the end of the iteration sequence $x = x^{(0)}$, $y = y^{(0)}$ and $Q = Q^{(0)}$, that is, the final system had a trivial solution. The matrix of the system is not singular and the only way to obtain the zero vector as a solution is by having the right hand side of the system of Eqs. 16 equal to zero.

If the sampling interval Δt is sufficiently small we can replace $\Delta t \sum_n$ with $\int_0^T dt$. The inverse of the variance matrix $\mathcal{M}^{-1}$ then becomes

$$\frac{2}{C_{tot}^2 e_s^2} \sum_k \begin{pmatrix} \int_0^T [\frac{\partial Q_k(t)}{\partial x}]^2 dt & \int_0^T [\frac{\partial Q_k(t)}{\partial x}][\frac{\partial Q_k(t)}{\partial y}] dt & \int_0^T [\frac{\partial Q_k(t)}{\partial x}][\frac{\partial Q_k(t)}{\partial Q}] dt \\ \int_0^T [\frac{\partial Q_k(t)}{\partial x}][\frac{\partial Q_k(t)}{\partial y}] dt & \int_0^T [\frac{\partial Q_k(t)}{\partial y}]^2 dt & \int_0^T [\frac{\partial Q_k(t)}{\partial y}][\frac{\partial Q_k(t)}{\partial Q}] dt \\ \int_0^T [\frac{\partial Q_k(t)}{\partial x}][\frac{\partial Q_k(t)}{\partial Q}] dt & \int_0^T [\frac{\partial Q_k(t)}{\partial y}][\frac{\partial Q_k(t)}{\partial Q}] dt & \int_0^T [\frac{\partial Q_k(t)}{\partial Q}]^2 dt \end{pmatrix} \quad (20)$$

Integrals in Matrix 20 can be numerically calculated and the matrix inverted. We present here only the most important parameters. The square root of the diagonal terms of the inverted matrix are the resolutions in x, y and Q respectively. Figures 10 and 11 show the position resolutions in both coordinates as a function of the position within the cell for two different cell geometries. The values of the resolutions in both directions are shown only for $y < 1.8mm$, that is for 90% of the cell's area. The resolution degrades for an incidence $y \leq d = 2mm$ as explained in the main section. To obtain the position information signals from the total of 3 consecutive cells, $k = -1, 0, 1$ were considered for $a = 2mm$ geometry (Figure 10) and the total of 5 cells, $k = -2, -1, 0, 1, 2$, were considered for $a = 1mm$ cell of Figure 11. All other parameters were the same for both simulations ($d = 2mm$, $e_s = 4nV/\sqrt{Hz}$, $C_{tot} = 5pF$, $T = 1.5\mu s$, $v = 3mm/\mu s$). The charge produced by the passing particle was assumed to be $12fC$, that is, the average charge produced by a minimum ionizing particle traversing $2cm$ of liquid argon. Values of all plotted σ are proportional to e_s and C_{tot} and inversely proportional to the signal charge Q as can be seen directly from the form of the inverse variance matrix $\mathcal{M}^{-1}$ in Equation 20 and Equation 11. Scaling with the other parameters is more complex.

The value of $e_s = 4nV/\sqrt{Hz}$ is modest. The first transistor of the preamplifier has to have a transconductance of only $0.7mS$. This value of transconductance can be reached with the drain current of $2.5 \times 20\mu A = 50\mu A$, $20\mu A$ being the "Maxwell's" current requirement, implying a modest power consumption of the front end electronics in the region of $0.2mW$/channel. The total power consumption of the vertex can be in 20 to 50W region.

The resolution shown is better than the mechanical stability of the physical layers of liquid argon within the inner tracker. If the background is not a problem we can save on the number of channels by increasing the cell size until the electronic resolution matches the mechanical tolerances.

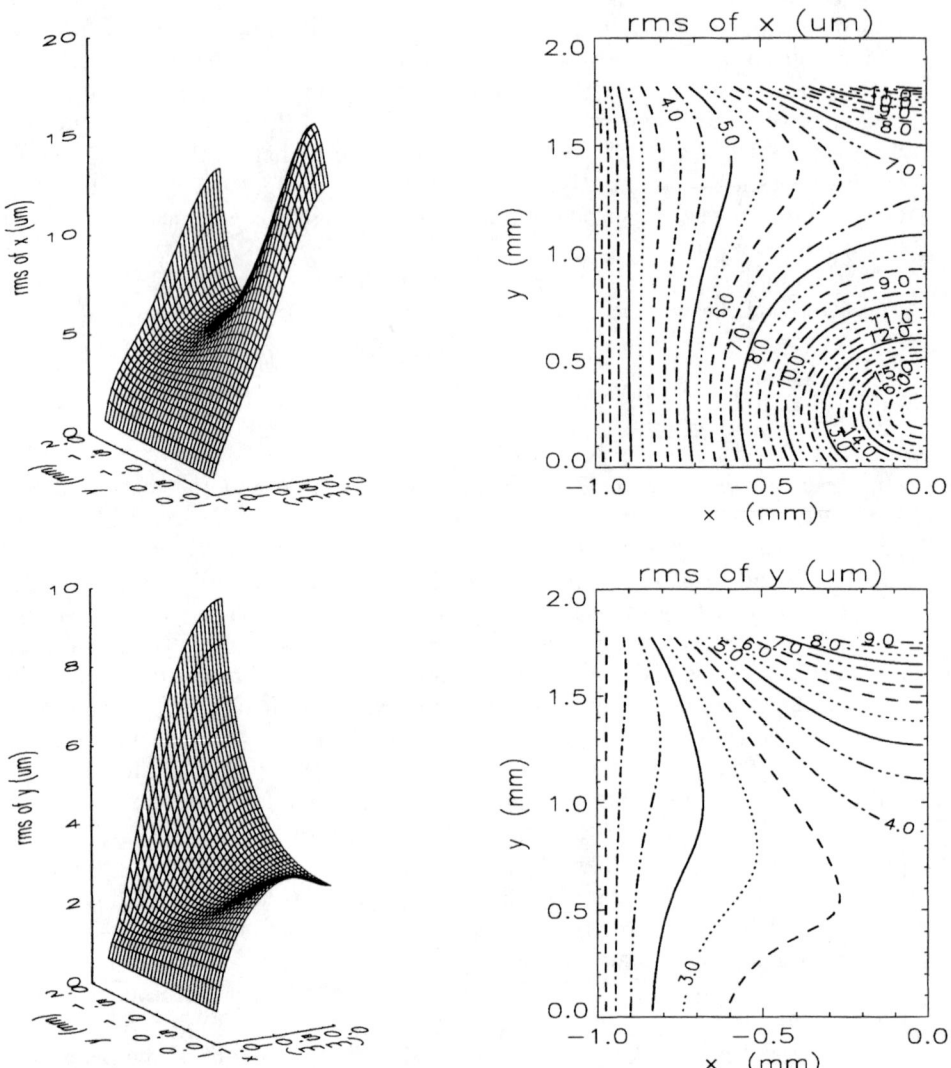

FIGURE 10. Position resolution along coordinate i) x obtained mainly by signal sharing (upper part) and ii) y obtained mainly from the drift time. Only 1/2 of the cell's size of $2mm$ in the x–direction is shown. The resolution is symmetrical around the y–axis ($x = 0$)

The optimal algorithm to obtain the position of a crossing charged particle may seem to be too involved to be practical. There are two answers to this objection.

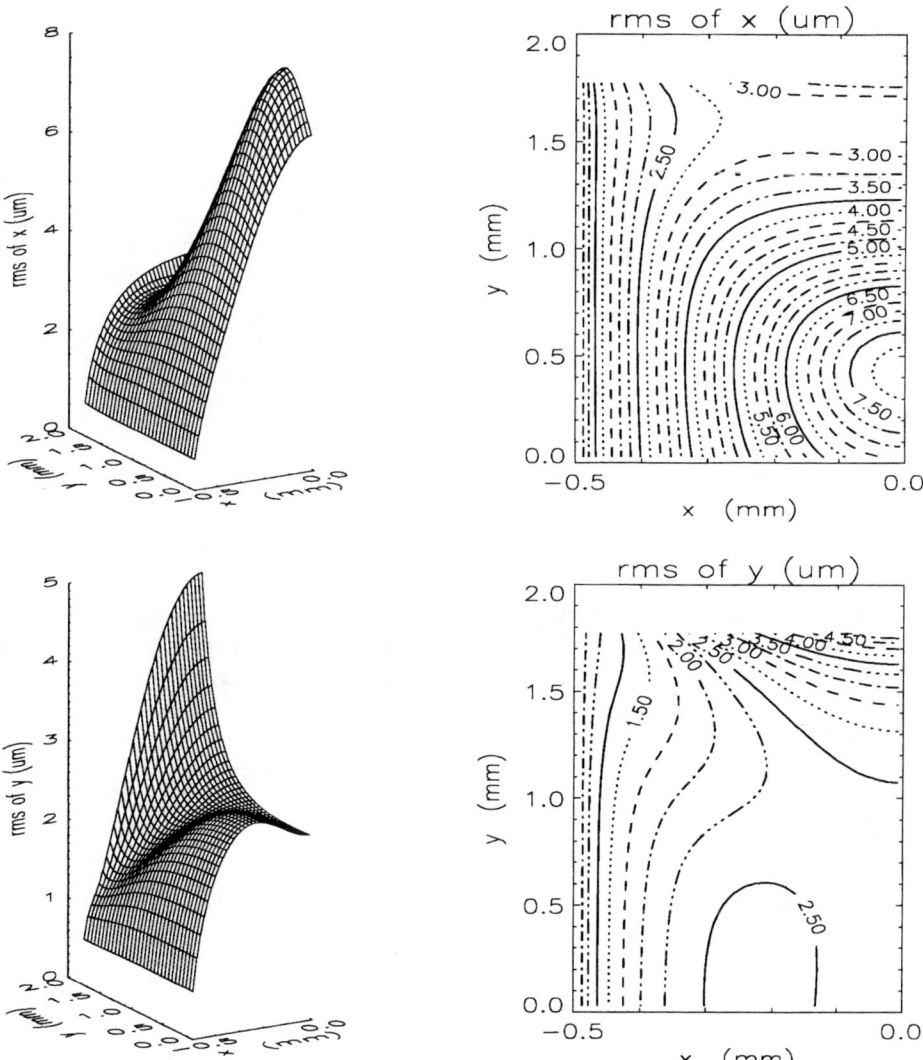

FIGURE 11. Position resolution along coordinate i) x obtained mainly by signal sharing (upper part) and ii) y obtained mainly from the drift time. Only 1/2 of the cell's size of $1mm$ in the x-direction is shown. The resolution is symmetrical around the y-axis ($x = 0$)

1. Given the increase of the power of processors with time, it may be practical, at the time of the detector construction, to put enough computing power right next to liquid argon layers to follow the optimum approach.

2. For a limiting case of $a \gg d$, that is, for one dimensional geometry a complete analytical calculation of maximum likelihood method can be carry through with two unknowns y and Q. The obtainable resolution in y is

$$\sigma_y = \frac{e_s \cdot C_{tot} \cdot d}{Q} \cdot \sqrt{\frac{3v}{2(d-y)}}. \tag{21}$$

A simple analysis of the precision attainable with a linear centroid finding filter gives the resolution of only $2/\sqrt{3} = 1.15$ times wider. A degradation of the position resolution by 15% is acceptable. Very likely, it is possible to develop a simple linear filter for the complete problem to determine both coordinates which is practical to implement and which does not degrade the resolution more than 15 to 30%.

Conclusions

In this contribution we have described a possible detector around an intersection region of a high energy $\mu^+\mu^-$ collider. We have considered the challenges coming from an intense background arriving at the same time as the very rare interactions of interest. The new technology for the vertex detector and inner tracker is based on long projective cells filled with liquid argon and pointing toward the interaction point. Such a detector is radiation hard and less sensitive to background. The performance of such a detector was analyzed in details. The results indicate that this technology would perform according the physics specification.

Moreover, this contribution proposes to instrument the two conical shape tungsten shields to improve the physics potential of the detection system and specifies a particular technology of achieving it.

A new option for the technology of the electromagnetic calorimeter was proposed. An improved muon system was shortly discussed.

REFERENCES

1. The $\mu^+\mu^-$ Collider collaboration, $\mu^+\mu^-$ *Collider–A Feasibility Study*, Snowmass DPF July 1996 Workshop, **BNL–52503, Fermi Lab–Conf.–96/092, LBNL–38946**, (1996); http://www.cap.bnl.gov/mumu/book.html
2. Ch. M. Ankenbrandt et al., (Muon Collider Collaboration) *Phys. Rev. Special Topics* **2**, 081001, (1999). This excellent review contains up to date list of references.
3. S. Geer and J. Chapman, in *Proceedings of 1996 DPF/DPB Summer Study on High-Energy Physics*; Fermilab Report No. **FERMILAB–CONF–96–375**.

4. Werner F. Schmidt, *Liquid state electronics of insulating liquids*, Boca Raton: CRC Press, 1997.
5. A. Castoldi at al., *Nucl. Instrum. and Meth.* **A399**, 227-243(1997).
6. M. I. Ferguson et al., *Nucl. Instrum. and Meth.* **A383**, 399-408 (1996).
7. GEM, Technical Design Report, April 30, 1993, GEM–TN–93–262, SSCL–SR–1219, Page 5–52.
8. John D. Jackson, *Classical Electrodynamics*, Second Edition, John Wiley & Sons, New York, (1975), Problem 1.12, Page 51.
9. G. Bertuccio et al., *Nucl. Instrum. and Meth.* **A322**, 271-279 (1992).
10. B. R. Martin, *Statistics for Physicist*, Academic Press, London and New York, (1971), Page 30.

Kaluza-Klein Physics at Muon Colliders

Thomas G Rizzo [1]

Stanford Linear Accelerator Center
Stanford CA 94309, USA

Abstract. We discuss the physics of Kaluza-Klein excitations of the Standard Model gauge bosons that can be explored by a high energy muon collider in the era after the LHC and TeV Linear Collider. We demonstrate that the muon collider is a necessary ingredient in the unraveling the properties of such states and, perhaps, proving their existence. The possibility of observing the resonances associated with the excited KK graviton states of the Randall-Sundrum model is also discussed.

INTRODUCTION

In theories with extra dimensions, $d \geq 1$, the gauge fields of the Standard Model(SM) will have Kaluza-Klein(KK) excitations if they are allowed to propagate in the bulk of the extra dimensions. If such a scenario is realized then, level by level, the masses of the excited states of the photon, Z, W and gluon would form highly degenerate towers. The possibility that the masses of the lowest lying of these states, of order the inverse size of the compactification radius $\sim 1/R$, could be as low as $\sim$ a few TeV or less leads to a very rich and exciting phenomenology at future and, possibly, existing colliders [1]. For the case of one extra dimension compactified on S^1/Z_2 the spectrum of the excited states is given by $M_n = n/R$ and the couplings of the excited modes relative to the corresponding zero mode to states remaining on the wall at the orbifold fixed points, such as the SM fermions, is simply $\sqrt{2}$ for all n. These masses and couplings are insensitive to the choice of compactification in the case of one extra dimension assuming the metric tensor factorizes, *i.e.*, the elements of the metric tensor on the wall are independent of the compactified co-ordinates.

If such KK states exist what is the lower bound on their mass? We already know from direct Z'/W' and dijet bump searches at the Tevatron from Run I that they must lie above $\simeq 0.85$ TeV [2]. A null result for a search made with data from Run II will push this limit to $\simeq 1.1$ TeV or so. To do better than this at present we must rely on the indirect effects associated with KK tower exchange in what

---

[1] Work supported by the Department of Energy, Contract DE-AC03-76SF00515

essentially involves a set of dimension-six contact interactions. Such limits rely upon a number of additional assumptions, in particular, that the effect of KK exchanges is the *only* new physics beyond the SM. The strongest and least model-dependent of these bounds arises from an analysis of charged current contact interactions at both HERA and the Tevatron by Cornet, Relano and Rico [3] who, in the case of one extra dimension, obtain a bound of $R^{-1} > 3.4$ TeV. Similar analyses have been carried out by a number of authors [4,5]; the best limit arises from an updated combined fit to the precision electroweak data [5] as presented at the 1999 summer conferences [6] and yields [7] $R^{-1} > 3.9$ TeV for the case of one extra dimension. From the previous discussion we can also draw a further conclusion for the case $d = 1$: the lower bound $M_1 > 3.9$ TeV is so strong that the *second* KK excitations, whose masses must now exceed 7.8 TeV due to the above scaling law, will be beyond the reach of the LHC. This leads to the important result that the LHC will *at most* only detect the first set of KK excitations for $d = 1$.

In all analyses that obtain indirect limits on M_1, one is actually constraining a dimensionless quantity such as

$$V = \sum_{n=1}^{\infty} \frac{g_n^2}{g_0^2} \frac{M_w^2}{M_n^2}, \qquad (1)$$

where, generalizing the case to d additional dimensions, g_n is the coupling and M_n the mass of the nth KK level labelled by the set of d integers n and M_w is the W boson mass which we employ as a typical weak scale factor. For $d = 1$ this sum is finite since $M_n = n/R$ and $g_n/g_0 = \sqrt{2}$ for $n > 1$; one immediately obtains $V = \frac{\pi^2}{3}(M_w/M_1)^2$ with M_1 being the mass of the first KK excitation. From the precision data one obtains a bound on V and then uses the above expression to obtain the corresponding bound on M_1. For $d > 1$, however, independently of how the extra dimensions are compactified, the above sum in V *diverges* and so it is not so straightforward to obtain a bound on M_1. We also recall that for $d > 1$ the mass spectrum and the relative coupling strength of any particular KK excitation now become dependent upon how the additional dimensions are compactified.

There are several ways one can deal with this divergence: (*i*) The simplest approach is to argue that as the states being summed in V get heavier they approach the mass of the string scale, M_s, above which we know little and some new theory presumably takes over. Thus we should just truncate the sum at some fixed maximum value $n_{max} \simeq M_s R$ so that masses KK masses above M_s do not contribute. (*ii*) A second possibility is to note that the wall on which the SM fermions reside is not completely rigid having a finite tension. The authors in Ref. [8] argue that this wall tension can act like an exponential suppression of the couplings of the higher KK states in the tower thus rendering the summation finite, *i.e.*, $g_n^2 \to g_n^2 e^{-(M_n/M_1)^2/n_{max}^2}$, where n_{max} now parametcrizes the strength of the exponential cut-off. (Antoniadis [7] has argued that such an exponential suppression can also arise from considerations of string scattering amplitudes at high energies.) For a fixed value of n_{max}, the exponential approach is found to be more effective and lead to a smaller sum than that obtained by simple truncation and thus to a weaker bound on M_1. (*iii*) A last scenario [9] is to note the possibility that the SM wall fermions may have a finite size in the extra dimensions which smear out and soften the couplings appearing in the sum to yield a finite result. In this case the suppression is also of the Gaussian variety.

We note that in all of the above approaches the value of the sum increases rapidly with d for a fixed value of the cut-off parameter n_{max}. For $d = 2(> 2)$ the sum behaves asymptotically as $\sim log\, n_{max}(\sim n_{max}^{d-2})$. This leads to the very important result that, for a fixed bound on V from experimental data, the corresponding bound on the mass of the lowest lying KK excitation rapidly strengthens with the number of extra dimension, d. Table I shows how the $d = 1$ lower bound of 3.9 TeV for the mass of M_1 changes as we consider different compactifications for $d > 1$. We see that in some cases the value of M_1 is so large it will be beyond the mass range accessible to the LHC as it is for all cases of the $d = 3$ example.

TABLE 1. Lower bound on the mass of the first KK state in TeV resulting from the constraint on V for the case of more than one dimension. 'T'['E'] labels the result obtained from the direct truncation (exponential suppression). Cases labeled by an asterisk will be observable at the LHC. $Z_2 \times Z_2$ and $Z_{3,6}$ correspond to compactifications in the case of $d = 2$ while $Z_2 \times Z_2 \times Z_2$ is for the case of $d = 3$.

	$Z_2 \times Z_2$		$Z_{3,6}$		$Z_2 \times Z_2 \times Z_2$	
n_{max}	T	E	T	E	T	E
2	5.69*	4.23*	6.63*	4.77*	8.65	8.01
3	6.64	4.87*	7.41	5.43*	11.7	10.8
4	7.20	5.28*	7.95	5.85*	13.7	13.0
5	7.69	5.58*	8.36	6.17*	15.7	14.9
10	8.89	6.42	9.61	7.05	23.2	22.0
20	9.95	7.16	10.2	7.83	33.5	31.8
50	11.2	8.04	12.1	8.75	53.5	50.9

SM KK STATES AT THE LHC AND LINEAR COLLIDERS

Let us return to the $d = 1$ case at the LHC where the degenerate KK states $\gamma^{(1)}$, $Z^{(1)}$, $W^{(1)}$ and $g^{(1)}$ are potentially visible. It has been shown [7] that for masses in excess of $\simeq 4$ TeV the $g^{(1)}$ resonance in dijets will be washed out due to its rather large width and the experimental jet energy resolution available at the LHC detectors. Furthermore, $\gamma^{(1)}$ and $Z^{(1)}$ will appear as a *single* resonance in Drell-Yan that cannot be resolved and looking very much like a single Z'. Thus if we are lucky the LHC will observe what appears to be a degenerate Z'/W'. How can we identify these states as KK excitations when we remember that the rest of the members of the tower are too massive to be produced? We remind the reader that many extended electroweak models [10] exist which predict a degenerate Z'/W'. Without further information, it would seem likely that this would become the most likely guess of what had been found.

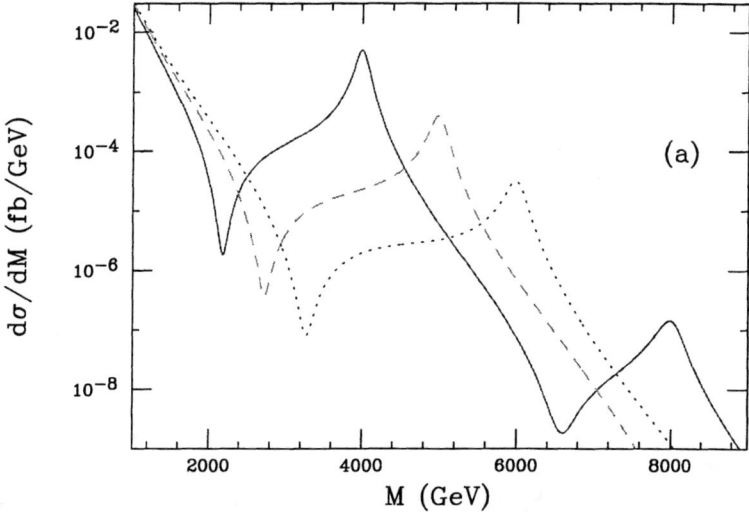

FIGURE 1. Cross section for Drell-Yan production of the degenerate neutral KK excitations $Z^{(n)}$ and $\gamma^{(n)}$ as a function of the dilepton invariant mass at the LHC assuming one extra dimension and naive coupling values with $1/R$=4(5, 6) TeV corresponding to the solid(dashed, dotted) curve. The second excitation is only shown for the case of $1/R = 4$ TeV.

To clarify this situation let us consider the results displayed in Figs. 1 for $d = 1$ where we show the production cross sections in the $\ell^+\ell^-$ channel with inverse compactification radii of 4, 5 and 6 TeV. In calculating these cross sections we have assumed that the KK excitations have their naive couplings and can only decay to the usual fermions of the SM. Additional decay modes can lead to appreciably lower cross sections so that we cannot use the peak heights to determine the degeneracy of the KK state. Note that in the 4 TeV case, which is essentially as small a mass as can be tolerated by the present data on precision measurements, the second KK excitation is visible in the plot. We see several things from these figures. First, we can easily estimate the total number of events in the resonance regions associated with each of the peaks assuming the canonical integrated luminosity of $100 fb^{-1}$ appropriate for the LHC; we find $\simeq 300(32, 3, 0.02)$ events corresponding to the

4(5,6,8) TeV resonances if we sum over both electron and muon final states and assume 100% leptonic identification efficiencies. Clearly the 6 and 8 TeV resonances will not be visible at the LHC (though a modest increase in luminosity by a factor of a few will allow the 6 TeV resonance to become visible) and we also verify our claim that only the first KK excitations will be observable. In the case of the 4 TeV resonance there is sufficient statistics that the KK mass will be well measured and one can also imagine measuring the forward-backward asymmetry, A_{FB}, if not the full angular distribution of the outgoing leptons, since the final state muon charges can be signed. Given sufficient statistics, a measurement of the angular distribution would demonstrate that the state is indeed spin-1 and not spin-0 or spin-2. However, for such a heavy resonance it is unlikely that much further information could be obtained about its couplings and other properties. In fact the conclusion of several years of Z' analyses [11] is that coupling information will be essentially impossible to obtain for Z'-like resonances with masses in excess of 1-2 TeV at the LHC due to low statistics. Furthermore, the lineshape of the 4 TeV resonance and the Drell-Yan spectrum anywhere close to the peak will be difficult to measure in detail due to both the limited statistics and energy smearing. Thus we will never know from LHC data alone whether the first KK resonance has been discovered or, instead, some extended gauge model scenario has been realized. To make further progress we need a lepton collider.

It is well-known that future e^+e^- linear colliders(LC) operating in the center of mass energy range $\sqrt{s} = 0.5 - 1.5$ TeV will be sensitive to indirect effects arising from the exchange of new Z' bosons with masses typically 6-7 times greater than $\sqrt{s}$ [11]. This sensitivity is even greater in the case of KK excitations since towers of both γ and Z exist all of which have couplings larger than their SM zero modes. Furthermore, analyses have shown that with enough statistics the couplings of the new Z' to the SM fermions can be extracted [12] in a rather precise manner, especially when the Z' mass is already approximately known from elsewhere, e.g., the LHC. (If the Z' mass is not known then measurements at several distinct values of $\sqrt{s}$ can be used to extract both the mass as well as the corresponding couplings [13].) In the present situation, we imagine that the LHC has discovered and determined the mass of a Z'-like resonance in the 4-6 TeV range. Can the LC tell us anything about this object?

The obvious step would be to use the LC to extract the couplings of the apparent resonance discovered by the LHC; we find that it is sufficient for our arguments below to do this solely for the leptonic channels. The idea is the following: we measure the deviations in the differential cross sections and angular dependent Left-Right polarization asymmetry, A_{LR}^ℓ, for the three lepton generations and combine those with τ polarization data. Assuming lepton universality(which would be observed in the LHC data anyway), that the resonance mass is well determined, and that the resonance is an ordinary Z' we perform a fit to the hypothetical Z' coupling to leptons, v_l, a_l. To be specific, let us consider the case of only one extra dimen-

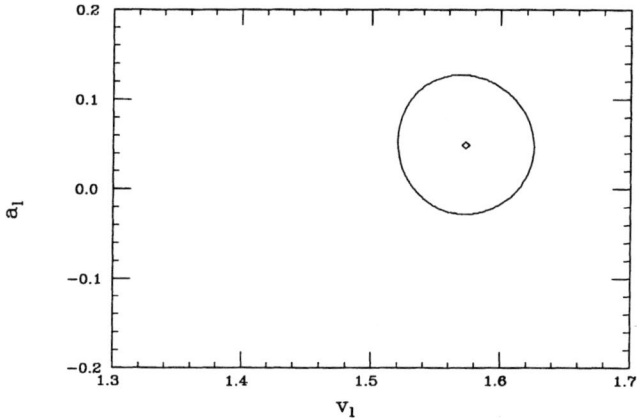

FIGURE 2. Fitted values of the parameters v_l and a_l following the procedures described in the text for a 4 TeV KK excitation at a 500 GeV e^+e^- collider. The contour described the 95% CL region with the best fit value as a diamond. The normalization is such that the corresponding SM Z boson's axial-vector coupling to the electron is -1/2.

sion with a 4 TeV KK excitation and employ a $\sqrt{s} = 500$ GeV collider with an integrated luminosity of 200 fb^{-1}. The result of performing this fit, including the effects of cuts and initial state radiation, is shown in Fig.2. Here we see that the coupling values are 'well determined' (i.e., the size of the 95% CL allowed region we find is quite small) by the fitting procedure as we would have expected from previous analyses of Z' couplings extractions at linear colliders [11–13].

The only problem with the fit shown in the figure is that the χ^2 is very large leading to a very small confidence level, i.e., $\chi^2/d.o.f = 95.06/58$ or CL=1.55 × 10^{-3}! (We note that this result is not very sensitive to the assumption of 90% beam polarization; 70% polarization leads to almost identical results.) For an ordinary Z' it has been shown that fits of much higher quality, based on confidence level values, are obtained by this same procedure. Increasing the integrated luminosity can be seen to only make matters worse. Fig.3 shows the results for the CL following the above approach as we vary both the luminosity and the mass of the first KK excitation at both 500 GeV and 1 TeV e^+e^- linear colliders. From this figure we see that the resulting CL is below $\simeq 10^{-3}$ for a first KK excitation with a mass of 4(5,6) TeV when the integrated luminosity at the 500 GeV collider is 200(500,900)fb^{-1} whereas at a 1 TeV for excitation masses of 5(6,7) TeV we require luminosities of 150(300,500)fb^{-1} to realize this same CL. Barring some unknown systematic effect the only conclusion that one could draw from such bad fits is that the hypothesis

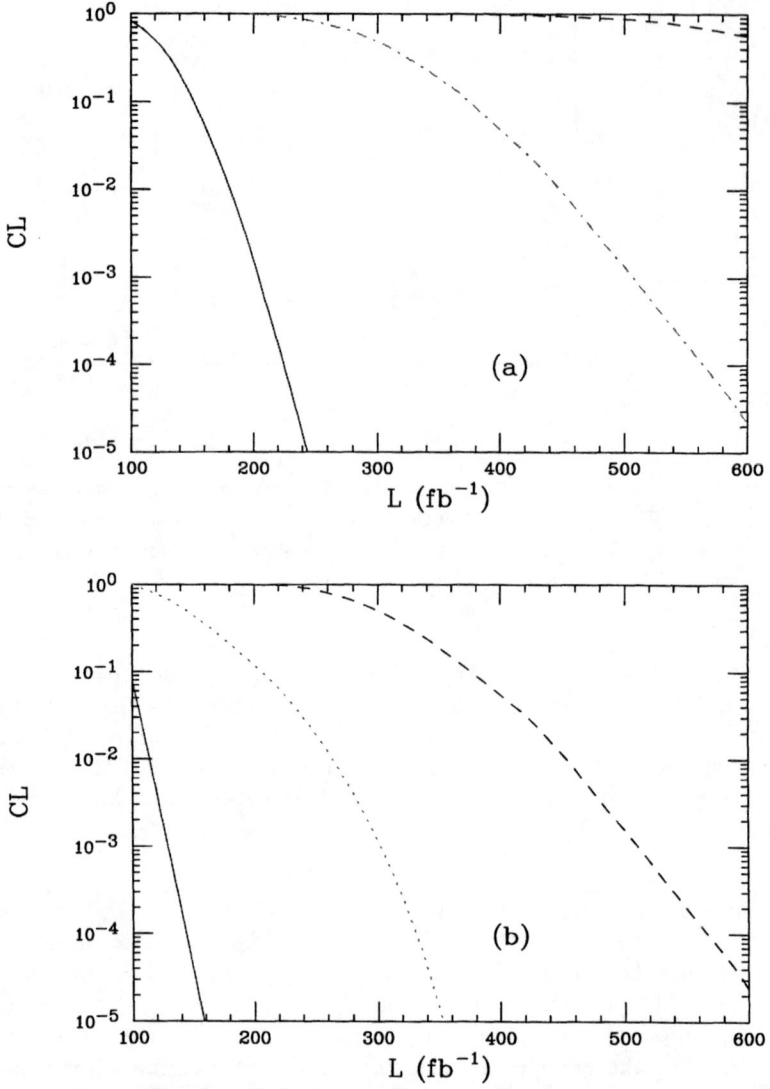

FIGURE 3. CL as a function of the integrated luminosity resulting from the coupling fits following from the analysis discussed in the text for both (a) a 500 GeV or a (b) 1 TeV e^+e^- collider. In (a) the solid(dash-dotted,dotted) curve corresponds to a first KK excitation mass of 4(5,6) TeV. In (b) the solid(dotted,dashed) curve corresponds to a first KK mass of 5(6,7) TeV.

of a single Z', and the existence of no other new physics, is simply *wrong*. If no other exotic states are observed below the first KK mass at the LHC, such as $\tilde{\nu}$ [14] or leptoquarks [15], this result would give very strong indirect evidence that something more unusual that a conventional Z' had been found but *cannot* prove that this is a KK state.

SM KK STATES AT MUON COLLIDERS

In order to be completely sure of the nature of the first KK excitation, we must produce it directly at a higher energy lepton collider and sit on and near the peak of the KK resonance. To reach this mass range will most likely require a Muon Collider. The first issue to address is the quality of the degeneracy of the $\gamma^{(1)}$ and $Z^{(1)}$ states. Based on the analyses in Ref. [4,5] we can get an idea of the maximum possible size of this fractional mass shift and we find it to be of order $\sim M_Z^4/M_{Z^{(1)}}^4$, an infinitesimal quantity for KK masses in the several TeV range. Thus even when mixing is included we find that the $\gamma^{(1)}$ and $Z^{(1)}$ states remain very highly degenerate so that even detailed lineshape measurements may not be able to distinguish the $\gamma^{(1)}/Z^{(1)}$ composite state from that of a Z'. We thus must turn to other parameters in order to separate these two cases.

Sitting on the resonance there are a very large number of quantities that can be measured: the mass and apparent total width, the peak cross section, various partial widths and asymmetries *etc*. From the Z-pole studies at SLC and LEP, we recall a few important tree-level results which we would expect to apply here as well provided our resonance is a simple Z'. First, we know that the value of $A_{LR} = [A_e = 2v_e a_e/(v_e^2 + a_e^2)]$, as measured on the Z by SLD, does not depend on the fermion flavor of the final state and second, that the relationship $A_{LR} \cdot A_{FB}^{pol}(f) = A_{FB}^f$ holds, where $A_{FB}^{pol}(f)$ is the polarized Forward-Backward asymmetry as measured for the Z at SLC and A_{FB}^f is the usual Forward-Backward asymmetry. The above relation is seen to be trivially satisfied on the Z(or on a Z') since $A_{FB}^{pol}(f) = \frac{3}{4}A_f$ and $A_{FB}^f = \frac{3}{4}A_e A_f$. Both of these relations are easily shown to fail in the present case of a 'dual' resonance though they will hold if only one particle is resonating.

A short exercise shows that in terms of the couplings to $\gamma^{(1)}$, which we will call v_1, a_1, and $Z^{(1)}$, now called v_2, a_2, these same observables can be written as

$$A_{FB}^f = \frac{3}{4}\frac{A_1}{D}$$
$$A_{FB}^{pol}(f) = \frac{3}{4}\frac{A_2}{D}$$
$$A_{LR}^f = \frac{A_3}{D}, \qquad (2)$$

where f labels the final state fermion and we have defined the coupling combinations

$$D = (v_1^2 + a_1^2)_e(v_1^2 + a_1^2)_f + R^2(v_2^2 + a_2^2)_e(v_2^2 + a_2^2)_f$$
$$+ 2R(v_1v_2 + a_1a_2)_e(v_1v_2 + a_1a_2)_f \qquad (3)$$
$$A_1 = (2v_1a_1)_e(2v_1a_1)_f + R^2(2v_2a_2)_e(2v_2a_2)_f + 2R(v_1a_2 + v_2a_1)_e(v_1a_2 + v_2a_1)_e$$
$$A_2 = (2v_1a_1)_f(v_1^2 + a_1^2)_e + R^2(2v_2a_2)_f(v_2^2 + a_2^2)_e + 2R(v_1a_2 + v_2a_1)_f(v_1v_2 + a_1a_2)_e$$
$$A_3 = (2v_1a_1)_e(v_1^2 + a_1^2)_f + R^2(2v_2a_2)_e(v_2^2 + a_2^2)_f + 2R(v_1a_2 + v_2a_1)_e(v_1v_2 + a_1a_2)_f,$$

with R being the ratio of the widths of the two KK states, $R = \Gamma_1/\Gamma_2$, and the $v_{1,2i}, a_{1,2i}$ are the appropriate couplings for electrons and fermions f. Note that when R gets either very large or very small we recover the usual 'single resonance' results. Examining these equations we immediately note that A_{LR}^f is now *flavor dependent* and that the relationship between observables is no longer satisfied:

$$A_{LR}^f \cdot A_{FB}^{pol}(f) \neq A_{FB}^f, \qquad (4)$$

which clearly tells us that we are actually producing more than one resonance.

Of course we need to verify that these single resonance relations are numerically badly broken before clear experimental signals for more than one resonance can be claimed. Statistics will not be a problem with any reasonable integrated luminosity since we are sitting on a resonance peak and certainly millions of events will be collected. With such large statistics only a small amount of beam polarization will be needed to obtain useful asymmetries. In principle, to be as model independent as possible in a numerical analysis, we should allow the widths Γ_i to be greater than or equal to their SM values as such heavy KK states may decay to SM SUSY partners as well as to presently unknown exotic states. Since the expressions above only depend upon the ratio of widths, we let $R = \lambda R_0$ where R_0 is the value obtained assuming that the KK states have only SM decay modes. We then treat λ as a free parameter in what follows and explore the range $1/5 \leq \lambda \leq 5$. Note that as we take $\lambda \to 0(\infty)$ we recover the limit corresponding to just a $\gamma^{(1)}(Z^{(1)})$ being present.

In Fig.4 we display the flavor dependence of A_{LR}^f as a functions of λ. Note that as $\lambda \to 0$ the asymmetries vanish since the $\gamma^{(1)}$ has only vector-like couplings. In the opposite limit, for extremely large λ, the $Z^{(1)}$ couplings dominate and a common value of A_{LR} will be obtained. It is quite clear, however, that over the range of reasonable values of λ, A_{LR}^f is quite obviously flavor dependent. We also show in Fig.4 the correlations between the observables $A_{FB}^{pol}(f)$ and $A_{FB}(f)$ which would be flavor independent if only a single resonance were present. From the figure we see that this is clearly not the case. Note that although λ is an *a priori* unknown parameter, once any one of the electroweak observables are measured the value of λ will be directly determined. Once λ is fixed, then the values of all of the other asymmetries, as well as the ratios of various partial decay widths, are all completely fixed for the KK resonance with uniquely predicted values. This means that we can

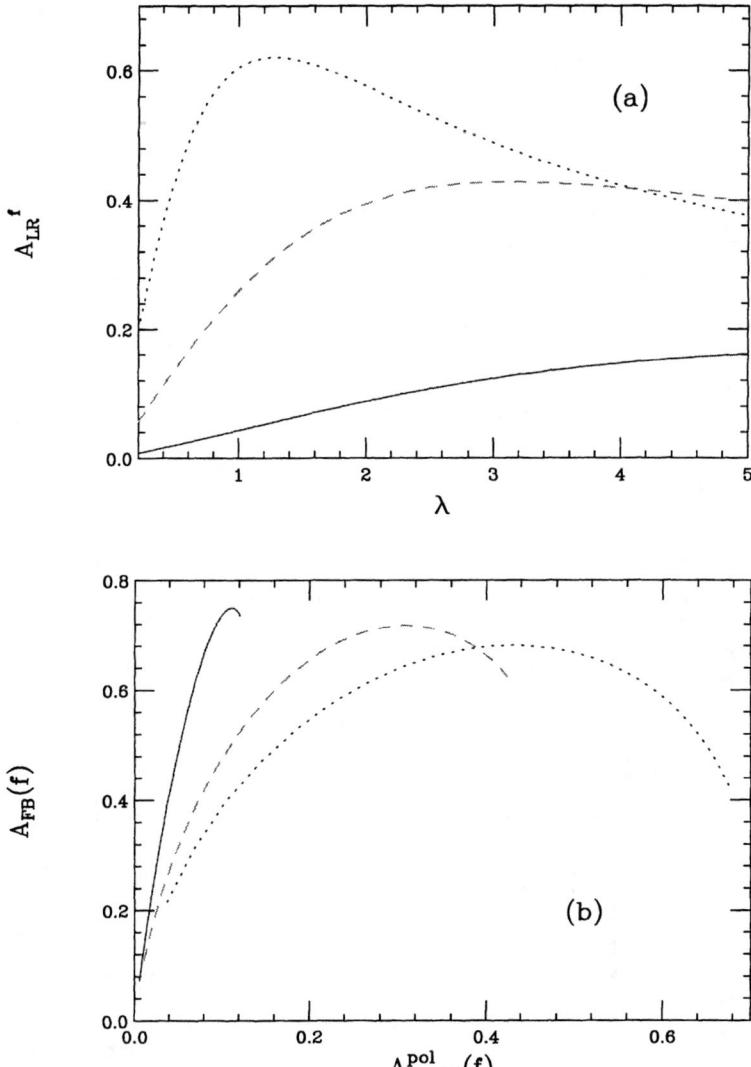

FIGURE 4. (a) A^f_{LR} as a function of the parameter λ for $f = \ell$(solid), $f = c$(dashed) and $f = b$(dots). (b) Correlations between on-peak observables for the same three cases as shown in (a). λ varies from 0.2 to 5 along each curve.

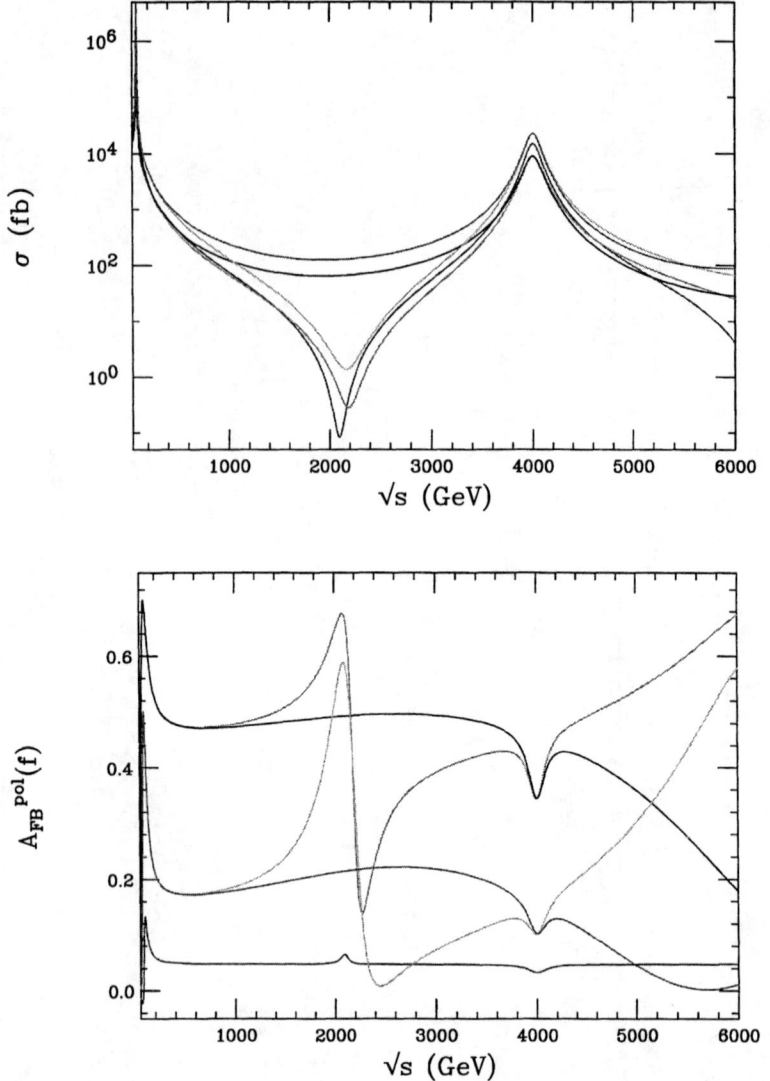

FIGURE 5. Cross sections and polarized A_{FB} for $\mu^+\mu^- \to e^+e^-\ b\bar{b}$ and $c\bar{c}$ as functions of energy in both the 'conventional' scenario and that of Arkani-Hamed and Schmaltz(AS) [9] where the quarks and leptons are separated in the extra dimension by a distance $D = \pi R$. The red curve applies for the μ final state in either model whereas the green(blue) and cyan(magenta) curves label the b and c final states for the 'conventional'(AS) scenario.

directly test the couplings of this apparent single resonance against what might be expected for a degenerate pair of KK excitations without any ambiguities.

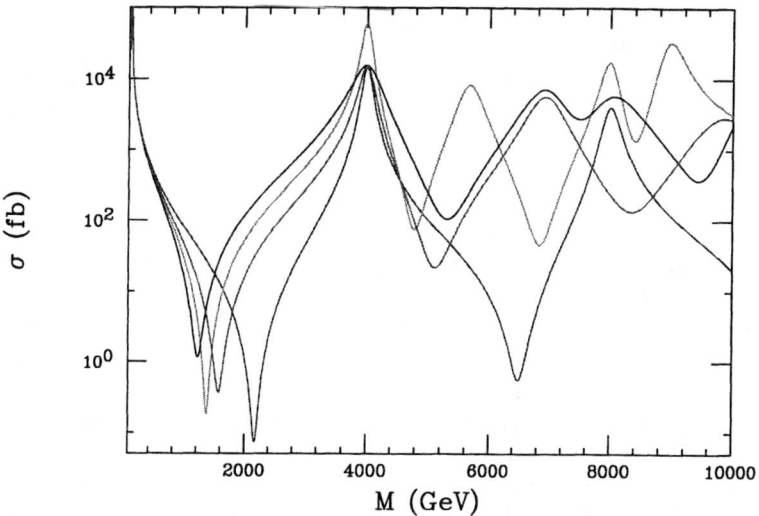

FIGURE 6. Same as Fig. 5a for the process $\mu^+\mu^- \to e^+e^-$ but now also including the models listed in Table 1 with $d = 2$ assuming $M_1 = 4$ TeV. The red(green,blue,purple) curve corresponds to the $S^1/Z_2(Z_2 \times Z_2, Z_{3,6}, S^2)$ compactifications.

In Figs. 5a and 5b we show that although on-resonance measurements of the electroweak observables, being quadratic in the $Z^{(1)}$ and $\gamma^{(1)}$ couplings, will not distinguish between the usual KK scenario and that of the Arkani-Hamed and Schmaltz(AS) (whose KK couplings to quarks are of opposite sign from the conventional assignments for odd KK levels since quarks and leptons are assumed to be separated by a distance $D = \pi R$ in their scenario) the data below the peak in the hadronic channel will easily allow such a separation. The cross section and asymmetries for $\mu^+\mu^- \to e^+e^-$ (or vice versa) is, of course, the same in both cases. Such data can be collected by using radiative returns if sufficient luminosity is available. The combination of on and near resonance measurements will thus completely determine the nature of the resonance.

We note that all of the above analysis will go through essentially unchanged in any qualitative way when we consider the case of the first KK excitation in a theory with more than one extra dimension as is shown in Fig.6. Here we see that the shape of the excitation curves for the $d = 1$ case and the $d > 1$ models listed in Table 1 will clearly allow the number of dimensions and the compactification scheme to be uniquely identified.

RANDALL-SUNDRUM GRAVITONS AT MUON COLLIDERS

The possibility of extra space-like dimensions with accessible physics near the TeV scale has recently opened a new window on the possible solutions to the hierarchy problem. Models designed to address this problem make use of our ignorance about gravity, in particular, the fact that gravity has yet to be probed at energy scales much above 10^{-3} eV in laboratory experiments. The prototype scenario in this class of theories is due to Arkani-Hamed, Dimopoulos and Dvali(ADD) [16] who use the volume associated with large extra dimensions to bring the d-dimensional Planck scale down to a few TeV. Here the hierarchy problem is recast into trying to understand the rather large ratio of the TeV Planck scale to the size of the extra dimensions which may be as large as a fraction of a millimeter. The phenomenological [17] implications of this model have been worked out by a large number of authors. An extrapolation of these analyses to the case of high energy muon colliders shows an enormous reach for this kind of physics.

More recently, Randall and Sundrum(RS) [18] have proposed a new scenario wherein the hierarchy is generated by an exponential function of the compactification radius, called a warp factor. Unlike the ADD model, they assume a 5-dimensional non-factorizable geometry, based on a slice of AdS_5 spacetime. Two 3-branes, one being 'visible' with the other being 'hidden', with opposite tensions rigidly reside at S_1/Z_2 orbifold fixed points, taken to be $\phi = 0, \pi$, where ϕ is the angular coordinate parameterizing the extra dimension. It is assumed that the extra-dimension bulk is only populated by gravity and that the SM lies on the brane with negative tension. The solution to Einstein's equations for this configuration, maintaining 4-dimensional Poincare invariance, is given by the 5-dimensional metric

$$ds^2 = e^{-2\sigma(\phi)}\eta_{\mu\nu}dx^\mu dx^\nu + r_c^2 d\phi^2 \,, \qquad (5)$$

where the Greek indices run over ordinary 4-dimensional spacetime, $\sigma(\phi) = kr_c|\phi|$ with r_c being the compactification radius of the extra dimension, and $0 \leq |\phi| \leq \pi$. Here k is a scale of order the Planck mass and relates the 5-dimensional Planck scale M to the cosmological constant. Examination of the action in the 4-dimensional effective theory in the RS scenario yields the relationship $\overline{M}_{Pl}^2 = M^3/k$ for the reduced effective 4-D Planck scale.

Assuming that we live on the 3-brane located at $|\phi| = \pi$, it is found that a field on this brane with the fundamental mass parameter m_0 will appear to have the physical mass $m = e^{-kr_c\pi}m_0$. TeV scales are thus generated from fundamental scales of order $\overline{M}_{Pl}$ via a geometrical exponential factor and the observed scale hierarchy is reproduced if $kr_c \simeq 11-12$. Hence, due to the exponential nature of the warp factor, no additional large hierarchies are generated.

A recent analysis [19] examined the phenomenological implications and constraints on the RS model that arise from the exchange of weak scale towers of gravitons. There it was shown that the masses of the KK graviton states are given by $m_n = kx_n e^{-kr_c\pi}$ where x_n are the roots of $J_1(x_n) = 0$, the ordinary Bessel function of order 1. It is important to note that these roots are *not* equally spaced, in contrast to most KK models with one extra dimension, due to the non-factorizable metric. Expanding the graviton field into the KK states one finds the interaction

$$\mathcal{L} = -\frac{1}{\overline{M}_{Pl}} T^{\alpha\beta}(x) h^{(0)}_{\alpha\beta}(x) - \frac{1}{\Lambda_\pi} T^{\alpha\beta}(x) \sum_{n=1}^{\infty} h^{(n)}_{\alpha\beta}(x). \qquad (6)$$

Here, $T^{\alpha\beta}$ is the stress energy tensor on the brane and we see that the zero mode separates from the sum and couples with the usual 4-dimensional strength, $\overline{M}_{Pl}^{-1}$; however, all the massive KK states are only suppressed by Λ_π^{-1}, where we find that $\Lambda_\pi = e^{-kr_c\pi} \overline{M}_{Pl}$, which is of order the weak scale.

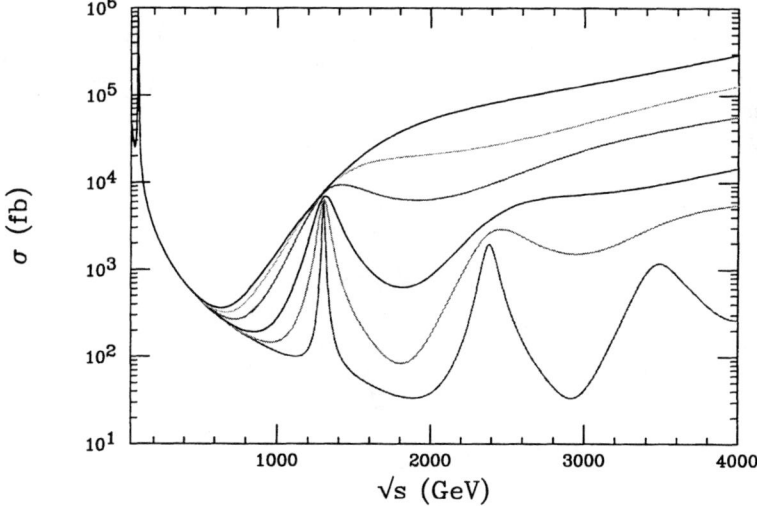

FIGURE 7. Cross section for $\mu^+\mu^- \to e^+e^-$ including the exchange of KK gravitons, taking the mass of the first mode to be 1.2 TeV, as a function of energy. From top to bottom the curves correspond to c=1.0, 0.7, 0.5, 0.3, 0.2, and 0.1.

This model has essentially 2 free parameters which we can take to be the mass of the first KK graviton mode and the ratio $c = k/\overline{M}_{Pl}$; the later quantity is restricted to be less than unity to maintain the self-consistency of the scenario (to prevent a radius of curvature smaller than the Planck scale in 5 dimensions) and if it is taken too small another hierarchy is formed. Figs.7 and 8 show the cross section and A_{FB} for the process $\mu^+\mu^- \to e^+e^-$ as a function of $\sqrt{s}$ in the presence of KK graviton resonances for several values of the parameter c. For large c one does not see the individual resonance structures (since the theory is strongly coupled and they are smeared together by their large widths which grow as $\sim c^2$) but only a very large shoulder somewhat similar to a contact interaction. For small c one sees the individual resonances with their widths growing rapidly with increasing mass as $\sim m_n^3$. Note that for large $\sqrt{s}$ where graviton exchange dominates the value of A_{FB} is driven to zero. Sitting on any of these KK resonances, in the case of small values of c, will immediately reveal the unique quartic angular distribution corresponding to spin-2 graviton exchange for the fermions in the final state $\sim 1 - 3\cos^2\theta + 4\cos^4\theta$.

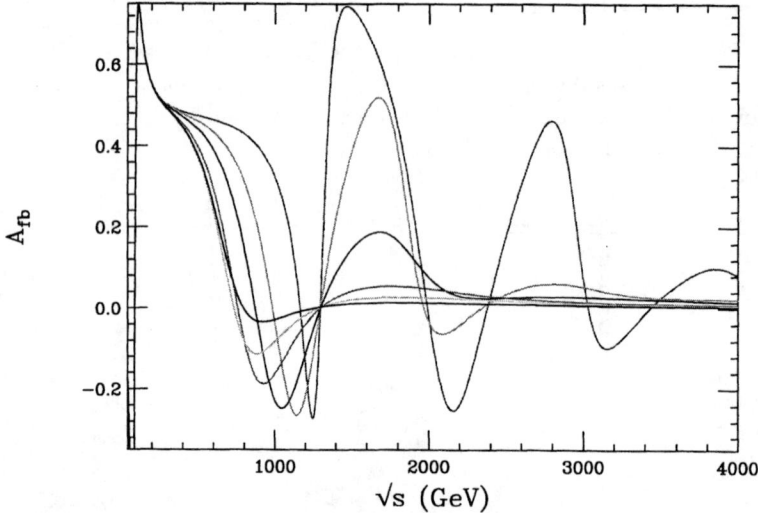

FIGURE 8. Same as the previous figure but now for the Forward-Backward asymmetry in the RS model. The color code is as in the previous figure.

CONCLUSIONS

Present data indicates that the masses of KK excitations of the SM gauge bosons must be rather heavy, e.g., > 3.9 TeV if $d = 1$. We have found that:

- With an integrated luminosity of 100 fb^{-1}, the LHC will be able to observe KK excitations in the mass range below $\simeq 6$ TeV but may not see any KK excitations when $d > 1$ since they are likely to be more massive. The LHC will not see the second set of KK resonances even when $d = 1$.

- The LHC cannot separate the KK states $\gamma^{(1)}$ from $Z^{(1)}$ which will appear together as a single resonance, nor can it obtain significant coupling constant information.

- The LHC cannot see the $g^{(1)}$ if its mass is in excess of ~ 4 TeV due to its large width and the energy resolution of the LHC detectors.

- The LHC cannot distinguish an extended electroweak model with a degenerate Z'/W' from a KK scenario. All we will know is the mass of these resonances.

- A LC with $\sqrt{s} = 0.5 - 1$ TeV will be sensitive to the existence of KK states with masses more than an order of magnitude larger than $\sqrt{s}$ for reasonable integrated luminosities $\simeq 100~fb^{-1}$.

- At a LC, the extraction of the couplings of an apparent Z', whose mass is known from measurements obtained at the LHC, can be performed in a straightforward manner with reasonable integrated luminosities. However, the Z' hypothesis will yield a poor fit to the data if the state in question is actually the combined $\gamma^{(1)}/Z^{(1)}$ KK excitation. The LC will not be able to identify this state as such–only prove it is not a Z'.

- A Muon Collider operating at or above the first KK resonance pole will identify it as a KK state provided polarized beams are available.

- Measurements of the KK excitation spectrum at Muon Colliders will be able to tell us both the number of extra dimensions and how they are compactified thus possibly revealing the basic underlying theory upon which the KK scenario is based.

- KK excitations of gravitons in the RS model can be studied in detail at both LC and Muon Colliders with Muon Colliders providing a much larger reach in explorable parameter space. These measurements can completely determine all of the parameters of this model.

Muon Colliders clearly offer a very important window into the physics of Kaluza-Klein excitations.

REFERENCES

1. I. Antoniadis, Phys. Lett. **B246**, 377 (1990); I. Antoniadis, C. Munoz and M. Quiros, Nucl. Phys. **B397**, 515 (1993); I. Antoniadis and K. Benalki, Phys. Lett. **B326**, 69 (1994); I. Antoniadis, K. Benalki and M. Quiros, Phys. Lett. **B331**, 313 (1994).
2. D0 Collaboration, S. Abachi et al.,Phys. Lett. **B385**, 471 (1996), Phys. Rev. Lett. **76**, 3271 (1996) and Phys. Rev. Lett. **82**, 29 (1999); CDF Collaboration, F. Abe et al., Phys. Rev. Lett. **77**, 5336 (1996), Phys. Rev. Lett. **74**, 2900 (1995) and Phys. Rev. Lett. **79**, 2191 (1997).
3. F. Cornet, M. Relano and J. Rico, hep-ph/9908299.
4. P. Nath and M. Yamaguchi, hep-ph/9902323 and hep-ph/9903298; M. Masip and A. Pomarol, hep-ph/9902467; W.J. Marciano, hep-ph/9903451; L. Hall and C. Kolda, Phys. Lett. **B459**, 213 (1999); R. Casalbuoni, S. DeCurtis and D. Dominici, hep-ph/9905568; R. Casalbuoni, S. DeCurtis, D. Dominici and R. Gatto, hep-ph/9907355; A. Strumia, hep-ph/9906266; C.D. Carone, hep-ph/9907362.
5. T.G. Rizzo and J.D. Wells, hep-ph/9906234.

6. J. Mnich, talk given at the *International Europhysics Conference on High Energy Physics(EPS99)*, 15-21 July 1999, Tampere, Finland; M. Swartz, M. Lancaster and D. Charlton talks given at the *XIX International Symposium on Lepton and Photon Interactions*, 9-14 August 1999, Stanford, California.
7. T.G. Rizzo, hep-ph/9909232; See also I. Antoniadis, K. Benalki and M. Quiros, hep-ph/9905311; P. Nath, Y. Yamada and M. Yamaguchi, hep-ph/9905415.
8. M. Bando, T. Kugo, T. Noguchi and K. Yoshioka, hep-ph/9906549. See also J. Hisano and N. Okada, hep-ph/9909555.
9. N. Arkani-Hamed and M. Schmaltz, hep-ph/9903417; N. Arkani-Hamed, Y. Grossman and M. Schmaltz, hep-ph/9909411.
10. For a discussion of a few of these models, see H. Georgi, E.E. Jenkins, and E.H. Simmons, Phys. Rev. Lett. **62**, 2789 (1989) and Nucl. Phys. **B331**, 541 (1990);V. Barger and T.G. Rizzo, Phys. Rev. **D41**, 946 (1990); T.G. Rizzo, Int. J. Mod. Phys. **A7**, 91 (1992); R.S. Chivukula, E.H. Simmons and J. Terning, Phys. Lett. **B346**, 284 (1995); A. Bagneid, T.K. Kuo, and N. Nakagawa, Int. J. Mod. Phys. **A2**, 1327 (1987) and Int. J. Mod. Phys. **A2**, 1351 (1987); D.J. Muller and S. Nandi, Phys. Lett. **B383**, 345 (1996); X.Li and E. Ma, Phys. Rev. Lett. **47**, 1788 (1981) and Phys. Rev. **D46**, 1905 (1992); E. Malkawi, T.Tait and C.-P. Yuan, Phys. Lett. **B385**, 304 (1996); E. Malkawi and C.-P. Yuan, hep-ph/9906215.
11. For a review of new gauge boson physics at colliders and details of the various models, see J.L. Hewett and T.G. Rizzo, Phys. Rep. **183**, 193 (1989); M. Cvetic and S. Godfrey, in *Electroweak Symmetry Breaking and Beyond the Standard Model*, ed. T. Barklow et al., (World Scientific, Singapore, 1995), hep-ph/9504216; T.G. Rizzo in *New Directions for High Energy Physics: Snowmass 1996*, ed. D.G. Cassel, L. Trindle Gennari and R.H. Siemann, (SLAC, 1997), hep-ph/9612440; A. Leike, hep-ph/9805494.
12. A. Djouadi, A. Lieke, T. Riemann, D. Schaile and C. Verzegnassi, Z. Phys. **C56**, 289 (1992); J. Hewett and T. Rizzo, in *Proceedings of the Workshop on Physics and Experiments with Linear e^+e^- Colliders*, September 1991, Saariselkä, Finland, R. Orava ed., (World Scientific, Singapore, 1992) Vol. II, p.489, ibid p.501; G. Montagna et al., Z. Phys. **C75**, 641 (1997); F. del Aguila and M. Cvetic, Phys. Rev. **D50**, 3158 (1994); F. del Aguila, M. Cvetic and P. Langacker Phys. Rev. **D52**, 37 (1995); A. Lieke, Z. Phys. **C62**, 265 (1994); D. Choudhury, F. Cuypers and A. Lieke, Phys. Lett. **B333**, 531 (1994); S. Riemann in *New Directions for High Energy Physics: Snowmass 1996*, ed. D.G. Cassel, L. Trindle Gennari and R.H. Siemann, (SLAC, 1997), hep-ph/9610513; A. Lieke and S. Riemann, Z. Phys. **C75**, 341 (1997); T.G. Rizzo, hep-ph/9604420.
13. T.G. Rizzo, Phys. Rev. **D55**, 5483 (1997).
14. T.G. Rizzo, Phys. Rev. **D59**, 113004 (1999).
15. For a review, see J.L. Hewett and T.G. Rizzo, Phys. Rev. **D56**, 5709 (1997) and Phys. Rev. **D58**, 055005 (1998).
16. N. Arkani-Hamed, S. Dimopoulos and G. Dvali, Phys. Lett. **B429**, 263 (1998) and Phys. Rev. **D59**, 086004 (1999); I. Antoniadis, N. Arkani-Hamed, S. Dimopoulos and G. Dvali, Phys. Lett. **B436**, 257 (1998.)
17. G.F. Giudice, R. Rattazzi and J.D. Wells, Nucl. Phys. **B544**, 3 (1999);

E.A. Mirabelli, M. Perelstein and M.E. Peskin, Phys. Rev. Lett. **82**, 2236 (1999); T. Han, J.D. Lykken and R. Zhang, Phys. Rev. **D59**, 105006 (1999); J.L. Hewett, Phys. Rev. Lett. **82**, 4765 (1999); T.G. Rizzo, Phys. Rev. **D59**, 115010 (1999).
18. L. Randall and R. Sundrum, hep-ph/9905221 and hep-th/9906182.
19. H. Davoudiasl, J.L. Hewett and T.G. Rizzo, hep-ph/9909255.

Might These Machines Be Affordable

Nicholas P. Samios

Physics Department, Brookhaven National Laboratory, Upton, NY 11973

Abstract. Experience in the construction of relatively large facilities is reviewed. The influence of the political climate and budget situation is discussed. All this with relation to the possibility and the strategy for the advent of future large accelerators such as the muon collider.

The scale, complexity and cost of accelerators have steadily increased over the years, even though the unit cost (for instance dollars per GeV) has been drastically reduced. Up to very recently the administrative and financial responsibility of constructing new machines has been the province of individual countries (or existing consortia such as CERN). As such the processes have been relatively straight forward with even the issue of outside and international participation being uniformly adopted and adhered to. Although the concept of a truly international accelerator was first discussed by V. Weiskopf in the '70's, and has been discussed many times and dismissed, the time may have come to more seriously discuss this possibility, certainly as the costs have escalated to over a billion dollars and has even brought into question the affordability of building such machines. In order to address this issue it is worthwhile to review recent activities in this area.

In this retrospective, there have been two large international endeavors that bear examination. The first is ITER, the International Thermonuclear Energy Reactor, whose aim was to engage in fusion research and development (R&D) and build a fusion reactor. This was a collaboration among Japan, U.S. and Western Europe and others, with extensive interactions, a division of effort on the R&D effort, a designed fusion machine and a process under way of deciding on a site. The cost was estimated to be ~8 B dollars with a 6-8 year construction time scale after a similar time period for these preliminary activities. The U.S. Congress then cancelled U.S. participation in this machine, even after the successful program at the Tokamac at Princeton. The second facility to be discussed is the Superconducting Super Collider (SSC). This accelerator was proposed by the U.S. high energy community and was a bold step forward in energy, 20 TEV x 20 TEV proton-proton collider with a high luminosity $10^{33}/cm^2/sec$. It also required a substantial R&D effort ~5 years with a projected construction period of ~8 years and a cost which started at $4.2B and reached $8.2B. The SSC began as a U.S. project but early on international participation was welcomed and promoted rather successfully. Japan, Russia, India, and other countries joined but were not truly partners nor were their monetary contributions major; in only one case, Japan, was there such a possibility and it failed to materialize. This machine also was cancelled by the U.S. Congress after spending $2B. The consequences of these actions are severe. In the fusion arena the U.S. is viewed as an unreliable partner; the U.S. program is in an R&D mode and there is uncertainty as to whether a fusion reactor will ever be built. The ending of the SSC has played havoc with the national if not the international high energy program. The canonical strategy involved the early realization of a high energy, high luminosity pp collider to provide a broad search for new physics such as a Higgs or evidence of super symmetry. This would set the energy scale for follow up lepton lepton colliders. The loss of the SSC has delayed the attainment of this logical program. The alternate and inferior LHC will become operational 6-8 years later than the SSC, in 2006-2008 time frame. As a result initiation of a lepton collider will probably be delayed or begun with less information and therefore be at greater risk in assuring that it would be a productive machine--not a nice situation. A common feature among these projects is their large cost and long time scale, with the consequence that they encompassed several administrations each one with its philosophy and agenda.

To calibrate the situation further one can examine more recent and ongoing relatively large scientific construction projects. Two such examples are the National Ignition Facility (NIF) at Livermore National Laboratory and the Spallation Neutron Source (SNS) at Oak Ridge National Laboratory. These are both national projects, both requiring substantial R&D and anticipated construction schedules of ~6-7 years and both have experienced some difficulty and criticism. The NIF project is well along in construction with an anticipated completion in 2003. However some technical and management uncertainties have recently emerged increasing the base cost of $1.2B by ~$350M and delaying the finish date by several years. The SNS project is just commencing but already has come under Congressional criticism concerning project management. Its estimated cost is $1.2B and with some delays in funding is expected to find its schedule expanded. Again we note that projects whose cost exceed $1B have a tendency to attract attention at any stage of their development. The recently proposed Japanese Hadron Facility (JHF) is of some interest. This is a joint effort by two Japanese Agencies, Monbusho and Science Technology Agency, is projected to be an International Research Center, with multi-purpose capabilities in nuclear-particle-materials and biological sciences. It will be interesting to follow the progress of this project since its estimated cost is ~$1.5B with a construction time of 5 years.

Before summarizing it is important to examine the status of the U.S. political climate especially with respect to the attitude towards science and budget. An important event took place in 1997; namely, the placement of budget caps as part of a deficit reduction program. This was done in order to deal with the national debt and it was forecast to last five years, with the major budget cuts taking place in the latter years. This was indeed adhered to in the early years '97-'99 and with the robust economy helped produce financial stability and a budget surplus. In fact budget surpluses are projected for the future, their magnitude dependent upon whether the budget caps are adhered to. The political situation is unclear but the budget trend (surpluses) is not. Although officially the FY '2000 U.S. budget kept the budget caps on there were some clever budget manipulations which in effect violated the agreement. Although there is still confusion, most informed opinion believes that the 2001 Clinton budget will break the budget caps with Congressional acquiescence. Furthermore, the fact that many individuals, industry, and Mr. Greenspan, attribute the vibrancy, innovation, and productivity of the U.S. economy to science and technology leads me to believe that science funding will benefit in the coming years.

Now what does this all mean. From an overall project cost point of view, one can say in general that those projects ranging in the $0.2-0.5B are doable and ongoing; $1-2B projects are possible and ongoing but receive extraordinary scrutiny; and those in the $10B range are unlikely to be approved in the next ~3 years. Those line items that take 6-8 years to complete also tend to encounter difficulties. As noted earlier they encompass more than one administration, possibly several Congresses where the majority changes and are prone to reductions when there are shortfalls in the total budget. In the particular case of muon colliders we are in the $5-10B category. In this case one would urge that the project be international in nature from as early as possible. This has to be truly international, not the present style of one country initiating the project and then asking others to join. I believe it will also be very important that the R&D program be well done and complete before a construction start: One of the foibles of some of present, and even past projects, is that the R&D had not all been accomplished which in turn caused delays and increased costs which in turn jeopardized the project. As noted earlier, I believe there will be funds for smaller projects and R&D funds for large projects in the immediate future because of the anticipated surpluses and emerging positive attitude towards science and technology. It would then be even possible to conceive of larger projects, such as a muon collider in the $5-10B range. Parenthetically the host country would be expected to fund at least half of the cost of the project, probably more like 65%, with the rest coming from the collaborating countries. Again I emphasize that it must be international in the core, with canonical advisory committees, design, construction groups, and experimental groups. I believe a new strategy paradigm will emerge in ~3 years, possibly more optimistic, however with unity and a strong physics case the possibility for constructing a relevant muon collider certainly exists but starting at least five years from 2,000.

Work supported by U.S. Department of Energy under contract DE-AC02-98CH10886.

REMARKS ON HIGH ENERGY MUON COLLIDER

A.Skrinsky

Budker Institute of Nuclear Physics
Novosibirsk, 630090, Russia

Abstract. Some aspects crucially important for very high energy muon collider are considered. While writing this sketchy article, we assumed the reference /1/ familiar to the reader.

Matching Of Final Cooling Sections

As was shown in /1/, the maximal luminosity $L_{\mu\mu}$ of very high energy muon collider is

$$L_{\mu\mu} = \frac{N_\mu^2}{4\pi} \cdot \frac{\gamma_{coll}^{\frac{3}{2}}}{\varepsilon_{neq6}^{\frac{1}{2}} \sigma_{longcoll}^{\frac{1}{2}}} \cdot \left(\frac{\Delta E_{max}}{E}\right)^{\frac{1}{2}} N_{life} f_0 .$$

Here N_μ – number of positive and negative muons collected in two corresponding bunches in each cycle; γ_{coll} – relativistic factor of muons at the collision stage; ε_{neq6} – 6-dimensional normalized (without π's) emittance reached at the final stage of ionization cooling; $\Delta E_{max}/E$ – relative energy spread at the collision stage limited either by experiment requirements, or by technical problems; $\sigma_{longcoll}$ – muon bunch length equal to the beta function at the collision region; N_{life} – number of useful turns at the collider with high average guiding field; f_0 – repetition frequency of generation/acceleration cycles.

Hence, one of the primary importance task is to reach the minimal ε_{neq6}. The most promising option is the use of current carrying lithium-berillium rods convoluted in helixes as moderators, interrupted by accelerating sections /1/. Apart of technical difficulties, there is a problem of proper matching of sequential moderation/acceleration sections (from the exit of one moderator section to the entrance of the next one), while the momentum spread at the cooling stage is inevitably very high – many percents /1/. Such a high spread leads to high chromatic aberrations. The focal lengths of more familiar individual lenses – short solenoids or quadrupole doublets – are proportional to the square of momentum of particles under focusing. But lithium and plasma lenses focal lengths are just proportional to the momentum, hence, their use should give possibility to reach low enough chromatic aberrations in matching sections much easier. One option of such matching is presented below (Fig.1).

	Li-Be helix	Li lens	Plasma lens		RF accelerating structure		Plasma lens	Li lens	Li-Be helix
pc, MeV/c	70		70		141	200		200	70
E$_{kin}$, MeV	21		21		71	121		121	21
β$_{tran}$, cm	0.7	3	95		84	190		3	1.1	0.7
L$_{frac}$, cm	85	1	1	8	20	125	35	11	2	1	85
H$_{max}$, T	20	0	2	0	0.3	0	0.5	0	2	0	20
R$_{curr}$, cm	0.4	1	4.5		7		7		1	0.4
I$_{foc}$, MA	0.4		0.9		0.1			0.2	0.9	0.4

Figure 1. Schematics of a matching section between 2 consecutive cooling sections.
Here: pc – current muon momentum; E$_{kin}$ – corresponding kinetic energy; β$_{tran}$ – current beta-function; L$_{frac}$ – lengths of section fractions; H$_{max}$ – magnetic field on the surface of the focusing element; R$_{curr}$ – radius of focusing element; I$_{foc}$ – peak current in the element (all the numbers are rough and need careful optimization).

Fig.2 illustrates chromatic aberration of similarly structured matching sections using plasma lenses and solenoidal ones.

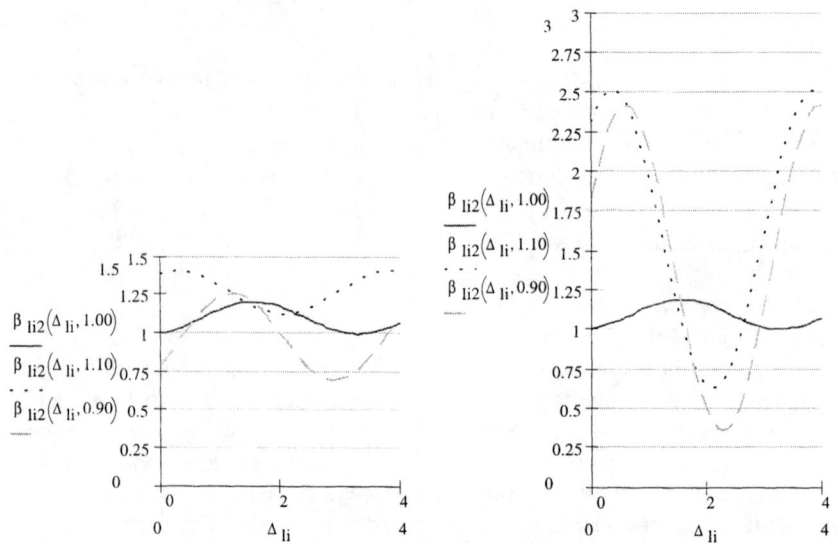

Figure 2. The beta-function at the first 4 cm in the second Li-Be helix for nominal muon momentum (1.00) and +/- 10% deviation for plasma lenses (left) and solenoidal lenses (right) cases (for the same corresponding focal lengths with optimization).

The short and strong lithium lenses at the exit and at the entrance of Li-Be helixes are necessary to make beta-function few times bigger than in the helixes and to ease low-aberration functioning of longer focal length plasma lenses. But to use them (instead of plasma lenses) at much higher beta-values inside accelerating structures is impossible – the multiple scattering results in unacceptable emittance growth.

At the Fig.3 an option for the same kind end matching section (at the exit of the cooling system) is presented. As is seen from Fig.4, resulting chromatic aberrations are acceptably small – quite comfortable for further acceleration and emittance gymnastics needed.

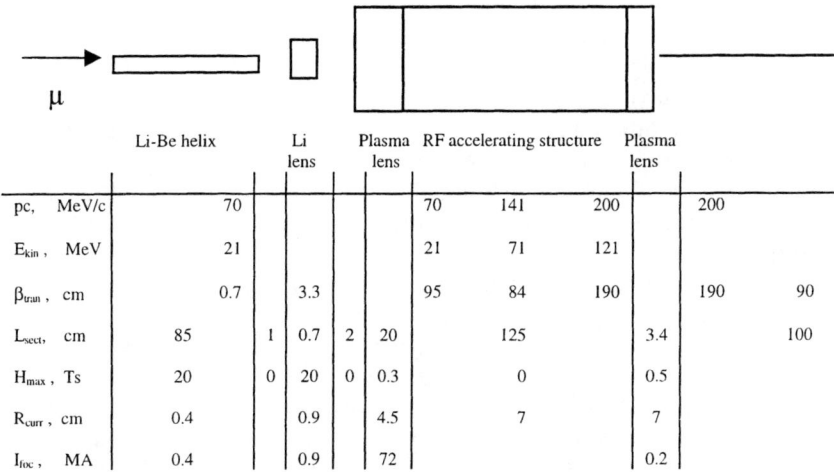

	Li-Be helix		Li lens			Plasma lens	RF accelerating structure		Plasma lens	
pc, MeV/c	70					70	141	200	200	
E_{kin}, MeV	21					21	71	121		
β_{tran}, cm	0.7		3.3			95	84	190	190	90
L_{sect}, cm	85	1	0.7	2	20		125	3.4		100
H_{max}, Ts	20	0	20	0	0.3		0	0.5		
R_{curr}, cm	0.4		0.9		4.5		7	7		
I_{foc}, MA	0.4		0.9		72			0.2		

Figure 3. The matching structure of the last cooling/accelerating section
(all the denotations the same as at Fig.1)

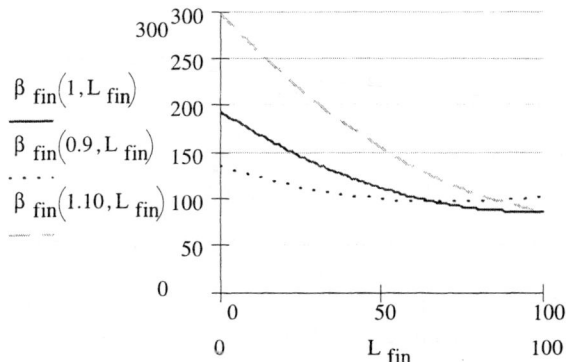

Figure 4. The beta-function upon the exit of last cooling/accelerating section
(the denominations the same as at Fig. 2).

On Beam-Beam Tune Shift Compensation

For higher energies, the beam-beam tune shift ξ (per interaction region) grows higher and higher, and for the "ultimate colliders" it reaches more than 0.5 and even more:

$$\xi_{coll} = \frac{r_\mu}{4\pi} \cdot N_\mu \sqrt{\frac{\gamma_{coll} \sigma_{longcoll} \Delta E_{coll} / E_{coll}}{\varepsilon_{neq6}}} \ .$$

Such a tune shift, in spite of short lifetime of muons, cannot be tolerated without damage for luminosity. Possible approach for preventing the damage is to compensate the shift with the liquid lithium jet of radius σ_{long}, crossing the interaction region. The electric fields are compensated perfectly, but equally needed magnetic fields compensation needs moving electrons of metal (or plasma). The scattering of these electrons leads to fast diffusion and to the growth of compensating current radius, hence to the fast disappearing of the useful effect. The better conditions for compensation would be if the "medium dense" plasma density 10^{19}-10^{20} e/cm^3 is used, what gives good compensation for few millimeters bunches length. The subject needs further study.

If such kind of compensation will be the real necessity, very severe background problem arises. Muons collide with target nuclei and produce lot of secondaries. Especially dangerous are neutral pions (decaying in photons) and neutrons. Charged particles of modest (up to few GeV/c) transversal momentum under action of longitudinal high magnetic field of the detector could be prevented from hitting the main sensitive parts of the detector (of course, non-detection of these particles limits experimental potential). But to deal with high angle neutrals is much more difficult.

The necessary switching to the lower plasma density eases the background problem, still remaining very worrying. The subject needs much more careful (and inventive) consideration.

On Collider Magnet

The decaying muons produce high energy electrons (many Megawatts!). Electrons emit fraction of their energy as high energy synchrotron radiation photons, which go to the outer wall of the magnet; electrons go to the inner wall. It is obligatory to prevent such power deposition in the cryogenic part of collider magnets. The putting inside a thick tungsten absorber would rise the size and the cost of the collider very substantially, and additionally will decrease its magnetic field, thus its luminosity.

The vertical size of muon beams at the main fraction of the collider is small – tens of microns (and lesser at higher energies). Thus, the non-cryogenic gap in the collider magnet could be very modest. The principal idea of "open median plane collider magnet" is presented at Fig.5.

Figure 5. Schematics of dipoles (left) and skew-quads (right) for "open median plane" collider.

The main problem is to prevent final electrons to hit the magnet before flying out because of vertical defocusing (or overfocusing) in quadrupole field.

Such structure of the magnet could help to remove and absorb somewhere far away out of detector the main fraction of decay products, thus to diminish very substantially this part of the background.

Let us have in mind a collider structure specificity for very high energies. At energies of 10 TeV (total) and higher muons live longer than acceleration cycles (usually assumed repetition rate is around 15 Hz). Hence, for higher energies and constant repetition rate more and more bunches shall live in the collider simultaneously. And for single ring collider practically unavoidably they would collide in parasitic regions. The only practical possibility to avoid problems with parasitic beam-beam collisions and to gain from "over-producing" of muons is to switch to double ring collider, keeping in actual operation the useful collision regions only.

ACKNOWLEDGEMENTS

The work benefited of many discussions with K.Lotov, V.Parkhomchuk, R.Palmer, G,Silvestrov, V.Telnov, A.Zholents, to whom I am very grateful. I am very grateful to B.King for his help and patience.

REFERENCES

1. Skrinsky A.N., "Towards Ultimate Luminosity Polarized Muon Collider (problems and prospects)", in *Proceedings of the Symposium on Physics Potential and Development of mu⁺mu⁻ Colliders,* San Francisco (1997); APS Proceedings **441**, pp. 249-264 (see there the other related references).

Limit on a horizontal emittance in high energy muon colliders due to synchrotron radiation

Valery Telnov

Institute of Nuclear Physics, 630090, Novosibirsk, Russia

Abstract. It is shown that at a 100 TeV muon collider the synchrotron radiation in the ring will determine the minimum horizontal emittance.

In all e^+e^- storage ring the horizontal emittance is determined by quantum nature of the synchrotron radiation. Emission of photons in regions with non-zero dispersion leads to growth of the horizontal emittance while the average energy loss leads to the cooling. As the result, some equilibrium emittance is reached. This is well known and corresponding formulae can be found elsewhere [1].

At muon colliders, the synchrotron radiation power is much smaller than that at e^+e^- storage rings due to higher particle mass: $P \propto (EB)^2/m^4$; however this suppression is compensated by the much higher energy and magnetic field.

In general, emittance in a storage ring depends on time in the following way

$$\epsilon_x = \epsilon_D(1 - e^{-t/\tau}) + \epsilon_0 e^{-t/\tau}, \tag{1}$$

where ϵ_0 is the initial emittance, ϵ_D is the equilibrium emittance, τ is the damping time. One can check (see also the B.King's table) that for the 2E=100 TeV muon collider $\gamma_\mu \tau_\mu \approx \tau_D \approx 1$ sec, so the emittance will be close to the equilibrium one.

The equilibrium *normalized* ($\epsilon_n = \gamma\epsilon$) horizontal emittance in a damping ring [1]

$$\epsilon_D = \frac{55\hbar}{32\sqrt{3}mc} \frac{\gamma^3}{J_x} \frac{\langle H \rangle}{\rho}, \tag{2}$$

where $J_x \approx 1$ is the damping partition number, $\langle H \rangle \propto \beta^3/\rho^2$ is the average value of the "H-function" characterizing a storage ring magnetic structure. One can see that for fixed energy and magnetic field $\epsilon_{nx} \propto 1/m.^4$

For an optimized FODO ring structure of the muon collider the minimum normalized horizontal emittance due to synchrotron radiation [1]

$$\epsilon_D \sim 100 \frac{E^3[\text{GeV}]}{N^3} \left(\frac{m_e}{m_\mu}\right)^4 \text{ cm rad}, \tag{3}$$

where N is the number of bending magnets in the ring. Extrapolating the number of magnets in the current designs to the 100 TeV energy as $N \propto \sqrt{E}$ (the usual energy dependence of the β-function) one can get $N \sim 1000$ for $2E = 100$ TeV. For this structure we obtain the normalized horizontal emittance at $t = \tau_D$

$$\epsilon_{nx} \approx 0.63\epsilon_D \approx 5 \times 10^{-3} \text{ cm rad}. \tag{4}$$

For comparison, in the B.King's table $\epsilon_{nx} = 8.7 \times 10^{-4}$ cm rad for "evolutionary" extrapolation and $\epsilon_{nx} = 2.1 \times 10^{-5}$ cm rad for "ultra-cold beams", which is smaller than above result by factors of 5 and 250, respectively.

This example shows that the considered effect is important for high energy muon colliders and this problem needs more accurate consideration. One should consider the more realistic case of the isochronous ring (which is necessary for muon colliders), including W-shielding which reduces the quad strength (and correspondingly N).

REFERENCES

1. H.Wiedemann, Particle Accelerator Physics, v.1, Springer-Verlag.

Problems and stoppers for $\gamma\gamma, \gamma\mu, \mu p$ colliders using very high energy muons

Valery Telnov

Institute of Nuclear Physics, 630090, Novosibirsk, Russia

Abstract.
It is well known that at linear e^+e^- (ee) colliders using laser backscattering one can obtain colliding $\gamma\gamma$, γe beams with energy and luminosity comparable to those in e^+e^- collisions. In this paper, it is explained why this can not be done at high energy muon colliders. Due to several physics reasons the $\gamma\gamma$ luminosity is suppressed here by a factor of 10^{14} ! Another option – γ's from a linear collider and muons from a muon collider – is also discussed (and has no sense either). Of course, one can study $\gamma^*\mu$ and $\gamma^*\gamma^*$ interactions at muon colliders in collisions with virtual photons as it is done now at e^+e^- storage rings. Muon-proton colliders are attractive only if the proton beam is cooled and has the same parameters as the muon beam, in which case $L_{\mu p} \sim L_{\mu\mu}$.

I INTRODUCTION

Firstly, I would like to explain the origin of this talk. Two weeks ago the chairman of our workshop Bruce King have sent me e:mail with the request to give a plenary talk on "prospects for very high energy $\gamma\gamma$ or $\gamma\mu$ colliders driven by the muon beams", he added that "even if it is impractical it would still be nice if you could give a brief explanation".

I have agreed to give such a talk but only without the word "prospects" in the title because I do not see any prospects here, only stoppers. Nevertheless, this physics is very interesting, and it is pleasure to me to tell briefly about high energy photon colliders based on e^+e^- (ee) linear colliders and explain why such photon colliders are completely impractical with muons.

The third combination of colliding particles, μp, is also discussed here very briefly.

II PHOTON COLLIDERS BASED ON LINEAR ee COLLIDERS

As you know, to explore the energy region beyond LEP-II, linear e^+e^- colliders (LC) in the range from a few hundred GeV to about 1.5 TeV and higher are under intense study around the world [1–4].

Beside e⁺e⁻ collisions, linear colliders provide a unique possibility for obtaining $\gamma\gamma$, γe colliding beams with energies and luminosities comparable to those in e⁺e⁻ collisions [5-9]. High energy photons for these collisions can be obtained using Compton scattering of laser light on high energy electrons. This idea is based on the following facts:

- Unlike the situation in storage rings, in linear colliders each beam is used only once.

- Using an optical laser with reasonable parameters (flash energy of 1 to 5 J) one can "convert" almost all electrons to high energy photons;

- The energy of scattered photons is close into the energy of initial electrons.

Each one of these items is vital for obtaining $\gamma\gamma$, γe collisions at energies and luminosities comparable to those in parental electron-electron collisions. [1]

The physics at high energy $\gamma\gamma,\gamma$e colliders is very rich and no less interesting than with pp or e⁺e⁻ collisions. This option has been included in the pre-conceptual design reports of LC projects [1-3], and work on full conceptual designs is under way. Reports on the present status of photon colliders can be found elsewhere [11,12].

Well, can we make similar photon colliders on the basis of muon colliders? What is the difference?

III $\gamma\gamma,\gamma\mu$ COLLIDERS BASED ON HIGH ENERGY $\mu\mu$ COLLIDERS

A Multi-pass collisions

At muon colliders two bunches are collided about 1000 times, which is one of the advantages over linear e⁺e⁻ colliders where beams are collided only once. However, if one tries to convert muons into high energy photons (by whatever means), the resulting $\gamma\gamma$ luminosity will be smaller than that in $\mu\mu$ collisions at least by a factor of 1000. This argument alone sufficient to give up the idea of $\gamma\gamma$ colliders based on high energy muon colliders. However, at this workshop F.Zimmermann proposed the idea of one pass muon colliders. So, I will continue enumeration of stoppers.

B Laser wave length

The required wave length follows from the kinematics of Compton scattering [6]. In the conversion region a laser photon with the energy ω_0 scatters at a small

---

[1] Here we do not discuss "photon colliders" based on collisions of virtual photons. This possibility always exists, however the luminositis and energies are considerably smaller than those in parental ee collisions, see sect.III F

collision angle (head-on) on a high energy electron(muon) with the energy E_0. The maximum energy of scattered photons (in direction of electrons) is given by

$$\omega_m = \frac{x}{x+1} E_0; \quad x = \frac{4 E_0 \omega_0}{m^2 c^4}, \tag{1}$$

where m is the mass of the charged particle. In order to obtain photons with the energy comparable to that of initial particles, say, 80 %, one needs $x \sim 4$, or the energy of laser photons

$$\omega_0 \sim m^2 c^4 / E_0. \tag{2}$$

The corresponding laser wave length is then

$$\lambda \sim 5 E_0 [\text{TeV}] \; \mu\text{m} \quad \text{for electron beams}; \tag{3}$$

$$\lambda \sim 0.12 E_0 [\text{TeV}] \; \text{nm} \quad \text{for muon beams}. \tag{4}$$

So, one can use optical lasers to make high energy photons by means of backward Compton scattering on electron beams, while at muon colliders one would have to use X-ray lasers!

C Flash energy

The probability of Compton scattering for an beam particle in the laser target $p \sim n \sigma_C l$, where n, l, σ_C are the density of the laser target, its length and the Compton cross section, respectively. The density $n \sim (A/lS)/\omega_0$, where A is the laser flash energy and S is the cross section of the laser beam which should be larger than that of the muon beam. [2] The Compton cross section for muon at $x = 4$ is about [6]

$$\sigma_C(x = 4) \sim \pi r_e^2 \left(\frac{m_e}{m_\mu}\right)^2, \tag{5}$$

where $r_e = e^2/m_e c^2$ is the classical radius of the electron.

From the above relations we get the required laser flash energy (for $p \sim 1$)

$$A \sim (S/\sigma_C) \omega_0 = \frac{S}{\pi r_e^2 E_0} \left(\frac{m_\mu}{m_e}\right)^4 m_e^2 c^4 = 1.5 \times 10^{-3} \frac{S[\mu\text{m}^2]}{E_0[\text{TeV}]} \left(\frac{m_\mu}{m_e}\right)^4 \; \text{Joule}. \tag{6}$$

At the muon collider with $E_0 = 50$ TeV, $S = 1$ μm^2 one needs the X-ray laser with the flash energy 10^5 J and the wave length of 6 nm (see eq. 2). This is certainly impossible. Beside this "technical" problem, there are even more fundamental stoppers for photon colliders based on muon beams, see below.

---

[2] In the case of the electron LC, where optical photons are used, the laser spot size is determined by diffraction: $a_\gamma \sim \sqrt{\lambda l/4\pi}$ which is several μm for LC electron beams [8]. At muon colliders, the required wave length is much shorter and diffraction can be neglected.

D e⁺e⁻ pair creation in the conversion region

Beside the Compton scattering at the conversion region, at muon colliders there is another competing process: e⁺e⁻ pair creation in collision of laser photons with the high energy muons, $\gamma\mu \to \mu e^+ e^-$. The ratio of the cross sections

$$\frac{\sigma_{\gamma\mu \to \mu e^+ e^-}}{\sigma_{\gamma\mu \to \gamma\mu}} \sim \frac{\frac{28\alpha r_e^2}{9} \ln \frac{4E_0 \omega_0}{m_e m_\mu c^4}}{\pi r_e^2 (m_e/m_\mu)^2} \sim 7 \times 10^{-3} \left(\frac{m_\mu}{m_e}\right)^2 \ln\left(\frac{m_\mu}{m_e} x\right) \sim 2000 \text{ at } x = 4. \quad (7)$$

So, high energy photons are produced with a very small probability, less than 1/1000 ! In all other cases muons lose their energy via creation of e⁺e⁻ pairs. This effect alone suppresses the attainable $\gamma\gamma$ luminosity at muon colliders by a factor of more than 10^6 !

E Coherent pair creation

OK, the yield of high energy photons from the conversion region is very small, but this is not the whole story. What happens to the "happy" photons at the interaction region? They will be "killed" by the process of coherent e⁺e⁻ pair creation in the field of the opposing muon beam. This process restricts the luminosity of photon colliders based on electron linear colliders [8–10].[3] The effective threshold of this process $\Upsilon = \frac{\omega}{m_e c^2} \frac{B}{B_0} \sim 1$, where ω is the photon energy, B is the beam field, $B_0 = \alpha e / r_e^2 \sim 4.4 \times 10^{13}$ Gauss. For the "evolutionary" $2E = 100$ TeV muon collider (see the B.King's tables) with $N = 0.8 \times 10^{12}$, $\sigma_z = 2.5$ mm, $\sigma_{x,y} = 0.2$ μm and a photon energy 40 TeV we have $\Upsilon \sim 180$. Using formulae given in ref. [8], one can find the probability of e⁺e⁻ pair creation during the bunch collision: it is very high, about 200. This means that only about 1% of high energy photons will survive in beam collisions and contribute to the $\gamma\gamma$ luminosity.

F Summary on $\gamma\gamma$, $\gamma\mu$ colliders based on high energy muons

1) The laser required for conversion of 50 TeV muons into high energy photons should have flash energy $A \sim 10^5$ J and wave length $\lambda \sim 5$ nm. This is impossible.

2) The achievable $\gamma\gamma$ luminosity

$$L_{\gamma\gamma}/L_{\mu\mu} \sim \frac{1}{1000} \times \left(\frac{1}{2000}\right)^2 \times \left(\frac{1}{100}\right)^2 = 2.5 \times 10^{-14} ! \quad (8)$$

Here the first factor is due to the one pass nature of photon colliders, the second one is due to the dominance of e⁺e⁻ creation at the conversion region (instead

---

[3] For an LC with the energy below about 1 TeV this effects is still not very important and one can obtain, in principle, $L_{\gamma\gamma} > L_{e^+e^-}$

of Compton scattering), and the third one is due to coherent pair creation at the interaction region. All clear. One can forget about $\gamma\gamma$ (and $\gamma\mu$ too) colliders based on high energy muon beams.

However, $\gamma\gamma,\gamma\mu$ interactions can be studied at muon colliders in collisions of virtual photons (without $\mu \to \gamma$ conversion). The luminosities in such collisions are [8]

$$L_{\gamma^*\gamma^*} \sim 10^{-2} L_{\mu\mu} \qquad W_{\gamma\gamma} > 0.1 \times 2E_0 \qquad (9)$$

$$L_{\gamma^*\gamma^*} \sim 10^{-4} L_{\mu\mu} \qquad W_{\gamma\gamma} > 0.5 \times 2E_0. \qquad (10)$$

$$L_{\gamma^*\mu} \sim 0.15 L_{\mu\mu} \qquad W_{\gamma\mu} > 0.1 \times 2E_0 \qquad (11)$$

$$L_{\gamma^*\mu} \sim 0.05 L_{\mu\mu} \qquad W_{\gamma\mu} > 0.5 \times 2E_0. \qquad (12)$$

IV $\gamma\mu$ COLLISIONS AT LC–MUON COLLIDERS

One can also consider $\gamma\mu$ colliders where high energy photons are produced at LC (on electrons) and then are collided with high energy muon beams. This option also has no sense for several reasons:
a) $N_e \sim 10^{-2} N_\mu$;
b) loss of photons at the IP due to coherent e^+e^- pair creation;
c) none of the LCs have the pulse structure of muon colliders (almost uniform in time), which results in a factor 100 times loss in luminosity.

All factors combined give $L_{\gamma\mu} < 10^{-5} L_{\mu\mu}$. Such $\gamma\mu$ collider has no sense; besides, $\gamma\mu$ collisions can be studied for free with much larger luminosities in $\gamma^*\mu$ collisions (see the end of the previous section).

V μp COLLIDERS

Let us first consider collisions of the LHC proton beams with muon beams of a 100 TeV muon collider. Without special measures, the luminosity in such collisions is lower than that in pp collisions at LHC due to larger distance between bunches at muon colliders (smaller collision rate).

$$L_{\mu p} \sim L_{pp} \times (\nu_\mu/\nu_p) \sim 10^{-3} L_{pp} \sim 10^{31} \text{ cm}^{-2}\text{s}^{-1}. \qquad (13)$$

This is too small for study of any good physics.

However, one can think about a special source of proton with several stages of electron cooling. If parameters of the proton beam are the same as those of the muon beam, then the luminosity at the 100 TeV μp collider $L_{\mu p} = L_{\mu\mu} \sim 10^{36}$ cm^{-2}s^{-1}. That is not easy to achieve, but such possibility is not excluded.

One of problems at such colliders is hadronic background. At $L_{\mu p} = 10^{36}$ and $\nu = 10^4$ the number of background γp reactions is about 5000/crossing. One can decrease backgrounds by increasing the collision rate (up to a factor of 5–10). It is not excluded that even with such backgrounds one can extract interesting physics. This option certainly makes sense, if a very high energy $\mu\mu$ collider is to be built. Its feasibility and potential problems should be studied in more detail.

REFERENCES

1. *Zeroth-Order Design Report for the Next Linear Collider* LBNL-PUB-5424, SLAC Report 474, May 1996.
2. *Conceptual Design of a 500 GeV Electron Positron Linear Collider with Integrated X-Ray Laser Facility* DESY 97-048, ECFA-97-182; R.Brinkmann et al., *Nucl. Instr. &Meth. A* **406** (1998) 13.
3. *JLC Design Study*, KEK-REP-97-1, April 1997; I. Watanabe et. al., KEK Report 97-17.
4. J.P. Delahaye et al., Acta Phys. Polon. B30:2029-2039, 1999; R. Bossart et al. CERN-PS-99-005-LP, Apr 1999.
5. I.Ginzburg, G.Kotkin, V.Serbo, V.Telnov,*Pizma ZhETF*, **34** (1981)514; *JETP Lett.* **34** (1982) 491 (Prepr. INP 81-50, Novosibirsk, Febr. 1981).
6. I.Ginzburg, G.Kotkin, V.Serbo, V.Telnov, *Nucl. Instr. & Meth.* **205** (1983) 47.
7. I.Ginzburg, G.Kotkin, S.Panfil, V.Serbo, V.Telnov, *Nucl. Instr.&Meth.* **219**(1984)5.
8. V.Telnov, *Nucl. Instr.&Meth.A* **294** (1990)72.
9. V.Telnov, *Nucl. Instr.&Meth.A* **355**(1995)3.
10. V.Telnov, *Proc. of ITP Workshop "Future High energy colliders"* Santa Barbara, USA, October 21-25, 1996, AIP Conf. Proc. No 397, p.259; e-print: physics/ 9706003.
11. V.Telnov, Proc. of the International Conference on the Structure and Interactions of the Photon (Photon 99), Freiburg, Germany, 23-27 May 1999, to be published in Nucl. Phys. Proc. Suppl. B, e-print: hep-ex/9908005.
12. V.Telnov, Proc. of World-Wide Study of Physics and Detectors for Future Linear Colliders (LCWS 99), Sitges, Barcelona, Spain, 28 Apr–5 May 1999, e-print: hep-ex/9910010, hep-ex/9910011.

Some problems in plasma suppression of beam-beam interactions at muon colliders

Valery Telnov

Institute of Nuclear Physics, 630090, Novosibirsk, Russia

Abstract.
The idea of plasma suppression of beam-beam effects at muon colliders is discussed. It is shown that one should take into account collisions in the plasma that were ignored before. Rough estimates show that this effect leads to a fast "recovery" of the beam magnetic field. For beam parameters characteristic for muon colliders the suppression of the magnetic component of the beam field (1/2 of the total force) is almost absent. It is also shown that the presence of the dense plasma (Li jet) at the interaction point leads to enormous hadronic background (due to photo-nuclear reactions) in the detector, about 10^7 particles per crossing at large angles which creates serious problems for experimentation.

I INTRODUCTION

One of main the problems for high energy muon colliders is the limitation of the luminosity due to beam-beam interactions. A measure of the beam-beam interaction is the tune-shift parameter ξ [1]. For round beams, $\xi = Nr_c/4\pi\epsilon_n$, where $r_c = e^2/mc^2$ is the classical radius of the beam particles, ϵ_n is the normalized transverse emittance. The value of ξ should be small enough ($\xi_{max} < 0.1$) - otherwise the beams are disrupted due to resonance diffusion. The maximum luminosity $L_{max} = N\gamma f \xi_{max}/r_c\beta$, where β is the β-function (usually $\beta \approx \sigma_z$), f is the collision rate. So, $L \propto \xi$. This effect puts a severe limit on the luminosity of muon colliders.

One of the possible solution of this problem is plasma suppression of beam-beam interactions [2–4]. In a sufficiently dense plasma one can expect that the induced charges and currents will decrease the beam fields and, consequently, the beam-beam effects. If plasma decreases the beam field by a factor K, one can use beams with K times smaller ϵ_n (to keep $\xi = \xi_{max}$), correspondingly the luminosity will be K times larger.

It is essential to reduce both electric and magnetic fields because in the vacuum their action on the opposing beam are equal in value and direction, the effective beam field is $|E|+|B|$. If the plasma density n_p is larger than the particle density of the colliding bunches n_b, the electric field of the beam will be suppressed by repelling (in the case of negatively charged bunch) or attracting (in the case of positively charge bunch) plasma electrons (ions are immobile). The nature of the magnetic field suppression is somewhat more complicated. In the linear approximation the resulting suppression of the beam-beam interaction [3]

$$\xi/\xi_0 \sim 4/(k_p \sigma_r)^2 \text{ for } k_p \sigma_r \gg 1, \qquad (1)$$

where $k_p = \omega_p/c$, $\omega_p^2 = 4\pi n_p e^2/m_e$, σ_r is the r.m.s. beam radius. A more accurate result, including nonlinear effects and finite plasma thickness, was obtained in ref. [4].

So, for a factor of 5 suppression of beam-beam interactions the plasma should satisfy the following requirements: a) $n_p > n_b$ and b) $k_p \sigma_r > 4$. For example, let us take parameters of the "evolutionary" 100 TeV muon collider (see the B.King's table) but with 5 times smaller ϵ_{nx}: $N = 0.8 \times 10^{12}$, $\sigma_r = \sqrt{2}\sigma_x = 0.13$ μm, $\sigma_z = 0.25$ cm. In this case $\xi = 0.5$ without suppresion, while the acceptable $\xi = 0.1$. The plasma density required for decreasing ξ by a factor of 5 is found from conditions a) and b), which give $n_p > 2.3 \times 10^{21}$ and $n_p > 2.5 \times 10^{22}$, respectively. As a source of plasma one can use a liquid Li jet [4] with electron density 1.5×10^{23} cm^{-3}. Such a target will be fully ionized by the muon beam and the return current. If this theory is correct, in the considered example one can increase the luminosity by a factor of 5. With other beam parameters one can expect even better results, up to 30 for the given beam diameter. All this sounds nice. However, there are two effects which create serious problems for this method, and, perhaps, close it:

- collisions in the plasma;
- hadronic background at large angles.

II COLLISIONS IN THE PLASMA.

In all papers on plasma suppression of the beam-beam interaction it was assumed that plasma is collisionless. This picture is not correct. In fully ionized plasma the electrons of the return current do not lose energy on ionization, do not lose energy in collisions with other electrons, because all electrons move with the same average velocity, and have very small energy loss in collisions with the ions; however, their longitudinal velocity is decreased due to the scattering on the ions.

The change of the longitudinal velocity in one scattering on the ion

$$\Delta v = -2v_0 \sin^2 \frac{\vartheta}{2} \approx -v_0 \frac{\vartheta^2}{2} \qquad (2)$$

The resulting friction in the plasma

$$\frac{d\vec{v}}{dt} = -\int v_0 \Delta v n_p d\upsilon = -\int v_0^2 \frac{\vartheta^2}{2} 8\pi n_p \left(\frac{e^2 Z}{m_e v_0^2}\right)^2 \frac{d\vartheta}{\vartheta^3} =$$

$$= -4\pi n_p \frac{\vec{v}}{v_0} \left(\frac{e^2 Z}{m_e v_0}\right)^2 \ln\frac{\vartheta_{max}}{\vartheta_{min}} = -4\pi n_p \frac{\vec{v}}{v_0} \left(\frac{e^2 Z}{m_e v_0}\right)^2 \ln\Lambda, \qquad (3)$$

where $\ln\Lambda = \ln(b_{max}/b_{min})$. The minimum value of the impact parameter b follows from the energy conservation $b_{min} = e^2 Z/mv_0^2$, and the maximum value of b is equal to the Debey length $\lambda_D = \sqrt{kT/m}/\omega_p$. For the considered plasma densities and electron velocities, $\ln\Lambda \approx 7 \sim 10$.

For estimation of the collision time one can take $dv = v$, which gives

$$\tau_{col} \sim \frac{v_0}{4\pi n_p \left(\frac{e^2 Z}{m_e v_0}\right)^2 \ln\Lambda}. \qquad (4)$$

The average velocity of electron in the return current $u \approx (n_b/n_p)c$. Although this velocity is not the same as v_0 due to transverse motion, let us the first assume $v_0 = u$. Then

$$\tau_{col} \sim (\frac{n_b}{n_p})^3 \frac{1}{4\pi c n_p Z^2 r_e^2 \ln\Lambda}. \qquad (5)$$

For the example given above $n_b/n_p \sim 1/60$, $Z = 3$, $n_p = 1.5 \times 10^{23}$, that gives $\tau_{col} \sim 10^{-17}$ sec, which is much smaller than the bunch collision time $\sigma_z/c \sim 0.25/3 \times 10^{10} \sim 10^{-11}$ sec.

So, the assumption of collisionless plasma is not valid. The plasma should be considered as a medium with some conductivity σ_c. Accurate calculation of conductivity is a complicated task because the electron drift in the longitudinal induction electric field with very small energy loss, only scatter. Due to the field their total kinetic energy continuosly grows, while the drift velocity is approximately constant: $u \sim c(n_b/n_p)$. Nevertheless, we can make some estimate.

Eq.3 is approximately valid even in the case of the "hot" return current if we will consider $\vec{v}$ as the drift velocity (from now on u) and v_0 as the "thermal" velocity, which is still unknown. The loss of the drift velocity given by Eq.3 is compensated by the induction electric field

$$|du/dt| = eE_{\|}/m_e. \qquad (6)$$

The conductivity is defined by equation

$$en_p u = \sigma_c E. \qquad (7)$$

Drift velocity is known:

$$u = c(n_b/n_p). \qquad (8)$$

Using Eqs 6,7,8, we get

$$\sigma_c = \frac{e^2 n_b c}{m_e |du/dt|} \qquad (9)$$

Using Eq.3, we obtain

$$\sigma_c = \frac{m_e v_0^3}{4\pi e^2 Z^2 \ln \Lambda} \tag{10}$$

Now we have to estimate v_0. The collision time is given by eq.4. Between two collisions the electron drift velocity is restored by the induction electric field and the total energy is increased by about $m_e u^2$, so as an estimate one can take the average kinetic energy to be equal to $m_e v_0^2 \sim N_{col} m_e u^2$. The maximum number of scatterings for the same electron can be estimated as the number of collisions after which its transverse displacement is equal to the beam radius (then this electron in the return current is replaced by the new one which comes from outside and is initially cool)

$$\tau_{col} v_0 \sqrt{N} \sim \sigma_r. \tag{11}$$

Using this arguments and Eq.10 we find

$$v_0 = c(4\pi n_b Z^2 \sigma_r r_e^2 \ln \Lambda)^{1/5} \tag{12}$$

For Z=3 (Li), $n_b = 2.5 \times 10^{21}$ (see the example above) and $\sigma_r = 0.15$ μm, $v_0 \sim 0.075c$.

So, the conductivity is found, see Eqs 10,12. Now we have to understand how the conductivity influences the plasma suppression of the bunch field. The return current is driven by the longitudinal electric field $E_\parallel$ that is caused by penetration of the beam magnetic field into the plasma. From Faraday law

$$E_\parallel \sim \frac{1}{c} \frac{d(B_\phi \sigma_r)}{dt}. \tag{13}$$

This electric field produces the return current equal approximately to the bunch current I (if compensation works)

$$\sigma_c E_\parallel \pi \sigma_r^2 \sim I. \tag{14}$$

Introdusing $B_\phi \sim 2I/c\sigma_r$ (the beam field when there is no beam field suppression) and Eqs 13,14, we obtain

$$\frac{dB_\phi}{B_\phi} \sim \frac{c^2 dt}{2\pi \sigma_r^2 \sigma_c}. \tag{15}$$

The relative value of the beam field which penetrates into the plasma during the time of the bunch collision is

$$\frac{\Delta B}{B} \sim \frac{c\sigma_z}{2\pi \sigma_r^2 \sigma_c} \sim \frac{2\sigma_z r_e Z^2 \ln \Lambda}{\sigma_r^2 (v_0/c)^3}, \tag{16}$$

where v_0 is given by Eq.12. For the example considered in this paper: $\sigma_z = 0.25$ cm, $Z = 3$, $\sigma_r = 0.13$ μm, $N = 0.8 \times 10^{12}$, $v_0/c \sim 0.075$, $\ln \Lambda = 7$ we get

$$\frac{\Delta B}{B} \sim 100 \text{ !!!} \tag{17}$$

In order to obtain plasma suppression by one order of magnitude we need $\Delta B/B \sim 0.1$. So, it seems that plasma does not help. Although my estimate is very approximate, it is very unlikely that a factor of 1000 is lost.

III BACKGROUNDS

A Photo-nuclear reactions

The total photo-nuclear cross section for lithium is [5]

$$\sigma_{\gamma Li} \sim 0.4 \times 10^{-27} \text{ cm}^2. \tag{18}$$

The number of virtual photon with the energy above 1 GeV per one muon at 100 TeV muon collider is

$$N_\gamma \sim \int \frac{2\alpha}{\pi} \ln\left(\frac{E}{\omega}\right) \frac{d\omega}{\omega} \sim 0.3 N_\mu. \tag{19}$$

The number of ph.n. reactions per bunch crossing generated by $2 \cdot 10^{12}$ muons (two beams) in $l = 0.5$ cm Li jet is

$$N_b = N_\gamma n_{Li} l \sigma_\gamma \sim 0.3 \times 2 \cdot 10^{12} \times 5 \cdot 10^{22} \times 0.5 \times 0.4 \cdot 10^{-27} = 0.6 \cdot 10^7 \text{ !} \tag{20}$$

Although most of the produced particles travel in the forward direction, each reaction produce gives approximately one particle $(\pi^\pm, \pi^0)$ at large angles, with $P \sim P_t \sim 300$ MeV. The total energy of these particles is greater than $2 \cdot 10^3$ TeV.

It is hard to imagine a detector which could work in conditions so terrible!

B e^+e^- production: $\mu Li \to \mu Li e^+ e^-$

The cross section of this reaction [6]

$$\sigma \approx \frac{28\alpha^2 r_e^2}{27\pi}(l^3 - 6.36 l^2)(Z_1 Z_2)^2, \tag{21}$$

where $l = \ln \frac{2(P_1 P_2)}{m_1 m_2} \approx \ln 2\gamma_\mu$. In the case of 50 TeV muons and the Li target, $\sigma = 1.8 \times 10^{-26}$ cm^2

The probability of e^+e^- pair creation by a muon in a 1 cm thick Li jet for 1000 crossings (as it is in the muon colliders) is about $1 - e^{-1}$ (one interaction length). In most cases the energy loss is not large but sufficient to knock the muon out of the 10^{-4} energy range which contributes to luminosity (muons with larger energy deviations are defocussed due to chromatic abberations of the final focus system). So, this effect will decrease the luminosity lifetime by about a factor of 2.

IV CONCLUSIONS

Suppression of the beam-beam effects by a dense plasma jet (Li) at the collision point is a very attractive idea. However, collisions in the plasma significantly change the picture. This effect was ignored before. Rough estimates show that this effect leads to fast "recovery" of the magnetic beam field, leaving it practically unsuppressed. This result should be checked by more accurate calculations.

Photo-nuclear reactions produce enormous hadronic backgrounds in the detector ($\sim 10^7$ particles/crossing at large angles), so the possibility of experimentation at such background conditions is practically impossible. Electro-production of e^+e^- pairs in Li jet leads to some decrease of the luminosity lifetime.

I would like to thank K.Lotov and A.Skrinsky for useful discussions and B.King for organization of the very fruitful workshop.

REFERENCES

1. H.Wiedemann, Particle Accelerator Physics, v.1.
2. D.H.Whittum, A.M.Sessler, J.J. Stewart and S.S.Yu, Part. Accel., v 34 (1990) 89.
3. G.V.Stupakov and P.Chen, Phys.ReV.Lett., 76 (1996) 3715.
4. K.V.Lotov, A.N.Skrinsky and A.V.Yashin, Budker INP 98-41.
5. Particle Data, Phys.Rev.D 54 (1996).
6. V.M. Budnev, I.F. Ginzburg, G.V. Meledin, V.G. Serbo, Phys. Rept. 15 (1974) 181.

Coherent e⁺e⁻ pair creation at high energy muon colliders

Valery Telnov

Institute of Nuclear Physics, 630090, Novosibirsk, Russia

Abstract.
It is shown that at muon colliders with the energy in the region of 100 TeV the process of coherent pair creation by the muon in the field of the opposing beam becomes important and imposes some limitations on collider parameters.

One of the main advantages of muon colliders is that the muon is much heavier than the electron and therefore the radiation (beamstrahlung) in beam collisions is suppressed. The relative energy loss during the beam collision $\Delta E/E \propto EB^2/m^4$.

However, there is another process in beam collisions which may be important for a high energy muon collider: it is the coherent e⁺e⁻ creation. In this process the e⁺e⁻ pair is created by a virtual photon in a strong field of the opposing muon beam $B \equiv |E| + |B|$. The process of coherent pair creation is very important for e⁺e⁻ linear colliders [1,2]. This process has large probability at

$$\kappa = (\omega/m_e c^2)(B/B_0) > 1, \quad B_0 = \alpha e/r_e^2 \sim 4.4 \times 10^{13} \text{ Gauss.} \qquad (1)$$

At a 100 TeV muon collider the energy and beam field are even higher than those at linear e⁺e⁻ colliders. So, one can expect that this process will be important for muon collider as well, because naively the cross section of this process depends only on E, B, m_e, but not on m_μ.

However, there is one effect in this process which makes a situation at electrons and muon beams very different. In e⁺e⁻ collisions, the maximum energy of virtual photons is almost equal to the electron energy, while at $\mu\mu$ colliders the maximum photon energy depends also on the mass of the produced system. This can be understood in the following way [3]. The minimum value of the photon mass, which corresponds to the case when the virtual photon has zero transverse momentum [4],

$$Q_{min}^2 = -q_{min}^2 = -(p-p')^2 \approx \frac{m^2 \omega^2}{E(E-\omega)}, \qquad (2)$$

where m is the mass of beam particles. Also, the cross section of e⁺e⁻ pair production is large only near the threshold $W^2 \sim 4m_e^2$. Besides, the cross section is negligible for $Q_{max}^2 > W^2$, i.e. $Q_{max}^2 \sim m_e^2$. As a result, from the inequality $Q_{min} < Q_{max}$ it follows

$$\omega < \gamma_\mu m_e c^2. \tag{3}$$

So, only photons with the energy $\omega < \gamma_\mu m_e c^2 \sim (1/200) E_\mu$ contribute to the process of coherent pair creation.

Nevertheless, at the 100 TeV $\mu\mu$ collider even such "low energy" photons can produce e^+e^- pairs. Indeed, for $N = 0.8 \times 10^{12}, \sigma_x = 2 \times 10^{-5}$ cm, $\sigma_z = 0.25$ cm, $E = 50$ TeV ("evolutionary" $\mu\mu(100)$ collider)

$$\kappa \sim \gamma_\mu \frac{B}{B_0} \sim 0.85. \tag{4}$$

Here I took $B \sim eN/\sigma_x \sigma_z$, which is close to the maximum effective beam field ($|B| + |E|$).

The probability of e^+e^- creation by the muon in the transverse magnetic field per unit length for $\kappa < 1$ [5]

$$W \sim \frac{0.013 \alpha^3 R^{5/2}}{r_e \gamma_\mu} e^{-2\sqrt{3}/R}, \tag{5}$$

where $R = \gamma_\mu(B/B_0)$.[1] For $R = 0.85$, $\sigma_z = 0.25$ cm, $E_\mu = 50$ TeV the probability of e^+e^- pair creation by the muon during its life (about 1000 beam collisions)

$$p \sim 1000 W \sigma_z \sim 0.1. \tag{6}$$

This is a large probability, the maximum that can be accepted. Further two times increase of the R value will lead to a one order decrease of the luminosity. Note, that in the process of the e^+e^- creation the muon loses about 1/200 of its energy that is much larger than the energy spread at muon colliders ($\sim 10^{-4}$), so the considered muon will no longer contribute to the luminosity (due to chromatic abberation).

Let us compare now coherent pair creation with beamstrahlung where the muon can also emit of sufficiently hard photon. We have seen that the probability of coherent pair creation is large when $\kappa \sim (EBe\hbar)/(m_\mu m_e^2 c^5) > 1$. In beamstrahlung, the muon is "lost" when the characteristic photon energy $E_\gamma/E \sim (EBe\hbar)/(m_\mu^3 c^5) = \kappa(m_e/m_\mu)^2 > \delta \sim 10^{-4}$ (see B.King's table of the 100 TeV "evolutionary" muon collider). One can see that the expression for a beamstrahlung does not contain m_e and is smaller by a factor of $(m_\mu/m_e)^2 \sim 4 \times 10^4$ than the characteristic parameter in coherent pair creation; however the upper limit on the beamstrahlung parameter is also smaller by a numerically similar factor. This means than both processes become important approximately at the same values of the muon energy and beam field. Which process is more important depends on the energy acceptance of the

---

[1] Here I distinguish R and κ because κ is approximately equal to $\gamma_\mu B/B_0$ while R is equal to this expession by definishion.

final focus system. For the considered parameters the coherent pair creation is more important.

The coherent e^+e^- pair creation in beam collisions is essential for the 100 TeV muon collider and imposes some limitations on design parameters.

REFERENCES

1. P. Chen, V. Telnov, *Phys. Rev. Letters*, **63** (1989)1796.
2. V.Telnov, *Nucl. Instr. &Meth.A* **294** (1990)72.
3. I.F. Ginzburg, IM-SO-RAN-96-2-202, hep-ph/9601273.
4. V.M. Budnev, I.F. Ginzburg, G.V. Meledin, V.G. Serbo, Phys. Rept. 15(1974)181.
5. V.N. Baier, V.M. Katkov, V.M. Strakhovenko, *Electromagnetic Processes at High Energies in Oriented Single Crystals*, World Scientific PC, Singapore, 1998.

FFAG Lattice Without Opposite Bends

Dejan Trbojevic*, Ernest D. Courant* and Al Garren[†]

*Brookhaven National Laboratory, Upton, NY 11973-5000, USA
[†]UCLA and LBNL, email: AAGarren@lbl.gov

INTRODUCTION

A future "neutrino factory" or Muon Collider requires fast muon acceleration before the storage ring. Several alternatives for fast muon acceleration have previously been considered. One of them is the FFAG (Fixed Field Alternating Gradient) synchrotron. The FFAG concept was developed in 1952 by K. R. Symon (ref. 1). The advantages of this design are the fixed magnetic field, large range of particle energy, simple RF; power supplies are simple, and there is no transition energy. But a drawback is that reverse bending magnets are included in the configuration; this increases the size and cost of the ring. Recently some modified FFAG lattice designs have been described where the amount of opposite bending was significantly reduced (ref. 2, ref. 3).

Muon acceleration needs to be very fast, because of the short muon lifetime. It is desirable to accelerate in as few turns as possible, subject to the limitations of RF capabilities. To accelerate muons from 5 GeV to 15 GeV, we may use as few as 10- 15 passes. This eases a lot of constraints on synchrotrons, like higher order betatron resonances; even the chromaticity should not be as important as normally. We present a first attempt to make an FFAG lattice without opposite bends to be used for muon acceleration within 5-15 GeV energy range.

MAJOR GOALS

The value of the dispersion function as well as the betatron functions through the synchrotron should be very small to permit accepting a large range of momentum of the circulating particles and keep within the aperture. The limitations on betatron functions are determined by the aperture size. If the aperture is assumed to be 70 mm (+-35 mm) then for the momentum offsets of:

$$-0.5 < \Delta p/p < 0.5,$$
the dispersion function ($\Delta x = D_x * \Delta p/p$) should be:

$$D_x < 70 \text{ mm},$$

between muon energies 5-15 GeV where the central energy is equal to 10 GeV.

The particle's momentum is assumed to accelerated by "distributed" accelerating cavities, as it travels through the synchrotron. The momentum increases from one "cell" to the next. We desire a lattice which allows particles within the momentum range of $-0.5 < \Delta p/p < 0.5$ to be stable without crossing integer tunes or resonances.. The necessity of dipoles with reversed field, as in the original FFAG design of K. R. Symon, is avoided by applying our previously reported FMC (flexible momentum compaction) method.

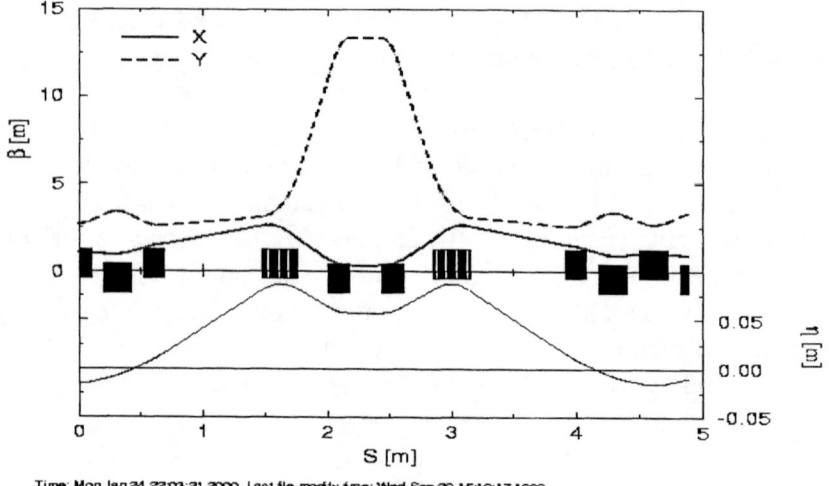

Figure 1 represents the betatron functions through one cell . Combined function magnets are presented as boxes.

THE FIRST EXAMPLE

A first attempt at a FFAG lattice without opposite bending is presented in fig. 1. The cell contains:

Major bending elements, including combined function magnets with small gradients (presented at the left and right side in figure 1), where all betatron functions, including dispersion, have small values. ($\beta_x < 0.3$ m, $\beta_y < 0.55$ m, $D_x > -0.03$ m).

A doublet QFS - QDS configuration (presented in the central region in fig. 1) consists of strong horizontally focussing QFS quadrupoles and a combined function magnet with vertical focusing. The maximum value of the dispersion function occurs in the middle of the of the focusing quadrupole QFS. The normalized dispersion space is defined by the Floquets' coordinate transformation as:

$$\zeta = \frac{D_x}{\sqrt{\beta_x}}, \quad \text{and} \quad \chi = D'\sqrt{\beta_x} + \frac{\alpha_x D_x}{\sqrt{\beta_x}}.$$

The top of fig. 2 represents the middle position of the betatron functions (as presented in fig. 1). Positions of the magnets are represented in fig. 2 by their names. At the bottom of fig. 2 is the position of the center of the combined function magnets with the largest bending angle. A vector from (0.0) to any point of this graph represent a square root of the amplitude function H.

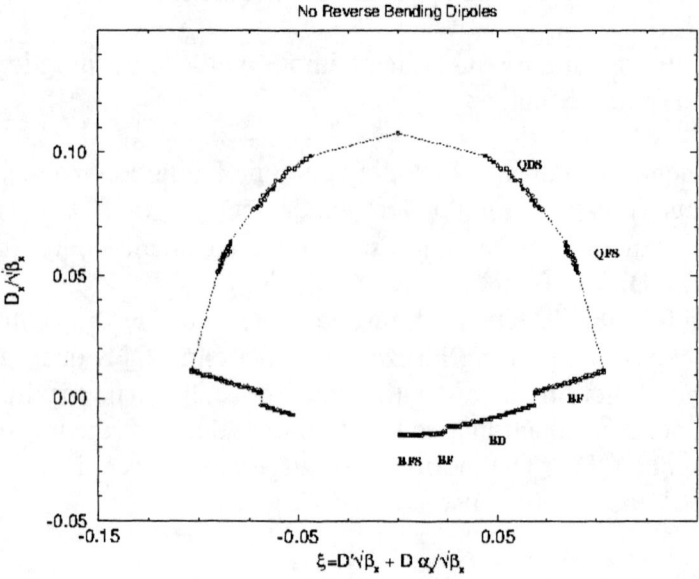

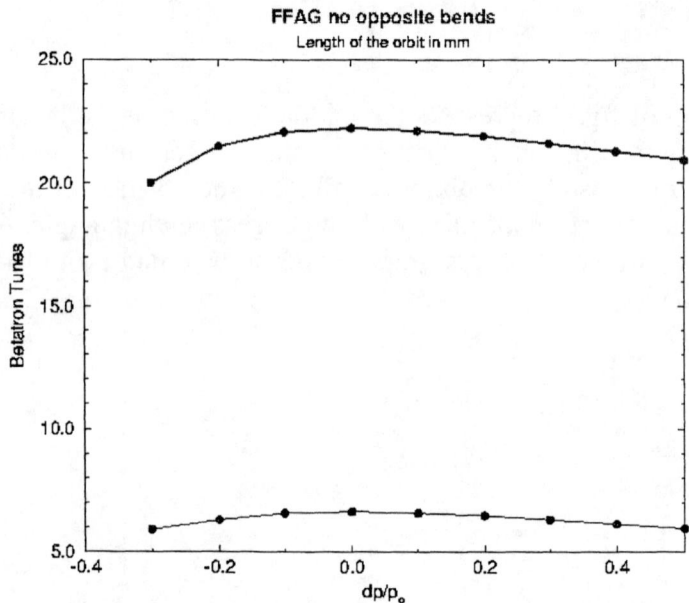

Figure 3 shows the tune dependence on momentum. At a momentum of $\Delta p/p = -0.3$ it is clear that the tune is very close to an integer value. As was noted above there will be only 10 –15 turns of the particles around the ring. If the particle starts with $0.7\, p_o$, where p_o represents the central value of momentum (which corresponds to energy of 10 GeV), immediately after few cells it would gain in momentum. The betatron tune for these particles would not represent unstable conditions.

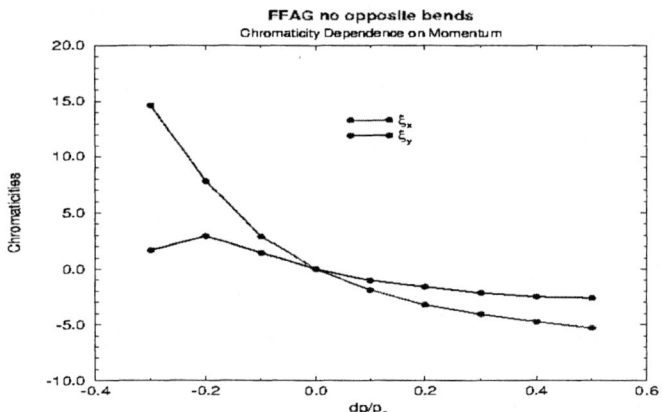

Figure 4 represents the chromaticity dependence on momentum of our first example. Again the particles with negative momentum offset of $\Delta p/p = -0.3$, will not go through unstable conditions as soon as they start orbiting around the synchrotron and gaining momentum kicks from the RF cavities.

CONCLUSIONS

The first example of an FFAG lattice without opposite bends shows promising results. It looks as if it may be possible to design an FFAG lattice which covering a momentum range $-0.5 < \Delta p/p < 0.5$, where even the particles with large momentum offsets have stable motion. The largest value of the difference in the path lengths of particles with different momenta, in this example, is of the order of 60 mm. This problem needs additional attention.

REFERENCES

1. K. R. Symon, "The FFAG Synchrotron – MARK I", MURA-KRS-6, November 12, 1954, pp. 1-19.
2. R. Ueno et. all, "Multi Orbit Synchrotron with FFAG Focusing For Acceleration of High Intensities Hadron Beams", PAC 99, New York, 1999, 2271-2273.
3. P. F. Meads, Jr., "A Compensated Dispersion-free Long Insertion for an FFAG Synchrotron", PAC 93, Proceedings of the 1993 Particle Accelerator Conference, Vol. 5 of 5, Washington D.C., pp. 3825-3827.

Muon Collider Workshop Summary

W. Willis
Columbia University

WHY THIS WORKSHOP NOW?

For most of us, the first reaction to the idea of a Workshop on muon colliders beyond ten TeV is that it must be premature, since we do not yet know how to make a reasonable muon storage ring at *any* energy. I have decided that the answer to that concern is a variant of "it is always later than you think." To illustrate my reasoning, I refer to a slide I showed in a 1981 talk at an ECFA talk at Oxford on "future accelerators," Figure 1. The emphasis was on research on novel accelerators and the far future. The organizers arranged a talk by Salam on the physics prospects, which of course was anticipated to be very optimistic. For balance, they asked me to provide a pessimistic view. Figure 1 shows my summary of the key physics results at the forefront of fundamental physics and the development of accelerators to ever-higher energies. The "Livingston Curve" shows the *effective* accelerator energy increasing exponentially with time. I also show the energy versus time for natural sources, radioactivity or cosmic rays. We see that this curve also has increased exponentially with time. I didn't have a name for it originally; I now I call it the "Auger Curve."

Since the beginning, there has been a competition between discoveries from observation of natural sources and from accelerator experiments. For example, x-rays were discovered with accelerators and nuclear decay was discovered with radioactive sources, at nearly the same time, just before the start of Figure 1. The electron and proton were discovered with accelerators; the nucleus was discovered with radioactive sources. The story continues with the positron and neutron and nuclear transmutations, muons and mesons and strange particles, all discovered from natural sources just a jump before accelerators could get to them. The antiproton broke the string. It received more publicity than seems appropriate for its relative significance, and it is a commonplace in our field that this was not only due to its physics significance but due to the relief that all the money spent on the accelerators and their experiments was starting to pay off in simple, easy to explain discoveries.

The last example at this time is the discovery of charmed particles by Niu. The evidence for the discovery was not really disputed at the time, but the same bias as in the case of the antiproton is perhaps visible in the treatment of the result by the particle physics establishment. Of course, the power of accelerators in following up on a mere discovery justifies that prejudice in the cases where that follow up is possible, but will it always be possible?

In the Oxford talk, I went on to relate the sad news that luminosity must go up together with the energy, if we continue to believe that fundamental physics is found at smaller distance scales. In 1981, that message was not familiar to part of the audience, and they

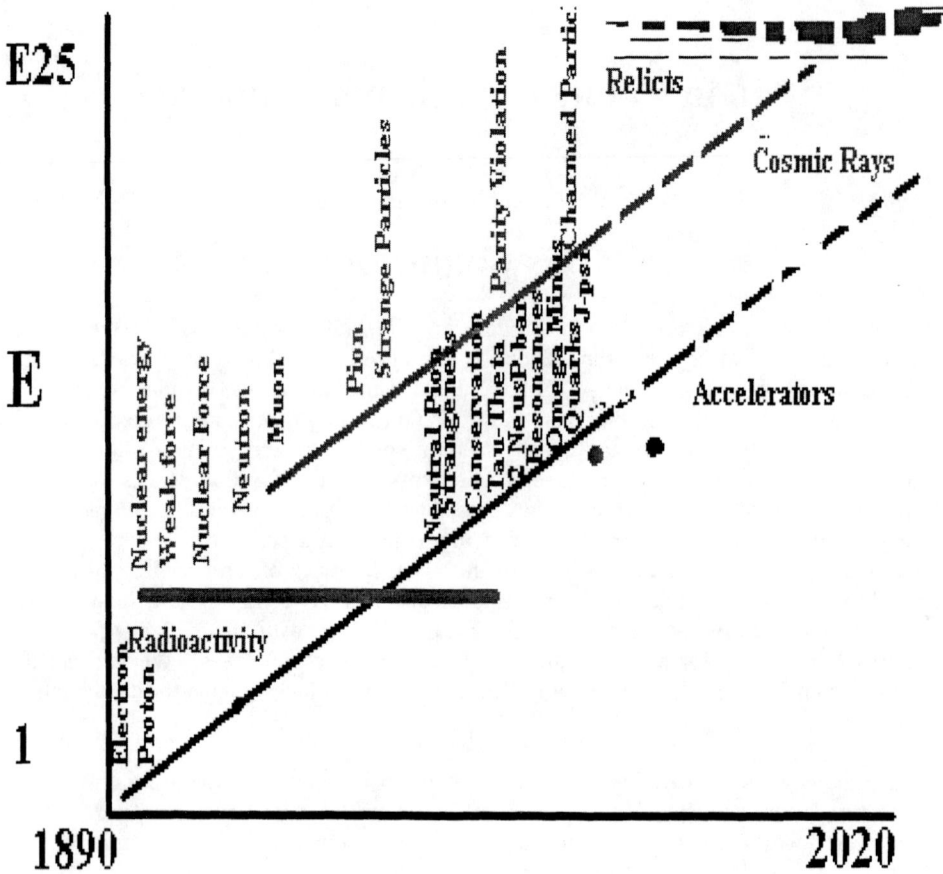

FIGURE 1.

not pleased with it. The fact that electrons, particularly, are disrupted as a beam and fragmented as particles by interaction with the opposing beam, in a way closely linked to the luminosity of the collisions, was also news to many of my auditors at that time. The free parameter is beam power, and that soon reaches unpalatable values. The practical limits for proton colliders as energy and luminosity go up together are well known. It was not difficult to sketch a pessimistic conclusion for the long-term future.

As addition to the "bad news," as far as accelerators are concerned (though not for physics!), I pointed out that *natural sources* are not necessarily finished. Indeed we have reason to expect to hear from them again. We know of new scales that certainly contain new physics. The most relevant is perhaps the Grand Unification Scale of 10^{15}GeV, and it is rather likely that there are new particles with this scale of mass. Some of these may still be around as "Relics." If so, there will be Relic Annihilation or decay at some level, and these events or their products will *someday* be observed. Since we do not expect ever to arrive at that energy with accelerators, when that day comes accelerators will no longer dominate the exploration of the frontier of fundamental physics. This event is shown on Figure 1, with a very large uncertainty on the time coordinate, of course.

My presentation acknowledged that there is still good news for accelerator-based physics, loopholes in the pessimistic arguments. **Resonances,** with a favorable width, perhaps lots of them, can drastically change the luminosity barrier at a high energy. Maybe **quantum mechanics** is wrong in some favorable way. Study of **large blobs** at very high energy density will benefit from high energy at a fixed geometrical cross section. (I had been promoting the search for quark-gluon plasma since 1979, following suggestions of altering the state of the vacuum I had heard in a talk by Ben Lee, and a very clear paper by T.D. Lee. This remark was a plug for that activity, mostly lost on my audience, though.) Especially, I listed **???**, that is, things turning up that are completely unanticipated. Every new accelerator *that has substantially raised the available collision energy* has made important discoveries.

I also listed the suggestion by Budker, made as least as early as the Kiev Rochester conference in 1970, that one could evade the problems of electron collisions by using muons. I received two comments privately after the talk. Salam said good talk, too pessimistic. Richter said muons are not the answer, since the problems with electrons are only helped with muons by $\log(m_\mu/m_e)$, which I think is surely too pessimistic!

Now some time has passed, and we can judge how events have transpired. As we feared, it is proving hard to stay on the Livingston curve. Three frontier accelerator projects have failed to complete in that time, ISABELLE, SSC and UNK. The LHC has not been completed as rapidly as one hoped ten years ago. We have not yet seen the approval of a linear collider. It does indeed take some optimism to see how we are going to get back on that curve. On the other hand, one is staying on the Auger curve much better than we expected. Indeed, this very fact constitutes, if verified, an important discovery, since it means that the fundamental limit of the Greisen cutoff has been surpassed. Any explanation would seem to involved new physics; the most popular being the decay of "nearby" cosmic strings to give super high energy particles. If this should be the case, we have arrived at the epoch foreseen on my old curve! It certainly illustrates how this epoch could have arrived. (Note in this model, the energy of the Relic decay is much higher than that of the observed particle, itself already very high. I had assumed that this would be the case when I drew the band in Figure 1.)

I cited this history and prophesy to illustrate that it is unwise for the field of accelerator physics for elementary particle physics, to waste too much time in activities that do not advance the energy frontier. In particular, it is not wise to spend very large resources, and inevitably consume a lot of time as well, on projects that do not advance the energy frontier. On the other hand, it has been shown unwise to over-reach by making too big a jump in energy beyond the reach of existing accelerators, if the cost thereby distorts the customary funding trend of the field.

On the basis of that reasoning, we can define the proper next step. The LHC fixes the energy reach of existing near-term projects. Its reach is just about 5 TeV in terms of the equivalent center of mass that can be observed. The size of the right increment in energy cannot be defined precisely, but a factor of about three has usually been found to be big enough, and more than that increases the risk. An increase of less than three starts to look frivolous in comparison to the resources involved. The exception would be in cases where there is a known threshold that is thought to be worthwhile. A case is the Cosmotron/Bevatron pair, where the center of mass difference was not so large, but the Bevatron was chosen to be capable of the production of antiprotons, and turned out to be well suited for the production of strong beams of K^- as well. Attempts to jump by larger factors seem laden with hubris. This defines the energy of the next frontier to be in the range 10-20TeV, and we may take 15GeV as a good number. The subject of this study, a 10TeV lepton collider, seems pretty well chosen from this point of view. Though we may not know how to get there, the recognition that this is our proper goal is helpful, since a heavy commitment to a path that cannot lead to such a facility is not money well spent. To identify paths that *could* lead to this end requires that study begin soon. This leads to this Workshop, and we should consider its results in the light of identifying these paths. Some intermediate goals are very likely necessary, and will no doubt produce good physics, but we should keep in mind, which aspects of them are on this path, and which aspects are diversions, and try to minimize the diversionary time and resources.

OUR PHYSICS WORKING GROUP DID VERY WELL BY US

We are proposing to spend very large resources to enter a new energy regime, and we must at the very least have a sound case that we expect to make discoveries of a magnitude proportionate to the cost. In practice this means that we are counting on the LHC delivering exciting discoveries that give strong promise of even most important phenomena in a well-defined energy range. Our Physics Working Group presented a number of attractive theories that are exactly of this character. If nature is as kind, we will be fortunate.

At the LHC we have an examples of the kind of physics motivation appropriate to the effort required to get the promised results. The arguments from existing electroweak data for new physics in the LHC range are very strong, though it is surely always possible to find models that can push the new phenomena just over the attainable threshold. More likely, something like SUSY will turn up. SUSY is a good example of how we can have experiment-friendly phenomena at very high energies, as seen in the amazing results of simulations of measurements at the LHC showing that one can make many high precision mass measurements despite unseen particles in the final state. There have been a lot of ideas lately that predict striking phenomena in the energy region near or somewhat above the top reach of the LHC. These were summarized in the report of the Physics Working Group. Here are some of my favorites:

- Kaluza-Klein excitations of Standard Model gauge bosons give huge cross sections on broad, multi-TeV peaks with big lepton decay rates. An interesting feature of this theory is that if LHC should see the lowest state, the next one would lie in the interval above 5 TeV and less than 15 TeV.

- Technicolor (Walking):
  - This is needed to avoid large flavor-changing neutral currents in technicolor, with spectral functions not saturated by low-lying resonances;
  - It displays a "tower" of vectors, rho, W etc extending up to more than 100 TeV ;
  - The lightest is around 1-2 TeV, so they sould be seen at LHC, and the WW and WWZ to be seen at the Muon Collider;
  - Good calorimetry is needed for these states.

- SUSY Messengers, gauge-modified SUSY models have them:
  - vector-like heavy "quarks", "charged leptons," "neutrinos" + scalar superpartners;
  - 10 TeV up, but get we can get an upper limit from LHC superpartner spectrum and neutralino decay length, which would then give us a target energy for the Muon Collider;
  - the charged leptons look like heavy muons;
  - the "quarks" hadronize;
  - the "neutrinos" show up as missing energy;
  - the scalars don't decay in the detector.

- Flavor changing neutral currents: —— but enough!

Lots of stuff! - - - a common feature is the use of LHC results to show that you need the new high energy and that it will be worth the effort to get it.

WHAT DO WE KNOW ABOUT A 15TEV MUON COLLIDER?

- (This is by no means a summary of the large amount of work presented.)
- Quite a bit is known, but not *nearly* enough.
- Proton drivers can produce and catch lots of muons in a short bunch, enough to get you in space charge and off-site radiation troubles.
- They can be cooled a lot by ionization, but how much and to what charge density needs *much* more study. From what we have heard here, it certainly seems too early to say that we can get small, highly bunched beams of 10^{12} muons.
- They can be stored for 1000 turns, and rings have been designed to hold impressive charges, at least if extreme low beta is not also required.
- It seems that they can be cooled optically in short times and to a very small emittance. We should give a ★ to Zholents et al, for progress on such an ambitious concept.

We were given a clear message that maintaining 1000 turns in a ring with extreme focusing at an Interaction Region is not simple:

Carol Johnstone: "an accelerator physicist's dream, but is it a workhorse?"

A number of ideas to ease this problem have been studied. It is suggested that we must go to axisymmetric focusing near the IR to avoid the huge betas in the matching sections for the quadrupole focused IR, and the sensitivity and non-linearities that arise for these conditions. We heard about two of these schemes:
  –Plasma lenses give strong first order focusing close to the IR (Cline).
  –Auxillary dense beams of electrons give first order focusing (Irwin).

We need more study of these.

An important question discussed in the Workshop was how much current is allowed in a conducting metal used as a degrader for ionization cooling? When is it not effective in shielding space charge? Parkhomchuk invoked his data on the limits of current density in two different rings holding protons and cooled by electrons. His data show that the electron cooling is not sensitive to the relative velocity of the electrons and protons over a wide range. He used this to argue that the limits on electron cooling are relevant to the ionization cooling parameters. He showed that his limits are explained in a simple plasma model. Having established some credibility for this model, he shows that some of the parameter sets for cooling of intense muon beams are well beyond the allowed limits. No doubt these arguments will receive careful evaluation in the muon collaboration.

- It seems possible that we have to choose: we can have *either* small beams *or* big charges.

This worry motivates ideas about working with fewer muons. There was quite a bit of discussion of beams with fewer muons, but in a tighter focus to keep the luminosity up. The problem with this is the disruption in the crossing of the beams that will prevent the beams from being captured and refocused back into the storage ring a thousand times.

  – One suggestion was to attempt a compensation of the electromagnetic forces by immersing the crossing point in a dense plasma, really in a wire of low-Z metal that becomes a plasma when the beams arrive. (Skrinsky)
    » Interesting discussions of the backgrounds generated in the detector, blowout of the positive ions during the beam crossing and radiation were given by Telnov.

NEW IDEAS AT THIS WORKSHOP

There was frustration with the problems of maintaining the muons in storage rings for 1000 turns while bedeviled by the sensitivity of the IR dynamics on the one hand and the disruption limit on the other. The latter limit forces the number of muons to be very large in terms of the space charge limits and the environmental radiation. The conundrum is sharpened when one considers the parameters of the ultra cool beams of the stochastic cooling design presented by Zholents. These considerations led Frank Zimmerman to take a look the option of acceleration in a ring and *single pass collisions*. This keeps one of the advantages of muons, the ability to stand acceleration in a ring, with large cost advantages over a linear acceleration. It renounces the other feature, the use of each muon in 1000 collisions. He points out that the loss is by no means a factor of 1000, since the disruption parameter can be increased greatly, perhaps by a factor of 100 or more. To get back the standard luminosity of 10^{35}, an increased repetition rate, by ten times in this example, still with a much smaller total number of muons.

The environmental radiation problem may be reduced still farther, since in the multi-pass collider, every muon decays, while in the single pass scheme, only a fraction of the muons decay in the accelerator, and even a much smaller fraction at full energy where they are troublesome.

The crucial difference from single pass electron collisions is that the muon collisions are not yet limited by beamstrahlung in the 10-20 TeV range.

This concept is attractive for the stochastically cooled beams, with small numbers (10^{9-10} muons) and tiny emittance, where getting luminosity with high disruption is allowed. With fewer muons and the other factors mentioned above, we could envision siting the collider at an existing laboratory. The stress on the proton driver and target are also reduced. The cost of the lasers in the cooling ring needs to be evaluated for these parameters.

With fewer muons and the other factors mentioned above, we could envision siting the collider at an existing laboratory. The stress on the proton driver and target are also reduced. The cost of the lasers in the cooling ring needs to be evaluated for these parameters.

THE WORKING GROUP ON EXPERIMENTS

There were new ideas here too. They address the two problems of muon colliders; first, the glow of soft electromagnetic radiation leaking in through the necessary gap in shielding looking at the interaction region. The origin is the flux of high-energy electrons from beam muon decay showering in the shielding. Second, the muons originating from Bethe–Heitler production of muons by decay electrons. These are roughly parallel to the beam, but at radial distances from the beam with a broad distribution extending up to meters. The cure to both is detectors that are to a significant degree blind to particle tracks that do not back to the interaction point, but this property is not in the standard menus of detectors. This is an interesting challenge for detector fans. The modest event rate, compared to the GigaHerz at the LHC, gives some space for innovating ideas.

- Rehak proposed a solution tracking charged particles by transverse drift in a projective volume filled with a sensitive liquid. This gives a large density of charge

- carriers, and the tracking accuracy could be good enough to meet the requirements of B identification.
- Heusch has emphasized the need for coverage at the smallest possible angle to the beam, in order to handle some exotic reactions such as those in e^+-e^+ collisions. Willis suggested to instrument the forward shielding cone with projective, fine-grained and extremely dense calorimetry. This will allow the extension of coverage to be extended to much smaller angles. This will not be effective for muons, but will allow the measurement of electrons, photons and hadrons with sufficiently large energy. The domain in which this is effective depends on an realistic model of the background muon flux and a very detailed model of the detector. Such a program will be carried out by Iulio Stumer.

REAL WORLD SCENARIOS

- We have learned that even though "$1G is not what it used to be" (Bunker Hunt, asked how it felt to lose $4G) it is not so easy to get **many** $G for a HEP project.
- A very dull story from the LHC would be a very poor start on an attempt to launch the next machine; fortunately we claim this is unlikely.
- Rizzo, for example, has given us a very fine scenario: a massive resonance (if not Kaluza-Klein, then Z′) with 10^4 events per fb on the mu-mu peak. If we *know* that, we could build a 10^{33-34} machine and upgrade later. Still going for 15 TeV!! That way we could get in business in a finite time.

CONCLUSIONS

- Some important new ideas came up
- Lots of good problems to work on were identified
- The physics ideas look very lively and promising
- The pace of study must be accelerated to settle questions that need to be settled *in time* to take advantage of the time window opened up by LHC discoveries, around 2010
- I think it particularly important to move on a demonstration of the optical stochastic cooling: this seems to open a window to such a machine at an existing laboratory
- The development of stochastic cooling, and very small emittance generally, is an activity unrelated to ν beams that require large numbers of muons. This means that collider R&D must retain a life apart from neutrino beams. This Workshop and others like it are needed in order to make clear to us what directions to follow. This answers the question with which I began: **Why Now**:

Final Focus Challenges for Muon Colliders at Highest Energies

F. Zimmermann
CERN, SL Division, Geneva, Switzerland

Abstract. This report discusses challenges in developing final-focus systems for muon colliders at 10–100 TeV [1]. The optics design is impeded by limited quadrupole gradients and the large beam emittances. Of interest are also spot-size dilutions accumulating over several turns. Tolerances on magnet vibration and field stability are comparable to those at future electron-positron linear colliders. While at 10 TeV nonlinear kinematic terms are still important, at 100 TeV synchrotron radiation may complicate the design. In view of the high charge per bunch and the multiple passes, wake fields and space-charge effects must be looked at carefully. For multi-TeV energies a single-pass muon collider is a promising option, since it poses no neutrino radiation hazard, can accommodate ultracold muon beams, and lends itself more easily to novel focusing techniques, such as plasma lenses or dynamic focusing, thereby avoiding the gradient limitations of conventional quadrupoles.

I   OVERVIEW

Table 1 compares the interaction-point (IP) parameters for a 10-TeV and a 100-TeV muon collider with their counterparts in a few existing or planned e^+e^- linear colliders. The IP beta functions, bunch length, and the interaction-point (IP) transverse spot sizes for a high-energy muon collider are comparable to those successfully obtained at the Stanford Linear Collider (SLC). All these values are quite relaxed compared with the corresponding numbers for a future e^+e^- collider. Moreover, the expected rms energy spread is extremely small, which softens constraints on the final-focus energy bandwidth.

However, four factors could complicate the final-focus design:

- the extremely high beam energy, in the face of limited quadrupole strengths, and large transverse emittances;

- a factor 20–75 higher bunch charge than SLC (factor 200–750 larger than CLIC);

- synchrotron radiation at 100 TeV;

- the multiple passes.

In the following sections we will discuss potential problems associated with each of these items.

TABLE 1. Comparison of final-focus and IP parameters for several e^+e^- and multi-TeV $\mu^+\mu^-$ colliders: the Stanford Linear Collider (SLC) is the first and so far only operating linear collider [2]; the Next Linear Collider (NLC) is the design for a future 1-TeV e^+e^- collider [3]; and CLIC [4] is a multi-TeV linear collider under study at CERN. The numbers for the 10 and 100-TeV muon colliders, μ-A and μ-B, correspond to the parameters provided by the workshop organizers.

parameter	symbol	SLC	NLC	CLIC	μ-A	μ-B
species		e^+e^-	e^+e^-	e^+e^-	$\mu^+\mu^-$	$\mu^+\mu^-$
cm energy [TeV]	E_{com}	0.1	1	3	10	100
Lorentz factor	γ	10^5	10^6	3×10^6	5×10^4	5×10^5
bunch population [10^{10}]	N_b	4	1	0.4	300	80
hor. emittance [μm]	$\gamma\epsilon_x$	50	4.5	0.68	38	8.7
vert. emittance [μm]	$\gamma\epsilon_y$	8	0.1	0.02	38	8.7
hor. beta [mm]	β_x^*	2.8	12	8	2.1	2.5
vert. beta [mm]	β_y^*	1.5	0.15	0.15	2.1	2.5
hor. spot size [nm]	σ_x^*	1700	235	43	1300	210
vert. spot size [nm]	σ_y^*	900	4	1.0	1300	210
bunch length [mm]	σ_z	1	0.12	0.03	2.2	2.5
rms energy spread [%]	σ_δ	0.1	0.3	0.3	0.06	0.011
pinch enhancement	H_D	2.0	1.45	2.24	1.08	1.11

II OPTICS

The primary difficulty in designing a final-focus optics is the limited quadrupole strength. Within the reach of present technology is a quadrupole gradient of 320 T/m [5]. Even if we optimistically extrapolate this to 500 T/m, at 100 TeV the corresponding normalized gradient is a meager $k \approx 3 \times 10^{-3}$ m^{-2}. This implies that the final quadrupole magnets must be long, and at least in one plane the beta functions can grow to large values, generating substantial chromaticity, which reduces the momentum bandwidth in that plane.

We used the automatic final-focus design program FFADA [6] and the general accelerator design code MAD [7] to develop a series of six test optics. Each of the final-focus models consists of three parts: a horizontal chromatic correction section (CCX), a vertical chromatic correction section (CCY) and a final transformer, usually consisting of two doublets. Each chromatic correction section comprises 4 bending sections and a $-I$ pair of sextupoles, positioned near the maximum dispersion points. The $-I$ transform between the sextupoles ensures that geometric aberrations and second-order dispersion induced by either sextupole cancel, and only chromaticity is generated. Residual aberrations, due to the chromatic break-

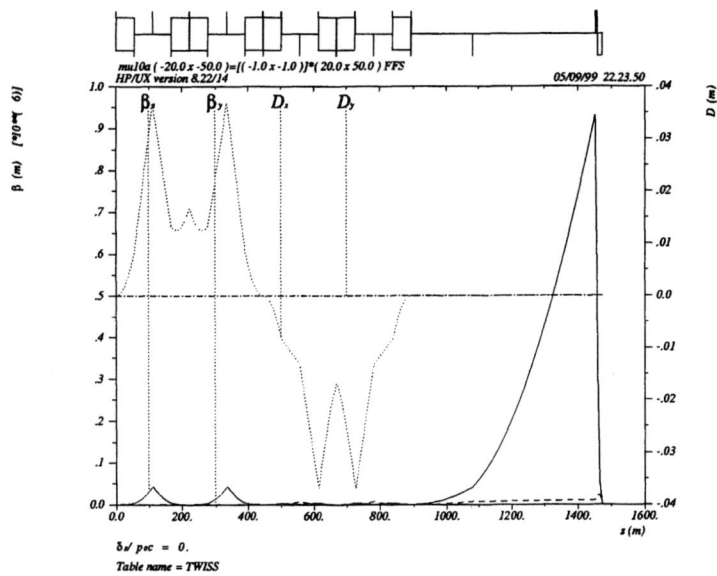

FIGURE 1. Twiss functions for a 10-TeV final focus ('10-TeV a') with a maximum quadrupole gradient of 320 T/m.

down of the $-I$, are of 4th order or higher. For the purpose of this exercise, the final transformer demagnification factor was chosen as 20 in the horizontal plane and 50 in the vertical, a free length of 2 m was assumed between the last quadrupole and the IP, and the final quadrupole is always vertically focusing, since inverting the polarity of the final doublet, keeping a constant demagnification ratio between the two planes, degrades the performance. Also, for the 10-TeV case, we studied an optics where we replaced the doublets in the final transformer by triplets, since a triplet configuration would be more natural for focusing round beams. No space has been assigned yet for sweeping dipoles, which may be needed for background suppression [8] .

Figure 1 depicts the beta functions and dispersion for a conventional final-focus system at 10 TeV. The total length of the system is about 1500 m. Counting both sides, the final focus would occupy about 20% of the total ring circumference (15 km), while for the lower-energy 100-GeV design [9] this fraction is 50%. Thus, the fraction of the collider ring assigned to the final focus decreases with beam energy, and at 100 TeV the total length of the final focus system of about 9 km amounts to only 9% of the collider circumference (100 km).

Table 2 summarizes the main parameters of six different optical models studied. Figures 2 and 3 compare the momentum acceptance for two of these optics. Table

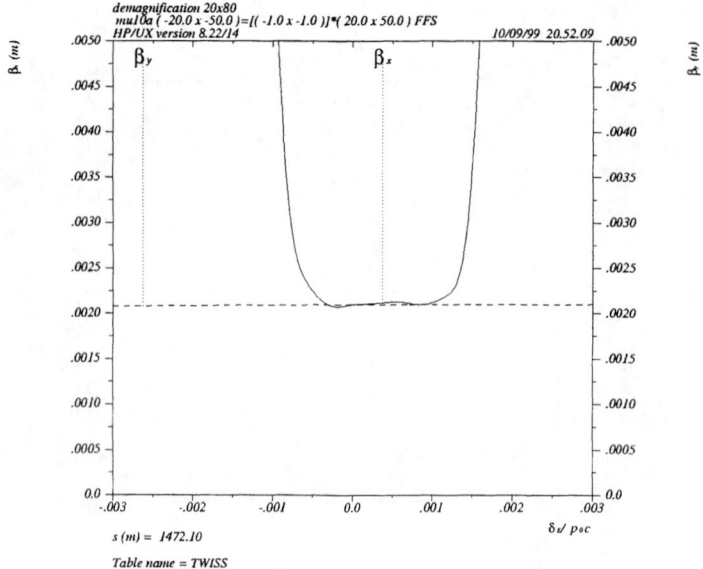

FIGURE 2. Momentum bandwidth for final-focus optics 10-TeV a, with a final-quadrupole gradient of 320 T/m. Shown is the beta function as a function of the relative momentum deviation.

and figures demonstrate that increasing the quadrupole gradient widens the acceptance. The best performance is obtained for the highest gradients, e.g., with optics '10-TeV b' at 10 TeV and '100-TeV c' at 100 TeV.

TABLE 2. Some parameters for the different final-focus optics considered. Listed are the name of the optics, the maximum field gradient in the final quadrupoles, the horizontal and vertical chromaticities, the final-focus length per side, and the total momentum acceptance.

optics	$(\partial B_y/\partial x)_{\max}$	ξ_x	ξ_y	L	$(\Delta p/p)_{\rm FW}$
10-TeV a	320 T/m	56000	3500	1472 m	0.15%
10-TeV b	3200 T/m	17000	1500	790 m	0.28%
10-TeV c	320 T/m	30000	10800	1660 m	0.22%
100-TeV a	3200 T/m	39500	7000	4500 m	0.14%
100-TeV b	3200 T/m	39500	7000	3000 m	0.12%
100-TeV c	7200 T/m	26000	4700	4500 m	0.21%

All examples for 100 TeV assume enormous final quadrupole gradients of either

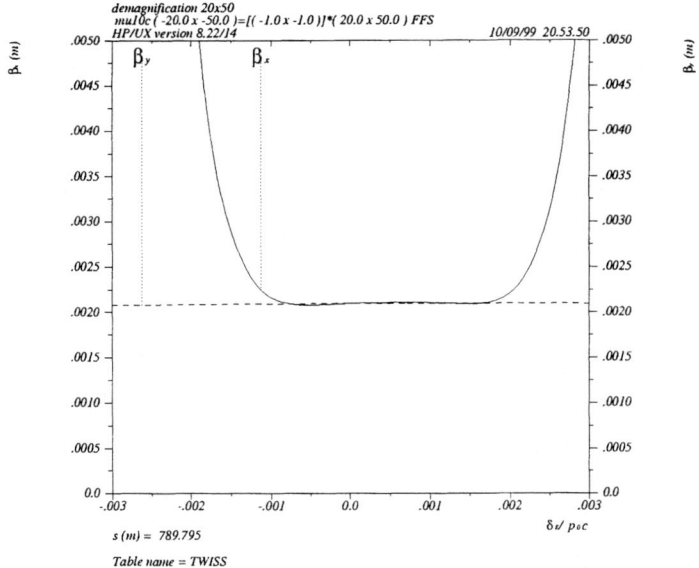

FIGURE 3. Momentum bandwidth for final-focus optics 10-TeV b, with a final-quadrupole gradient of 3200 T/m. Shown is the beta function as a function of the relative momentum deviation.

3200 T/m or 7200 T/m, without which we could not find a satisfactory optics solution. Not only at 100 TeV, but already for the 10-TeV systems the required quadrupole gradients are incompatible with the desired beam stay clear assuming present magnet technology [10]. For example, with a peak beta function of almost 1000 km in optics '10-TeV a' and using the beam parameters of Table 1, the maximum rms beam size inside the last two quadrupoles is 2.8 cm. If we require at least 2σ beam stay clear and another 3 cm space for a tungsten liner, the inner radius of the final quadrupoles must be 9 cm, which is 3–4 times larger than the maximum aperture in present high-gradient superconducting quadrupole designs [5].

Therefore, we must assume one of three possibilities: either a breakthrough in magnet technology, or the use of novel high-gradient focusing methods, or emittances which are substantially smaller than those listed in Table 1. The first option is too speculative. Dynamic focusing or plasma lenses, which come to mind for the second option, are not well adapted to a multi-turn application. It remains the third option, namely to demand much smaller transverse emittances. Reduced emittances would also alleviate problems with higher-order optical aberrations which are discussed next.

Although for all six test optics the bandwidth in the horizontal plane is not large,

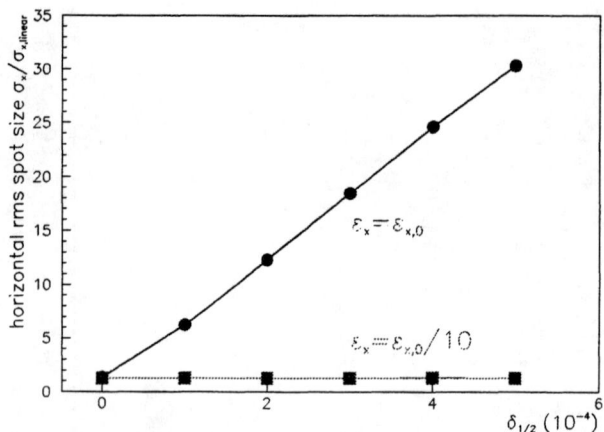

FIGURE 4. Simulated horizontal rms IP spot size as a function of the half width of a flat energy distribution, for optics '10-TeV b'. Results are shown for the nominal and for a ten times decreased emittance. The vertical axis shows the spot size divided by its value expected from the linear optics.

in some cases it appears barely sufficient for the extremely small energy spread of the high-energy muon beam. However, tracking shows that the situation is worse. None of the presented systems achieves a satisfactory performance in the horizontal plane, due to higher-order chromo-geometric aberrations. So the horizontal spot size for optics '10-TeV b' with a ±0.042% flat energy distribution is 35 μm, instead of the expected 1.3 μm.

Figure 4 shows the dependence of the horizontal blow-up on the energy spread and on the transverse emittance, demonstrating that it is caused by higher-order chromo-geometric terms. Without energy spread, the horizontal spot size shrinks to 1.78 μm. We have not identified the cause of the residual small blow-up for zero energy spread (imperfections of the phase advances between sextupoles are one possibility). Figure 5 illustrates that this blow up vanishes for a 10-times smaller emittance. For nominal emittances the vertical spot size is still close to the ideal value.

The chromaticity ξ listed in Table 2 is defined in the linear collider sense. It quantifies the increase of the IP spot size due to an rms momentum spread δ_{rms} according to

$$\frac{\Delta\sigma}{\sigma_0} = \xi\delta_{rms}. \qquad (1)$$

The actual spot size is obtained as the quadratic sum of the linear spot size σ_0 and the chromatic contribution $\Delta\sigma$:

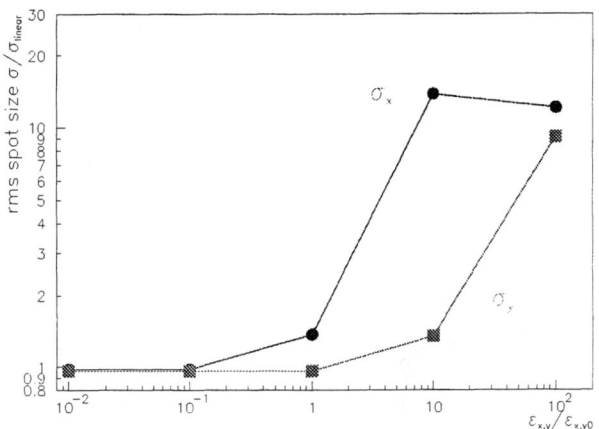

FIGURE 5. Simulated rms IP spot sizes with zero energy spread as a function of the transverse emittances, for optics '10-TeV b'. The vertical axis shows the spot size divided by its value expected from the linear optics; depicted along the horizontal axis is the emittance normalized to the design emittance.

$$\sigma = \left(\sigma_0^2 + (\Delta\sigma)^2\right)^{1/2}. \qquad (2)$$

The sextupoles in the chromatic correction section are adjusted so that the total chromaticity ξ is zero. Therefore, to first order, the incoming energy spread has no effect on the IP spot size. However, if a certain amount of energy spread $(\Delta\delta)_{\text{rms}}$ is generated between the sextupoles and the final doublet, it will interact only with the uncompensated chromaticity of the final doublet and increase the spot size. Such energy spread can be generated by longitudinal wake fields, space charge, or synchrotron radiation.

III WAKE FIELDS AND SPACE CHARGE

The higher bunch charge, compared with most other accelerators, motivates a careful look at wake fields and space charge effects. Transverse wake fields amplify any initial orbit fluctuation, and, thereby, cause the beams to only partially overlap at the collision point, especially on later turns. Longitudinal wake fields and space charge impair the chromatic correction, if the energy spread which they induce between sextupoles and the final quadrupoles becomes comparable to the inverse chromaticity $1/\xi_{x,y}$. If the error in the chromatic correction accumulates over n turns, as it probably will, the tolerance on the induced energy spread is $1/(\xi_{x,y}n)$. Using the parameters of Table 2 and $n = 1000$ turns the tolerable energy spread induced between sextupoles and final doublet is only a few times 10^{-8}! Already tiny changes in the particle energies can spoil the chromatic correction, especially if they accrue over multiple turns.

The discussion of the longitudinal wake fields follows that for the NLC [11]. At the muon collider, the bunch length is long compared to the characteristic distance s_0 defined by

$$s_0 = \left(\frac{cb^2}{2\pi\sigma_c}\right)^{1/3} \quad \text{(cgs units)}, \tag{3}$$

where b is the chamber radius, c the speed of light, and σ_c the conductivity. For an aluminium chamber ($\sigma_c \approx 3.2 \times 10^{17}$ s^{-1}) with $b = 1$ cm radius: $s_0 = 25$ µm.

Following Refs. [11,12], the change in rms energy spread due to the resistive-wall wake field reads

$$\frac{d\sigma_E}{ds} = 0.205 \frac{Ne^2}{b^2}\left(\frac{s_0}{\sigma_z}\right)^{3/2} \quad \text{(cgs units)}. \tag{4}$$

With aluminium, and in more practical units, the induced relative energy spread is

$$(\Delta\delta)_{\rm rms} = 3.6 \frac{N[10^{12}]}{E[{\rm eV}]b[{\rm cm}](\sigma_z[{\rm cm}])^{3/2}} \Delta s[{\rm cm}], \tag{5}$$

where Δs is the length of the beamline segment considered. For an aluminium beam pipe with 1 cm radius the induced energy spread is $(\Delta\delta)_{\rm rms} \approx 3 \times 10^{-8}$ at 10 TeV assuming a length $\Delta s \approx 1.5$ km, and it is $(\Delta\delta)_{\rm rms} \approx 2 \times 10^{-9}$ at 100 TeV over $\Delta s \approx 4.5$ km. This is not far from the tolerable limit.

To suppress the enormous background from muon decay, a large number of collimators in the final focus may be necessary [8]. Geometric wake fields from these collimators also generate an rms energy spread, and can impair the chromatic correction. For an untapered collimator of radius a in a beam pipe of radius b the induced rms energy spread is [11,13–15]

$$(\Delta\delta)_{\rm rms} \approx 0.444 \times \frac{Nr_\mu}{\gamma\sigma_z} \ln\frac{b}{a}. \tag{6}$$

For example, consider a collimator at 1.5 cm in a 2-cm beam pipe. At 10 TeV the rms energy spread is $(\Delta\delta)_{\rm rms} \approx 7 \times 10^{-8}$, and at 100 TeV, $(\Delta\delta)_{\rm rms} \approx 1.6 \times 10^{-9}$.

The 10-TeV value is already significant. In addition, the contributions of different collimators to $(\Delta\delta)_{\rm rms}$ add linearly. Therefore, if there are about 100 untapered collimators between sextupoles and final doublet, their geometric wake field can drastically degrade the chromatic correction and increase the IP spot size.

The effect can be reduced if the collimators are tapered. According to Ref. [16], the rms energy spread caused by a tapered collimator in the high-frequency limit is

$$(\Delta\delta)_{\rm rms} = 0.44 \times \frac{Nr_\mu}{\gamma\sigma_z}(1-\tilde{\eta}_1)\ln\frac{b}{a}, \qquad (7)$$

where $\tilde{\eta} = \min(1.0, \eta_1)$ with

$$\eta_1 = \frac{g\sigma_z}{(b-a)^2} \qquad (8)$$

and g the length of the taper. If the taper is sufficiently shallow, then $\tilde{\eta}_1 = 1$, and in the approximation used zero additional energy spread is generated. For example, with a bunch length of 2.2 mm and a change in the chamber radius $(b-a)$ of 5 mm, the condition $\tilde{\eta}_1 = 1$ is fulfilled, if the taper is slightly longer than 1 cm.

Also the longitudinal space-charge force induces an rms energy spread that can affect the chromatic correction [17]. Ignoring the dependence on the transverse coordinates, the difference in energy spread between two locations '1' and '2' can be estimated as

$$(\Delta\delta)_{\rm rms} \approx 0.28 \times \frac{2Nr_\mu}{\sqrt{2\pi}\sigma_z\gamma}\ln\left(\frac{(\sigma_x+\sigma_y)_2 b_1}{(\sigma_x+\sigma_y)_1 b_2}\right), \qquad (9)$$

where b_1 and b_2 denote the beam pipe radius at the two locations. Assuming that the beam-pipe radius is the same, the change in rms energy spread between the sextupoles and the final doublet is determined by the ratio of the sum of horizontal and vertical beta functions at these two locations. If no care is taken, this can easily be a factor of 20 (see Fig. 1), which would amount to $(\Delta\delta)_{\rm rms} \approx 2.6 \times 10^{-7}$. This would be acceptable for a single turn, but it will degrade the chromatic correction over a few hundred turns. Detailed simulations with the exact space-charge field could be performed, e.g., using a modified version of MAD as in Ref. [17].

We now discuss the transverse wake fields. The centroid deflection due to the resistive-wall wake field is given by [3]

$$\Delta y' = 2\frac{Nr_\mu}{\gamma\sigma_z}\frac{L}{b^3} <f_R> \sqrt{\lambda\sigma_z}\,(\Delta y), \qquad (10)$$

where (Δy) denotes the offset of the beam in the vacuum chamber, b the beam-pipe radius, $<f_R> = 0.82$ a factor arising from averaging the wake field over the longitudinal bunch distribution, and $\lambda = 0.045$, 0.15, and 1.2 nm for Cu, W, Ti [3]. For a tungsten beam pipe of 2-cm radius extending over a length of 100 m, the jitter enhancement is $\Delta y/y \approx 5 \times 10^{-6}\,\beta[{\rm m}]$ for the 10-TeV parameters, where β denotes the beta function at the location of the wake. So, with beta functions reaching values 1000 km the transverse resistive-wall wake could be significant.

Finally, we estimate the transverse geometric wake field from a collimator (or mask), assuming that the bunch length is short compared with all radial apertures. If the collimator is not tapered, the centroid wake field kick is [18,19]

$$\Delta y' = \frac{2Nr_\mu}{\gamma}\left(\frac{1}{a^2}-\frac{1}{b^2}\right)y, \qquad (11)$$

where a and b are the radius of the collimator and the beam pipe, respectively, and y the centroid offset. For 10-TeV and considering a transition between 1 cm and 1.5 cm radius, it evaluates to $\Delta y' \approx 10^{-5} y[m]$. This is two times larger than the deflection from the resistive wall wake.

Fortunately, as in the longitudinal case, a taper can again reduce the size of the geometric wake field. With a taper of length g, the wake-field kick reads [3]

$$\Delta y' = \frac{N r_\mu}{\gamma \sigma_z} \frac{2(b-a)^2}{abg} <f_G> y \qquad (12)$$

with $<f_G> = 0.282$.

In conclusion, wake fields and space charge will impose strong constraints on optics design, beam-pipe apertures, and beam parameters.

IV SYNCHROTRON RADIATION AND BEAMSTRAHLUNG

Since muons are 200 times heavier than electrons, at 100 TeV the synchrotron radiation in the final-focus bending magnets becomes comparable to that at a 500-GeV e^+e^- collider. The rms energy spread induced by synchrotron radiation over a bending length l_B with a total bend angle θ is [20]

$$\delta_{\rm rms} = \left(\frac{55}{24\sqrt{3}} r_\mu \lambda_\mu \gamma^5 \frac{\theta^3}{l_B^2} \right)^{1/2}, \qquad (13)$$

where $\gamma = E_\mu/(m_\mu c^2)$ denotes the Lorentz factor, r_μ the classical muon radius, and λ_μ the muon Compton wavelength. For example, assuming a total bending angle of only 2.5 mrad over a length of 2000 m at 50 TeV beam energy, we find $\delta_{\rm rms} = 2 \times 10^{-9}$, which is insignificant, since the synchrotron radiation is incoherent and the energy spread increases as the square root of the number of turns. As the opposite extreme, suppose next that the strength of the dipole magnets is chosen equal to 10 T, e.g., in order to optimize the momentum bandwidth. In this case, the total bending angle over a 2-km length is 118 mrad, and the rms energy spread evaluates to $\delta_{rms} \approx 6 \times 10^{-7}$, which after 1000 turns becomes 2×10^{-5}. This is now comparable to the inverse chromaticity and will affect the IP spot size.

The synchrotron radiation in the last quadrupoles (Oide effect) is not an issue, because the IP beta function is more than 10 times larger than for electron-positron colliders. Also beamstrahlung, i.e., synchrotron radiation in the field of the opposing beam is still small. Two parameters characterizing the beamstrahlung are Υ and n_γ. Apart from a factor 2/3, the Υ parameter is equal to the critical energy of the beamstrahlung in units of the incident beam energy. Its average value over the collision is [21]

$$\Upsilon = \frac{5}{6} \frac{r_\mu^2 \gamma N_b}{\alpha \sigma_z \sigma_x (1 + \sigma_y/\sigma_x)}, \qquad (14)$$

where α denotes the fine-structure constant, and r_μ the classical muon radius. For the 100-TeV parameters (see Table 1) we find $\Upsilon \approx 3 \times 10^{-5}$. The number of beamstrahlung photons emitted per muon is [21]

$$n_\gamma \approx \frac{2\alpha r_\mu N_b}{\sigma_x(1+\sigma_y/\sigma_x)}, \qquad (15)$$

which evaluates to $n_\gamma \approx 0.4$. This is not completely insignificant. Finally, the average beamstrahlung energy loss per collision reads $\delta_B \approx \Upsilon n_\gamma/2$, and for 100 TeV it is $\delta_B \approx 7 \times 10^{-6}$. After 100 turns the average energy loss is equal to the rms energy spread.

In conclusion, at 100 TeV synchrotron radiation and beamstrahlung start to become noticeable and their effect should be included in the final-focus design.

V KINEMATIC TERMS

The dynamic aperture of the muon collider ring may be limited by nonlinear kinematic terms [22]. The general trajectory equations in a magnetic field read [23]:

$$x'' = \frac{e}{p}\sqrt{1+z'^2+x'^2}\left[x'B_s - (1+z'^2)B_x + x'z'B_z\right] \qquad (16)$$

$$z'' = -\frac{e}{p}\sqrt{1+z'^2+x'^2}\left[z'B_s - (1+z'^2)B_z + x'z'B_x\right] \qquad (17)$$

with $p = \gamma m v$ the particle momentum.

To examine the importance of this effect, we have tracked a distribution of 100 on-momentum particles through the final focus. We varied the initial emittances, and performed simulations with and without the nonlinear kinematic terms. Results are illustrated in Fig. 6. For the nominal emittance at 10 TeV, the effect of the nonlinear contributions is small. They increase the rms spot size by less than 1%. However, if we track a distribution with a 25 times larger emittance (in both planes) the rms spot size at the IP is ten times larger with kinematic terms than without. This suggests that the kinematic terms may indeed affect the dynamic aperture.

At 100 TeV, for optics '100-TeV a', the same relative blow up by a factor 10 is obtained for a 250-fold increase in the nominal emittance. Thus, the kinematic terms appear to lose importance towards higher energies.

VI MULTI-TURN EFFECTS

Table 3 lists some general muon-collider parameters at 10 and 100 TeV. At 10 TeV the muons are stored in the collider ring for 740 turns; at 100 TeV they are stored for 380 turns. At the end of the 10-TeV store the luminosity has decreased by 50%, and the two beam currents by about 30%. Optimum luminosity is achieved

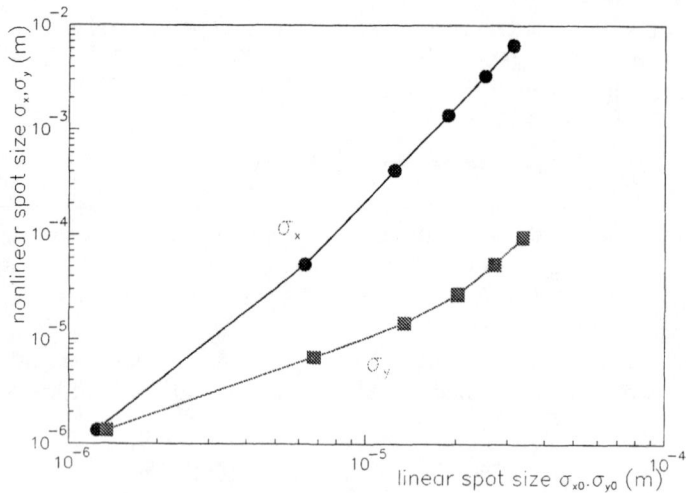

FIGURE 6. Spot size including kinematic terms as a function of the linear spot size, for the optics '10-TeV a' with a constant $\beta^*_{x,y}$ and varying emittance.

TABLE 3. A few collider parameters at 10 and 100 TeV, as provided by the workshop organizers.

energy	10 TeV	100 TeV
circumference	15 km	100 km
repetition rate	27 Hz	7.9 Hz

if new bunches are injected into empty rf buckets, while the old bunches continue to collide until no particles are left.

Even if the design IP spot size is reached on the first turn, it will be difficult to maintain this spot size on subsequent turns. Figure 7 shows a preliminary simulation result for optics '10-TeV b', which illustrates a significant blow-up on later turns. The horizontal blow-up for zero energy spread could have its origin in imperfections of the $-I$ or of the phase advance between the sextupoles and the final doublet, and/or in a small mismatch between turns. A larger concern is the enormous spot-size increase for a nonzero (but small) energy spread, which we attribute to higher-order chromogeometric aberrations.

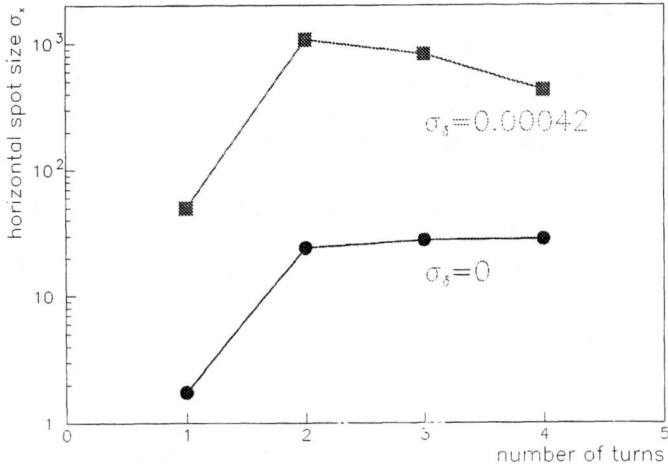
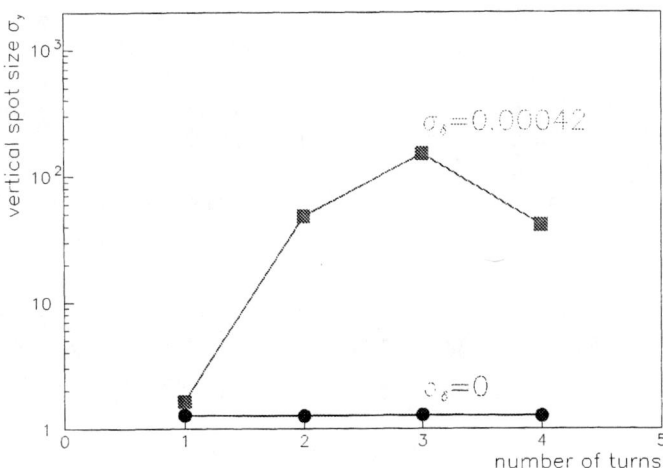

FIGURE 7. Simulated horizontal (top) and vertical (bottom) rms IP spot sizes as a function of turn number, with and without beam energy spread, for optics '10-TeV b' and choosing a fractional tune of 0.31 in both transverse planes.

VII TOLERANCES AND TUNING

Single-pass tolerances can be calculated with the FFADA code [6]. Figure 8 shows tolerances on transverse magnet position and quadrupole field strengths, for the optics '10-TeV a'. According to the calculations, the final quadrupoles must be stabilized at the 20-nm level. Without feedback the multi-turn tolerances are significantly tighter than the single-pass tolerances. Thus, a turn-by-turn orbit feedback may be indispensible.

At the SLC, the different aberrations, such as dispersion, waist, and linear coupling, were corrected at regular time intervals. These corrections were performed by scanning and adjusting combinations of quadrupole and skew quadrupole magnets (so-called multiknobs), such that the IP spot sizes were minimized or the luminosity maximized. The spot sizes were determined from beam-beam deflection scans; luminosity and beamstrahlung could be measured with dedicated monitors.

At a muon collider, both beams would be affected by changes in the quadrupole settings, which will complicate the tuning procedures, unless only beam observations from the first turn are used. For independent control of the two beams, it may be necessary to use electrostatic quadrupole magnets, a challenge at these beam energies. Moreover, the multiknobs should not change any of the betatron tunes. There may also be a need or desire to match the optics from turn to turn, which perhaps could be done with pulsed quadrupole magnets. Different from a single pass collider, also the outgoing aberrations must be matched, which doubles the number of tuning scans required.

Tuning will thus become more difficult. However, the many turns also open up, in principle, the possibility to perform several complete tuning scans during one cycle.

VIII BACKGROUND, HEAT LOAD AND COLLIMATION ISSUES

There are two sources of background: (1) muon decay [8], and (2) beam halo [24]. Only the second source exists in electron-positron colliders. Over a distance of 1 km 10^8 muons are lost at 10 TeV, and 2×10^6 at 100 TeV.

As a means to reduce the background and the heat load of superconducting quadrupoles due to the decaying muons, it was proposed to install sweeping magnets. For a 2-TeV collider, these sweeping dipoles cover about 40% of the final-focus length and have a strength of 8.5 T. There may even be a need for sweeping magnets between the IP and the first (last) quadrupoles, which might lengthen l^* and enhance the chromaticity of the system.

A second approach, for reduced heat load, is to add a several cm thick tungsten liner inside the quadrupoles. This of course lowers the quadrupole gradient and, like the sweeping dipoles, it will further aggravate the problem of higher order chromaticity.

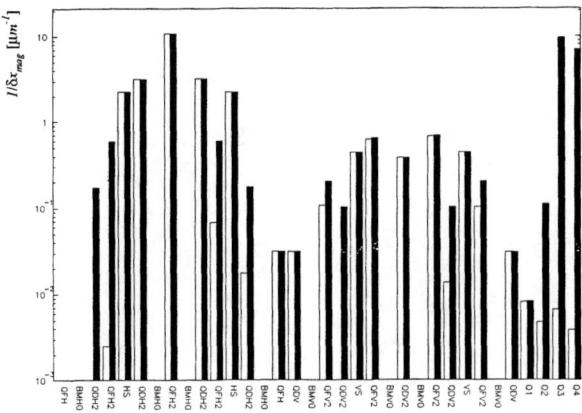

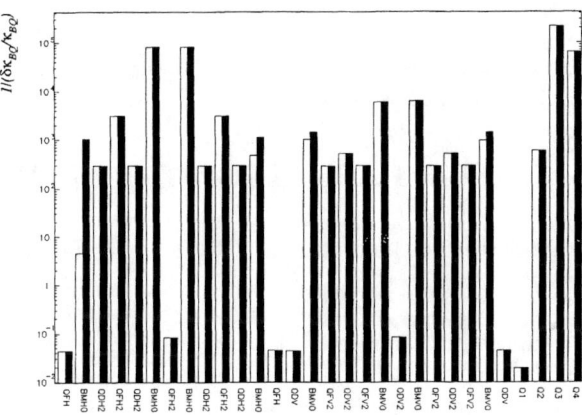

FIGURE 8. Single-pass tolerances for quadrupole, bending and sextupole magnets in optics '10-TeV a'. Shown are tolerances on magnet position (top) and field strength (bottom). Each value displayed corresponds to a luminosity loss of 2%. The full bars represent pulse-to-pulse 'jitter' tolerances, due to both induced orbit motion and beam size increase at the interaction point. This orbit jitter can be corrected within a few pulses using a fast orbit feedback. The open bars are 'drift' tolerances referring only to increases in the IP spot size. Since the beam size tuning will be performed less frequently, these tolerances must be met over a longer time scale, *e.g.*, minutes.

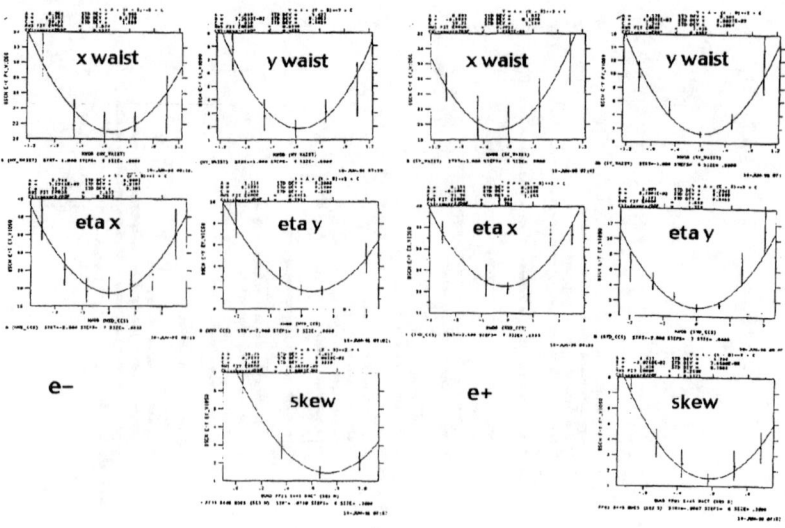

FIGURE 9. Typical tuning scans at the SLC interaction point. Dispersion, waist and coupling were corrected evey few hours. To this end, orthogonal multiknobs were varied, and for each multiknob setting the rms beam sizes determined via beam-beam deflection scans. The knobs were set to the minimum of the fitted parabola.

IX SINGLE-PASS MUON COLLIDER

The design of a muon ring collider at multi-TeV energies faces severe, perhaps insurmountable problems:

- The neutrino radiation is likely to limit a ring collider to energies below a few TeV. The radiation hazard arises because the neutrino cross sections increase almost linearly with energy, while the angular divergence of the emitted neutrinos decreases as $1/\gamma$. As a net result the neutrino radiation dose increases as the 3rd power of energy [25], and at multi-TeV energies easily exceeds the US Federal limit [26].

- The beam has to survive hundreds of passes through a final-focus system more challenging than that of the SLC, retaining the same constant emittance. This appears non-trivial, as already the extracted beam at the SLC showed large emittance degradation even in the absence of collisions.

- If optical stochastic cooling [27–29], or other future techniques reduce the muon beam emittances by several orders of magnitude, the luminosity of a ring collider is limited by the beam-beam tune shift.

- Synchrotron radiation in the collider ring will likely increase the nominal beam emittance [30], casting doubt on the value of ultracold muon beams for high-energy muon ring colliders.

- Conventional quadrupole magnets appear to be too weak for a satisfactory optics design; advanced high-gradient schemes like plasma lenses or dynamic focusing are not well suited for a multi-pass system.

Similarly, several difficulties lie in the way of an electron-positron linear collider at multi-TeV energies. The most dramatic is the effect of beamstrahlung, the coherent pair creation at high Υ, and the associated degradation of the luminosity spectrum and large background.

A single-pass muon collider (SPMC) solves all the above problems: Because of the larger muon mass, the beamstrahlung at 10 TeV or 100 TeV is still contained. There is no need to preserve the emittances after the collision, and the beam can be dispersed onto a dump (downwards, or upwards), thereby reducing the density of neutrino radiation by orders of magnitude. Note that, as an option, the beams could still be accelerated in a ring [31], from which they might then be extracted, focused to a small spot size, and collided only once.

TABLE 4. Example parameters for single-pass muon colliders at 10, 100 and 1000 TeV.

parameter	symbol	SPMC-0	SPMC-I	SPMC-II	SPMC-III
cm energy [TeV]	E_{cm}	3	10	100	1000
luminosity [10^{35} cm^{-2} s^{-1}]	L	1.2	2.1	7.2	5.4
beam energy [TeV]	E_b	1.5	5	50	500
muons/bunch [10^{12}]	N_b	5	3	0.8	0.2
bunches/train	n_b	1	1	1	1
repetition rate [Hz]	f_{rep}	160	27	7.9	3.2
normalized tr. emittances [μm]	$\gamma\epsilon_{x,y}$	15	2	0.5	0.25
6-dim. normalized emittance [10^{-12} m^3]	$\gamma^3\epsilon_{6d}$	16	1.5	0.23	0.30
rms energy spread	δ_{rms}	1%	1%	1%	1%
rms bunch length [mm]	σ_z	0.5	0.8	0.2	0.1
relativistic Lorentz factor [10^4]	γ	1.41	4.7	47	473
IP beta functions [mm]	$\beta^*_{x,y}$	0.5	0.8	0.2	0.1
IP spot sizes [nm]	$\sigma_{x,y}$	730	184	14.5	2.3
beamstrahlung energy loss	δ_B	7×10^{-7}	8×10^{-6}	4×10^{-3}	0.14
Upsilon parameter	Υ	2×10^{-6}	1.0×10^{-5}	1.4×10^{-3}	0.04
beamstrahlung photons/lepton	N_γ	0.71	1.67	5.61	8.43
luminosity enhancement factor	H_D	2.00	3.67	3.77	2.83

Table 4 shows example parameters for single-pass muon colliders at 3, 10, 100 and 1000 TeV, with typical luminosities of a few 10^{35} cm^{-2} s^{-1}. The 6-dimensional emittance for 3 TeV is a factor 5 smaller than the workshop strawman-design value in Table 1. At higher energies the 6-dimensional emittance is much further reduced. For these beam energies the small IP beta functions cannot be achieved with conventional magnets, but must be based on more exotic techniques, such as a plasma lens [32,33] or dynamic focusing [34–36]. The IP beta functions listed in the table correspond to the minimum values achievable with a plasma lens, as estimated in Eq. (20) below. The average muon current at 1000 TeV is chosen a factor 10 smaller than at 100 TeV, in order to account for the muon decay and to limit the beam power. The 3-TeV case is special since here conventional quadrupoles can still be used to focus the beams to the desired spot sizes. The 3-TeV IP beta functions of 0.5 mm assumed in both planes are similar to those proposed for a 500-GeV $\gamma\gamma$ collider [37]. Since at the 3-TeV muon collider beamstrahlung is insignificant, it will provide a much purer luminosity spectrum than a multi-TeV e^+e^- collider.

To focus 50-TeV muon beams, advanced focusing techniques are indispensible. One option is a plasma lens. We denote the beam size at the plasma lens by σ_r, the beta function at the lens by β_r, the rms transverse beam emittance by ϵ, the plasma density by n_p and the distance between plasma lens and collision point by l^*. If the plasma density is smaller than the beam density, the focusing strength of the plasma lens is [33]

$$K = \frac{2\pi r_\mu}{\gamma} n_p \qquad (18)$$

and the reduction of the focusing field by the plasma return current is small as long as [33]

$$k_p \sigma_r \gg 1, \qquad (19)$$

where $k_p = \sqrt{4\pi r_e n_p}$ is the plasma wave number. Combining Eqs. (18) and (19) and using the approximation $l^* \approx 1/K^{1/2}$ we obtain a lower bound on the IP beta function achievable with a plasma lens:

$$\beta^* \geq 2\frac{r_e}{r_\mu}(\gamma\epsilon_{x,y}). \qquad (20)$$

This bound depends only on the mass ratio of electrons and muons and on the normalized beam emittance. Emittances and beta functions in Table 4 were chosen in accordance with Eq. (20). If the normalized emittance can be reduced further, e.g., via optical stochastic cooling [29], the IP beta function may be decreased in proportion! The luminosity scales as

$$L \propto \frac{1}{(\gamma\epsilon_x)(\gamma\epsilon_y)}. \qquad (21)$$

Another technique that has been proposed for electron-positron colliders is dynamic focusing [34–36]. For example, with a demagnification factor $\xi = l^*/\beta^*$ of 33 the focusing by the lens beam would reduce the betafunction at the interaction point from about 2 cm to 20 μm. The required lens-beam charge is [35]

$$N_q \approx \frac{3\gamma\epsilon_{x,y}}{r_\mu}\xi, \qquad (22)$$

which amounts to $N_q \approx 2.7 \times 10^{14}$ for a muon collider at 10 TeV, and $N_q \approx 6 \times 10^{13}$ at 100 TeV. Due to its dependence on normalized emittance and particle mass, this charge is much higher than it would be for an electron beam. Also dynamic focusing becomes easier for reduced beam emittances.

X CONCLUSIONS

Already for a single pass, the design of a final-focus optics for a multi-TeV muon collider is nontrivial, due to the limited strength of conventional quadrupoles and the large geometric emittances. The optics design must rely on novel focusing methods, better cooling of the muon beam, or advances in the design and construction of high-field magnets. The multiple passes imply many additional challenges concerning spot-size stability, tolerances, and tuning. A single-pass collider is a promising option which avoids these last (and many other) problems, is much better adapted to advanced focusing techniques, and could achieve a luminosity of several 10^{35} cm^{-2} s^{-1}, with normalized emittances $\gamma\epsilon_{x,y} = 2$ μm at 10 TeV, and 0.5 μm at 100 TeV. The assumed 6-dimensional emittances are still many orders of magnitude larger than the ultimate emittance that might be attained by optical stochastic cooling [29], leaving room for substantially higher luminosity. If plasma lenses are employed, the luminosity scales inversely with the product of the normalized transverse emittances.

ACKNOWLEDGEMENTS

I am grateful to B. King for inviting me to the workshop and for continuous encouragement. I thank C. Johnstone, E. Keil and V. Telnov for helpful discussions and informations. Support by K. Hübner, J. Gareyte, and F. Ruggiero is appreciated.

REFERENCES

1. B. King, "Studies for Muon Collider Parameters at Center-of-Mass Energies of 10 TeV and 100 TeV", presented at IEEE PAC99, New York (1999).
2. SLAC Linear Collider Conceptual Design Report, SLAC-R-0229 (1980).
3. The NLC Design Group, "Zeroth Order Design Report for the Next Linear Collider", presented at the 1996 DPF/DPB Summer Study on New Directions in High Energy Physics (Snowmass '96).
4. J.P. Delahaye and I. Wilson for the CLIC Study Team, "CLIC, a Multi-TeV e$^+$e$^-$ Linear Collider", CERN/PS 99-062.

5. S. Caspi, K. Chow, A.F. Lietzke, A.D. McInturff, M. Morrison, R.M. Scanlan, G. Ambrosio, G. Bellomo, F. Broggi, L. Rossi, "Design of a Nb$_3$Sn High Gradient Low-Beta Quadrupole Magnet" 15th International Conference on Magnet Technology, Bejing, China (1997).
6. O. Napoly and B. Dunham, "FFADA: Computer design of final focus systems for linear colliders," Contributed to 4th European Particle Accelerator Conference (EPAC 94), London, England, 27 Jun - 1 Jul 1994.
7. H. Grote, F.C. Iselin, "The MAD Program", CERN/SL/90/13 (1990).
8. C. Johnstone and N.V. Mokhov, "Optimization of a Muon Collider Interaction Region with Respect to Detector Backgrounds and the Heat Load to the Cryogenic Systems", presented at the 1996 DPF/DPB Summer Study on New Directions in High Energy Physics (Snowmass '96).
9. C.M. Ankenbrandt et al., "Status of Muon Collider Research and Development and Future Plans", PRST-AB 2, no. 8., (1999).
10. E. Keil, presentation at this workshop and private communication (1999).
11. T. Raubenheimer and F. Zimmermann, "Longitudinal Wake Fields and Chromatic Spot-Size Dilution in the NLC Final Focus", SLAC NLC Note 23 (1996).
12. A. Piwinski, DESY Report 72/72 (1972).
13. V.E. Balakin and A.V. Novokhatsky, Proc. 12-th Int. Conf. High Energy Accel., Batavia, p. 117 (1984).
14. L. Palumbo, Part. Acc. 25, pp. 201–216 (1990).
15. S.A. Kheifets, IEEE Trans. Mcrowave Theory Technique MTT-35, 753-760 (1987).
16. S.A. Heifets and S.A. Kheifets, "Coupling Impedance in Modern Accelerators", Rev. Mod. Physics, Vol. 63, no. 3 (1991).
17. F. Zimmermann and T. Raubenheimer, "Longitudinal Space Charge in Final-Focus Systems for Linear Colliders", NIM A 390, p. 279 (1997).
18. E. Gianfelice, L. Palumbo, IEEE Tr-NS, 37, 2, p. 1084 (1990).
19. F. Zimmermann, K.L.F. Bane, C.K. Ng, "Collimator Wake Fields in the SLC Final Focus", Proc. of EPAC96, Sitges, p. 504 (1996).
20. Sands, M., 1979, "The Physics of Electron Storage Rings", SLAC–121.
21. Chen, P., and K. Yokoya, 1992, "Beam-Beam Phenomena in Linear Colliders", in 'Frontiers of Particle Beams: Intensity Limitations', Lecture Notes in Physics 400, Springer-Verlag.
22. W. Wan, "Tune-Shift with Amplitude due to Nonlinear Kinematic Effect", 1999 IEEE PAC, New York (1999).
23. K. Steffen, "High-Energy Beam Optics", Interscience Publishers (1965).
24. A. Drozhdin, N. Mokhov, C. Johnstone, W. Wan, and A. Garren, "Scraping Beam Halo in $\mu^+\mu^-$ Colliders", 4th International Conference on Physics Potential and Development of Muon Colliders", San Francisco, California (1997).
25. B. King, "Potential Hazards from Neutrino Radiation at Muon Colliders", presented at IEEE PAC99, New York (1999).
26. B. King, "Neutrino Physics at a Muon Collider", Proc. of the Workshop on Physics at the First Muon Collider and Front End of a Muon Collider, Fermilab (1997).
27. A. Mikhalichenko, M. Zolotorev, "Optical Stochastic Cooling", Phys. Rev. Lett. 71, p. 4146 (1993).

28. M.S. Zolotorev and A.A. Zholents, "Transit time method of optical stochastic cooling", Phys. Rev. E50, 3087 (1994).
29. A. Zholents, contribution to this workshop.
30. V. Telnov, contribution to this workshop.
31. C. Johnstone, remark at this workshop.
32. P. Chen, "A Possible Final Focusing Mechanism for Linear Colliders", Part. Acc. 20, 171 (1987).
33. W. Barletta et al., "Plasma Lens Experiments at the Final Focus Test Beam", Proc. of the 6th workshop on Advanced Accelerator Concepts, Lake Geneva (1994).
34. J. Irwin, "Dynamic Focusing Schemes for Linear Colliders", Proc. IEEE PAC97, Vancouver, B.C., (1997).
35. J. Irwin, R. Helm, K. Thompson, D. Schulte, "Further Developments in Dynamic Focusing", Proc. IEEE PAC99, New York (1999)
36. J. Irwin, contribution to this workshop.
37. C. Adolphsen et al., "Zeroth Order Design Report for the Next Linear Collider", LBNL-PUB-5424, SLAC Report 474, submitted to Snowmass 96 (1996).

LIST OF PARTICIPANTS

Anderson, Gregory (Northwestern U.)
Berg, Scott (BNL)
Berger, Mike (Indiana U.)
Berz, Martin (Mich. State U.)
Cline, David (UCLA)
Courant, Ernest (BNL)
Dawson, Sally (BNL)
Firestone, Alexander (NSF)
Garren, Al (UCLA)
Goldberg, Marvin (NSF)
Hanson, Gail (Indiana U.)
Harrison, Mike (BNL)
Heusch, Clem (SLAC)
Irwin, John (SLAC)
Johnson, Colin (CERN)
Johnstone, Carol (Fermilab)
Kahn, Stephen (BNL)
Keil, Eberhard (CERN)
King, Bruce (BNL)
Kirk, Tom (BNL)
Kustom, Robert (Argonne Lab.)
Lane, Ken (Boston U.)
Lebrun, Paul (Fermilab)
Lotov, Konstantin (BINP)
Lykken, Joe (Fermilab)
Makino, Kyoko (Mich. State U.)
McDonald, Kirk (princeton)
Mills, Fred (Fermilab)
Nagamine, Kanetada (KEK/RIKEN)
Norton, Peter (RAL)
Padamsee, Hasan (Cornell U.)
Parkhomchuk, Vassily (INP, Novosibirsk)
Parsa, Zohreh (BNL)
Paul, Peter (BNL)
Rehak, Pavel (BNL)
Rizzo, Tom (SLAC)
Samios, Nicholas (BNL)
Sessler, Andrew (LBL)
Skrinsky, Alexander (INP, Novosibirsk)
Telnov, Valery (INP, Novosibirsk)
Willis, Bill (Columbia U.)
Wilson, Mark (DOE)
Zadorozhny, Vladimir (NAS, Ukraine)
Zholents, Alexander (LBL)
Zimmermann, Frank (CERN)
Zisman, Michael (LBL)

AUTHOR INDEX

B

Benary, O., 1
Berg, J. S., 13
Berger, M. S., 32
Berz, M., 38, 217

C

Cline, D., 260
Courant, E. D., 333

D

Dawson, S., 48

E

Erdélyi, B., 38

G

Garren, A., 66, 333
Gatti, E., 260

H

Heusch, C. A., 68, 260

K

Kahn, S., 1, 260
Keil, E., 80
King, B. J., 86, 122, 142, 165, 260
Kirk, T., 260

L

Lane, K., 181
Lebrun, P., 190

Lotov, K. V., 201
Lykken, J. D., 208

M

Makino, K., 38, 217

N

Nagamine, K., 228
Norton, P., 260

P

Parkhomchuk, V. V., 236
Parsa, Z., 239, 249

R

Radeka, V., 260
Rehak, P., 260
Rizzo, T. G., 290

S

Samios, N. P., 260, 308
Skrinsky, A., 311
Stumer, I., 1

T

Tcherniatine, V., 260
Telnov, V., 316, 318, 324, 330
Trbojevic, D., 333

W

Willis, W., 260, 339

Z

Zadorozhny, V., 249
Zimmermann, F., 347